Other volumes in this series:

Volume 1	**Power circuit breaker theory and design** C.H. Flurscheim (Editor)
Volume 4	**Industrial microwave heating** A.C. Metaxas and R.J. Meredith
Volume 7	**Insulators for high voltages** J.S.T. Looms
Volume 8	**Variable frequency AC motor drive systems** D. Finney
Volume 10	**SF6 switchgear** H.M. Ryan and G.R. Jones
Volume 11	**Conduction and induction heating** E.J. Davies
Volume 13	**Statistical techniques for high voltage engineering** W. Hauschild and W. Mosch
Volume 14	**Uninterruptable power supplies** J. Platts and J.D. St Aubyn (Editors)
Volume 15	**Digital protection for power systems** A.T. Johns and S.K. Salman
Volume 16	**Electricity economics and planning** T.W. Berrie
Volume 18	**Vacuum switchgear** A. Greenwood
Volume 19	**Electrical safety: a guide to causes and prevention of hazards** J. Maxwell Adams
Volume 21	**Electricity distribution network design, 2nd edition** E. Lakervi and E.J. Holmes
Volume 22	**Artificial intelligence techniques in power systems** K. Warwick, A.O. Ekwue and R. Aggarwal (Editors)
Volume 24	**Power system commissioning and maintenance practice** K. Harker
Volume 25	**Engineers' handbook of industrial microwave heating** R.J. Meredith
Volume 26	**Small electric motors** H. Moczala *et al.*
Volume 27	**AC-DC power system analysis** J. Arrill and B.C. Smith
Volume 29	**High voltage direct current transmission, 2nd edition** J. Arrillaga
Volume 30	**Flexible AC Transmission Systems (FACTS)** Y-H. Song (Editor)
Volume 31	**Embedded generation** N. Jenkins *et al.*
Volume 32	**High voltage engineering and testing, 2nd edition** H.M. Ryan (Editor)
Volume 33	**Overvoltage protection of low-voltage systems, revised edition** P. Hasse
Volume 34	**The lightning flash** V. Cooray
Volume 35	**Control techniques drives and controls handbook** W. Drury (Editor)
Volume 36	**Voltage quality in electrical power systems** J. Schlabbach *et al.*
Volume 37	**Electrical steels for rotating machines** P. Beckley
Volume 38	**The electric car: development and future of battery, hybrid and fuel-cell cars** M. Westbrook
Volume 39	**Power systems electromagnetic transients simulation** J. Arrillaga and N. Watson
Volume 40	**Advances in high voltage engineering** M. Haddad and D. Warne
Volume 41	**Electrical operation of electrostatic precipitators** K. Parker
Volume 43	**Thermal power plant simulation and control** D. Flynn
Volume 44	**Economic evaluation of projects in the electricity supply industry** H. Khatib
Volume 45	**Propulsion systems for hybrid vehicles** J. Miller
Volume 46	**Distribution switchgear** S. Stewart
Volume 47	**Protection of electricity distribution networks, 2nd edition** J. Gers and E. Holmes
Volume 48	**Wood pole overhead lines** B. Wareing
Volume 49	**Electric fuses, 3rd edition** A. Wright and G. Newbery
Volume 51	**Short circuit currents** J. Schlabbach
Volume 52	**Nuclear power** J. Wood
Volume 905	**Power system protection, 4 volumes**

IET POWER ENGINEERING SERIES 11

Series Editors: Prof. A.T. Johns
J.R. Platts
G. Ratcliffe

Conduction and Induction Heating

Conduction and Induction Heating

E.J. Davies

The Institution of Engineering and Technology

Published by The Institution of Engineering and Technology, London, United Kingdom

First published 1990
Reprinted 2007

The Institution of Engineering and Technology
Michael Faraday House
Six Hills Way, Stevenage
Herts, SG1 2AY, United Kingdom

www.theiet.org

British Library Cataloguing in Publication Data
Davies, E.J.
Conduction and induction heating.
1. Heat engineering
I. Title II. Series
621.402

ISBN (10 digit) 0 86341 174 6
ISBN (13 digit) 978-0-86341-174-8

Printed in the UK by Antony Rowe Ltd, Eastbourne
Reprinted in the UK by Lightning Source UK Ltd, Milton Keynes

For Margaret, who makes it all possible

Chance favours the prepared mind

Pasteur

The philosopher may be delighted with the extent of his views; the artificer with the readiness of his hands; but let the one remember that, without mechanical performance, profound speculation is but an idle dream, and the other that, without theoretical prediction, dexterity is little more than brute instinct.

Samuel Johnson

Bessel functions are beautiful functions in spite of their many applications

Anon. British Mathematician

Those who fall in love with practice without science are like a sailor who steers a ship without a helm or compass, and who never can be certain where he is going.

Leonardo

It is easier to write ten volumes on theoretical principles than to put one principle into practice.

Tolstoy

Contents

Part II Induction heating

Foreword

John Davies and I have known each other over a number of years through the activities of the British National Committee for Electroheat.

John Davies has had a long association with Electroheat and its application to Industrial processes. As a Professor of Electrical Engineering at the University of Aston he initiated project and research studies on aspects of conduction and induction heating which in turn influenced his staff to introduce the subjects to other establishments as they progressed. Thus today there are several educational centres where the continuing investigation of these subjects is due to John Davies' inspiration and the work exemplified in this book.

Several organisations have benefited from the work undertaken and presented by John Davies. He has been a member of the British National Committee for Electroheat for nearly twenty years and has been successively Chairman of their Education and Training Committee, Technology Committee and Induction Heating Group. He has participated in congresses and other activities of the International Union of Electroheat (UIE). As a Fellow of the Institution of Electrical Engineers he has held several posts on the Institutions' Boards, Divisions and Groups and published several papers in the Institutions' journals. He gave the IEE Hunter Memorial Lecture in 1981 on Electroheat for Industry.

The Induction Heating Handbook, written by John Davies with Peter Simpson and published in 1979, has proved to be a valuable reference on the subject. This latest book provides much updated and additional data which will be invaluable to designers, constructors and users of the many types of equipment using conduction and induction heating.

In particular, the final part of this comprehensive book puts theory into practice and considers the practical applications. Sample calculations and details of some of the equipment available are presented, and tables of useful constants and parameters needed for calculations are given in the appendices.

I believe this book should sit on the shelves of all those associated with Electroheat and the promotion of energy efficiency.

It gives me special pleasure, as a friend, to sign this foreword during my year in office as President of the Institution of Electrical Engineers and to have the opportunity to note the important work carried out by John Davies in the service of electrical engineering through his work in University, Industry and the Institution of Electrical Engineers.

Dr J.C. Smith

Preface

In his recent best-seller, Stephen Hawking says that he was advised that every equation halved the sales. He got round this by subterfuge, but his solution was not open to me. With over 100 equations, Hawking's rule would mean that my sales would be decreased by 2^{100}! Tudbury managed to present the subject in a simplified manner but his technique would not suit my message or style. My view is that this topic is essentially mathematical and that this is the only way to real understanding.

Another viewpoint which was put to me very strongly was that most design is done by computer nowadays, and so a new book should reflect this by including calculating programs in the text. All the new curves and many of the old ones in this book have been calculated or checked by computer, so no one can say that I reject the computer. The best practical programs are solidly based on theory, and so someone has to understand the equations: it is that understanding which I am trying to bring out in this book. Having said that, the reader does not need to use this mathematics every time he does a calculation. It is for fundamental understanding that I am aiming.

As far as I know, this is the first time that conduction and induction have been treated together, but the topics will be seen to be inextricably linked, the same theory occuring in both but for different quantities.

In this book I have tried to record the unchanging basics of the subject. In the last three decades, supply systems have changed dramatically and we can be sure that they will go on changing as new semiconductor devices are produced. A book based on this rapidly becomes dated and the topic is best treated separately in a format which allows frequent change and which is written by someone more skilled in the electronic art than myself.

Heat flow is an essential part of this study and is treated at some length in Part III, starting with the simplifying assumption that the heat is produced in a very shallow surface layer, and going on to analyse the actual heat patterns.

The fourth part deals with the practice. This changes slowly, and so its inclusion is justifiable if one does not have too many photographs, because the external appearance of the apparatus changes faster than the internals – very often these changes are merely cosmetic. Also, photographs of equipment seldom tell the reader anything about its functioning.

Professor E.J. Davies
Sutton Coldfield and Park House, Natland, Cumbria, England.
1987–1989

Acknowledgments

If I have seen further (than you and Descartes)
it is by standing upon the shoulders of Giants.

Sir Isaac Newton
Letter to Robert Hooke 1675

If Newton saw fit to emphasise his debt to his precursors, how much more it behoves us lesser mortals to do so. Electroheat is a subject with its foundations in Fourier, both for heat transfer and the principles of the diffusion equation.

Then Heaviside made his contribution, as he did to so much of electromagnetism; in addition to the particular papers quoted, we owe to this 'Giant' the form of vector analysis now taken for granted and the operational calculus on which many of the solutions quoted in this book are based (and for which modern mathematics tries to rob him of credit, perhaps as revenge for his cruel mockery of the mathematicians of his time). He discovered Poynting's energy-flow theorem independently and analysed the result more thoroughly. To Heaviside, too, we indirectly owe the SI integrated system; while it is not exactly the solution that he advocated, it was his trenchant attack on the irrationality of the old systems that prepared the way for the ultimate solution that unites electrical and mechanical quantities. Even the nomenclature we use owes much to Heaviside. The reader will have noticed by now that I have a respect not far short of idolatry for Heaviside! His giant shoulders form a broad base on which to build our edifice.

At that time, it was not practical to heat anything by induction, because the power sources and PF correction capacitors were not available, and so the subject lay fallow.

On the theoretical side, I have found McLachlan on Bessel functions and Carslaw & Jaeger on heat flow extremely useful. Neither of these sources is 'original' in the sense that Heaviside undoubtedly was, but they are both excellent pieces of work which link pure mathematics with practice. I have drawn on their wisdom extensively, while trying to extend their results in the practical direction.

Turning to practitioner – theoreticians, we have Dreyfus of Sweden, who combined great analytical skill with an excellent understanding of what was occurring in an object being heated. Following in his footsteps at ASEA, Dr Sundberg has been very helpful to me. He gave me a copy of his ASEA internal publication 'Elektrougnar och induktiva omrörare', which improved my non-existent Swedish at the same time as helping my technology. His doctoral thesis (in English) is also helpful, but specialised.

In 1960, a book of special interest to me was published, which combined theory and practice as well as drawing on the excellent papers from Westinghouse by Baker, with whom its author had obviously worked. This was Simpson's 'Induction heating – coil and system design', which soon became the standard text, but sadly it was allowed to go out-of-print by his publisher.

By 1978, I felt ready to write a book of my own to replace the out-of-print Simpson and, having been offered a sabbatical semester, decided to do something about it. Among other aims, I wanted a book with more theory, so that it could be used for teaching. I approached Peter Simpson for permission to use parts of his work and discovered that he had quite independently started to rewrite his original book. He kindly suggested that we collaborate, which I gratefully accepted; the result was Davies & Simpson (1979), which was well received. This, too, was allowed to go out-of-print, with the copyright reverting to us.

In the 1970s, the Electricity Council embarked on an ambitious programme of Electroproduction education into which Davies & Simpson was integrated. In 1986, I wrote 'Heat transfer for induction heating', which was largely based on Davies & Simpson but given a teaching format. On leaving Aston University in 1983, I contracted with the Electricity Council to write an original Teaching Monograph on 'Direct resistance heating'. This was published in 1987. In the process of writing, it became increasingly obvious that conduction heating and induction heating were not only similar, but had a duality relationship; this was mentioned in the monograph but could not be explored for space reasons. It was also apparent that conduction heating was underexploited in industry and that perhaps the topic needed rethinking in the light of the semiconductor revolution. Hence this book. I am extremely grateful to Professor David Langman, Brian Booth and John Williams of the Electricity Council for all the help and encouragement that they have given me throughout. In particular, Part I of this book uses both the text and some of the artwork from the monograph and I am indebted to the Electricity Council, who had copyrighted my material for my protection, for permission to use this. I must also thank Lewis Tozer, who edited both works with great care.

When this book was being planned, I had to decide whether to produce an expanded and rearranged Davies & Simpson or something substantially different. After much heartsearching, I decided on the latter course because I felt that the electronics side needed a book of its own, which could be updated in line with the rapid changes that take place in that subject, whereas the power side changes more slowly. I am extremely grateful to Peter Simpson for agreeing to the use of material from our previous collaboration; in particular, he has allowed me to draw heavily on chapters 2, 3 and 4, which form the basis of chapters 22, 23 and 24 of this book.

I have had considerable help from my old friends Tony Baines of Radyne and Norman Kerruish. I had helpful suggestions from Don Martin, Bob Ireson, Bill Wilkinson and Mike Evans. Ian Harvey and Colin Edgerley of the Electricity Council Research Centre at Capenhurst have given me unstintingly their expertise.

My son-in-law, Christopher Tyler, has helped me with many computer programs.

Mike Thelwell of the British National Committee for Electroheat has given me much advice and encouragement, as well as practical help.

Finally, my thanks to Dr Katie Petty-Saphon, Janet Aitchison and Jill Priest at PPL and especially to Brian Goodale, my copy editor, who have helped me in so many ways.

List of symbols

A	area, m^2
b	thickness of slab heated from one side, m
b	inner radius of tube, m
$2b$	thickness of slab heated from both sides, m
B	magnetic flux density, T
c	mass specific heat, J/kg K
$(c\gamma)$	volume specific heat, J/m^3 K
C	constant in radiation equation
d	outside diameter of cylinders and tubes, m
E	electric field intensity, V/m
f	frequency, Hz
f_c	critical frequency, Hz
H	magnetic field strength, A/m
h	height, m
I	current, A
I_0, J_0, K_0	Bessel functions
j	$\sqrt{-1}$
J	current density, A/m^2
J	rms value of surface J, A/m^2
k	thermal conductivity, W/m K
k	auxiliary quantity $= \sqrt{(2j\alpha^2)}$
L, l	length, m
	water quantity in litres/min
m	as suffix, denotes mean value
m	auxiliary quantity $= \sqrt{(2)}\alpha$
$M_0(\theta_0)$, $M_1(\theta_1)$, $N_0(\phi_0)$, $N_1(\phi_1)$	polar Bessel functions
O	higher orders (HO)
p	power loss constant for induction, numeric
p_r	power loss constant for DRH, numeric
P	power, W
P_w	workpiece power, W
P_v	volume power density, W/m^3

$\mathscr{P}$	power loss/unit surface area, W/m^2
$\mathscr{P}_s$	electromagnetic power-density surface, W/m^2
$\mathscr{P}_n$	net power-density after subtracting radiation, W/m^2
$\mathscr{P}_R$	radiation power density, W/m^2
q	reactive power loss constant for induction, numeric
q_r	reactive power loss constant for DRH, numeric
Q	energy, J
$\mathscr{Q}$	water quantity, m^3/s
Q	reactive power/unit surface area, var/m^2
Q_w	reactive power, var
r	radius, m
R	outer radius of cylinder or tube, m
R	resistance, Ω
$\bar{R}$	resistance per metre length, Ω/m
R_{ac}	effective resistance of slab, cylinder, etc., when carrying alternating current, Ω
R_{dc}	resistance of slab, cylinder, etc., when carrying direct current, Ω
R'	resistance per unit surface area, i.e. 1 m long × 1 m wide, Ω
s	surface value, used as suffix
t	thickness of slab $(=2b)$, or of tube wall, m
t	time, s
t_0	time to heat billet/slab, without radiation
t_R	time to heat billet/slab, including radiation
T	absolute temperature, K
T_s	absolute surface temperature, K
T_a	absolute temperature of receiving surface, K
T_f	final temperature, K
$T_{\mathscr{P}}$	artificial temperature, representing power input, K
v	velocity, m/s
V	potential difference, V
w	width, m
X	reactance, Ω
X_{ac}	effective reactance of slab, cylinder, etc., when carrying alternating current, Ω
X'	reactance per unit surface area
Z	impedance, Ω
Z'	impedance of unit length and width, Ω

Greek

α	auxiliary quantity = 1/depth of penetration = $1/\delta$, m^{-1}

α_n	roots of auxiliary equation $J_1(\alpha) = 0$
β_0	roots of auxiliary equation $J_1(\beta) = 0$
γ	density, kg/m^3
δ	depth of penetration $= \sqrt{(2\rho/\mu\omega)}$, m
ε	emissivity constant of surface, dimensionless
θ	as suffix, variable value of quantity with temperature
θ	peripheral direction (tubes and cylinders)
θ	temperature, K or °C
θ	temperature difference, K or °C
θ_b	back temperature rise (single sided heating), K
θ_c	centre temperature, K or °C
θ_d	differential temperature
θ_m	mean temperature rise, K or °C
θ_s	surface temperature, K or °C
Θ_1	$T_1/T_{\mathscr{P}}$
Θ_2	$T_2/T_{\mathscr{P}}$
μ_0	magnetic constant, H/m
μ_r	relative μ, numeric
ρ	resistivity, Ωm
τ	normalised time, numeric $= \left[\frac{kt}{(c\gamma)b^2} \text{ or } \frac{kt}{(c\gamma)R^2}\right]$
$\cos\phi$	power factor
Φ	magnetic flux, Wb
ω	angular frequency, rad/s
η	efficiency, %
η	viscosity, m^2/s

Remarks

Simple suffices, e.g. J_z, J in the z direction. Various suffices which are specialised are defined in the text, e.g. Fig. 22.7.
Compound suffices, e.g. H_{xy}, H in x direction as a function of y.
Example 15.3 uses suffix 1 below Curie and suffix 2 above Curie for $\mathscr{P}$, θ, k, $c\gamma$ etc.
Soaking conditions (Chapter 16) uses barred symbols, e.g. $\bar{\theta}_m$, for the temperature conditions prevailing when the power is switched off at the start of soaking.

Part I

Direct resistance heating

Chapter 1

Fundamentals

1.1 Introduction

The special feature of conduction and induction heating is that the heat is produced inside the material by currents flowing in it. From this, it follows that the material must be a conductor, as currents do not flow in insulators, by definition. Dielectrics are also heated from inside, but not by Joule effect (i.e. not by currents): the mechanism is completely different. It is natural to discuss conduction and induction heating in the same book because, as we shall see, they are closely related. As far as the author knows, this common theory has not been emphasised before, but we shall see that it leads to interesting relationships and that ideas developed for one can be used for the other. The order of words in the title is not chance; although conduction heating is much less common than induction heating, for reasons that we shall explore, there are considerable advantages in developing the fundamental ideas from conduction heating. In part this is because we are all schooled in Ohm's law and so have an innate understanding of conduction, whereas induction is not as 'natural'.

Conduction heating is commonly known as *direct resistance heating*. The 'resistance heating' part of this term is obvious; it is 'direct', as opposed to 'indirect', because the heating occurs in the material. Indirect resistance heating is extensively used in industry, where the heat is produced in a resistor and transferred to the workpiece by convection or radiation. The word 'direct' implies that the current flows in the workpiece itself. It has nothing to do with direct or alternating currents! For convenience, we shall use DRH for direct resistance heating.

At first sight, the concept of heating a metal by passing current through it is attractively simple. All the energy supplied to the workpiece is stored in the material as heat, raising its temperature, assuming no convection or radiation losses. In practice, this simplicity is marred by restrictions on shape and the problem of contacts, but it can still be stated dogmatically that *where it is appropriate* DRH is the best method available. The 'ifs and buts' of that statement will be developed later.

1.2 Simple electrical equations

The basic ideas of DRH are delightfully simple, using only the most elementary concepts. We shall see later that they have to be modified when we use alternating current (AC), but for the moment we can restrict ourselves to direct current (DC). Consider a metallic conductor, of resistivity ρ Ωm, with a uniform cross-section (which can be any shape) of area A m^2 and of length l m. Then the resistance R is given by

$$R = \rho l/A \qquad \text{ohms} \tag{1.1}$$

Assume that electrodes are fixed to the ends of the conductor – this will be considered in more detail later – and that V volts are applied to the electrodes. The current I is given by

$$I = V/R \qquad \text{amperes} \tag{1.2}$$

and the power P dissipated in the resistor, by the Joule effect, is

$$P = I^2R = V^2/R \qquad \text{watts} \tag{1.3}$$

Equations 1.1 to 1.3 use the circuit approach to the heating, but we shall find it useful to express those facts in their field forms. The field quantities involved are the electric field intensity $\boldsymbol{E}$ V/m and the current density $\boldsymbol{J}$ A/m^2.

From first principles, $\boldsymbol{E} = \rho\boldsymbol{J}$, but this can easily be obtained from eqns 1.2 and 1.1:

$$I = V/R = V/(\rho l/A)$$
$$\rho I/A = V/l$$
$$\rho J = E$$

This general form of Ohm's law has been derived here from the circuit Ohm's law of eqn 1.2, but it is absolutely general. It is one of the Maxwell equations. The directions of $\boldsymbol{E}$ and $\boldsymbol{J}$ are both axial in this specific example. The sameness of direction is also universally true, so the equation can be written in vector quantities. In the example, ρ, $\boldsymbol{E}$ and $\boldsymbol{J}$ are constant in the whole conductor, but the equation is equally true if they are variable. Having established that we are dealing with vector quantities, it is convenient to work with the scalar quantities E and J.

The field equivalent of eqn 1.3 is obtained by rearranging:

$$P = I^2R = \rho l I^2/A = \rho(lA)\,(I/A)^2$$
$$= \rho J^2 \text{ (volume of conductor)}$$

or

$$\text{loss/volume} = \rho J^2 \qquad \text{W/m}^3 \tag{1.4}$$

In the example, the loss is uniformly distributed through the workpiece, but the expression is true even when ρ and J are both changing. It is convenient to give this 'volume loss' a new symbol related to P, say P_v; so

$$P_v = \rho J^2 \quad \text{W/m}^3 \tag{1.5}$$

1.3 Simple electrothermal equations

Since we are aiming to heat the conductor electrically, we need to think about the property of the material that is concerned with heat storage. Unfortunately, the quantity usually defined is *mass* specific heat, with the symbol c and units of J/kg K, i.e. the quantity of heat needed to raise the temperature of one kilogram of the material through one kelvin. But, in electroheat, we are normally thinking in terms of volume rather than mass, so we need the *volume* specific heat, with units of J/m^{3}K. This has not been given a special symbol, but since it is easily derived from c by multiplying by density (kg/m^3), we can use a compound symbol ($c\gamma$), which will usually appear in a bracketed form. (The normal BS symbol for density is ρ but, as this would cause confusion in electrothermal work with resistivity, γ has been substituted. BS also allow δ for density, but we shall see later that this would be confused with depth of penetration.)

Consider our original conductor, length l, area A, being heated with P_v W/m^3 for t seconds. We have

$$\text{energy in} = P_v\,(\text{volume})\,t \quad \text{joules} \tag{1.6}$$

$$\text{energy stored} = \theta_m\,(\text{volume})(c\gamma) \quad \text{joules} \tag{1.7}$$

With no losses, energy in is equal to energy stored, and therefore

$$P_v t = \theta_m (c\gamma)$$

where θ_m is the mean temperature-*rise* of the material. In this example, the word 'mean' is redundant, since the whole volume is at the same temperature, but we shall need the concept later.

Rearranging gives the rate-of-rise of temperature:

$$\theta_m / t = P_v / (c\gamma) \tag{1.8}$$

This equation is valid for all 'volume-heating' concepts and applies equally to dielectric heating.

Example 1.1

Mild-steel rods, 12 mm diameter and 8 m long, are to be heated in pairs by DRH from 20°C to 720°C by short-circuiting one pair of ends and feeding the two remaining ends from a DC source (Fig. 1.1). Assuming that the volume specific heat ($c\gamma$) has the constant value of 5 MJ/m^{3}K, and that ρ has the constant value of 0.5 $\mu\Omega$ m (neither assumption being true; see Section 1.4), calculate the DC

Fig. 1.1 *Example 1.1*

supply needed:

(a) to heat the rods in 1 minute, using circuit concepts
(b) to heat the rods in 1 minute, using field concepts.

Solution

(a)

$$\begin{aligned}\text{energy stored} &= \theta_m \,(\text{volume})(c\gamma)\\ &= (700)\,[2 \times 8 \times (\pi/4) \times (12/1000)^2]\,(5 \times 10^6)\\ &= 6.33 \quad \text{MJ}\\ \text{resistance} &= \rho l/A\\ &= (0.5/10^6)\,[16/(113.1 \times 10^{-6})]\\ &= 0.071 \quad \Omega\end{aligned}$$

But energy $= I^2Rt$. Therefore

$$\begin{aligned}I^2 &= (6.33 \times 10^6)/(0.071)\,(60)\\ I &= 1220 \quad \text{A}\\ V &= IR = 86 \quad \text{V}\\ P &= VI = 105 \quad \text{kW}\end{aligned}$$

(b) From eqn 1.8,

$$\begin{aligned}P_v &= (c\gamma)\theta_m/t\\ &= (5 \times 10^6)\,(700/60) = 58.3 \times 10^6 \quad \text{W/m}^3\end{aligned}$$

From eqn 1.5,

$$\begin{aligned}P_v &= \rho J^2\\ J^2 &= P_v/\rho\\ &= (58.3/10^6)/(0.5 \times 10^{-6})\\ J &= 10.8 \times 10^6 \quad \text{A/m}^2\end{aligned}$$

$$I = J\,(\text{section})$$
$$= (10.8 \times 10^6)\,(\pi/4)\,(12/1000)^2$$
$$= 1220 \quad \text{A}$$

But $E = \rho J$. Therefore

$$E = (0.5/10^6)\,(10.8 \times 10^6)$$
$$= 5.4 \quad \text{V/m}$$
$$V = El$$
$$= 5.4 \times 16 = 86.4 \quad \text{V}$$

The power can be obtained from VI, as before, or by multiplying P_v by the total volume, i.e.

$$P = (58.3 \times 10^6)\,(\pi/4)\,(12/1000)^2\,(16) = 105 \quad \text{kW}$$

This simple example shows that both concepts lead to the same answers, with roughly the same amount of work. Note that the 10^6 factors frequently cancel. There would be some argument for working in Ω mm and using mm^2 for sections, but here it is preferable to keep to SI and cancel powers of 10; in practice, one would work in millimetres.

The example was chosen to show that DRH gives workable numbers if the workpiece is long and thin. An 80 V, 1200 A, 100 kW transformer is practical, without difficulties in connecting it to the load. The next example shows how the numbers change if the load is shorter and fatter.

Example 1.2
As previous example, except that the workpiece is 0.25 m long and 100 mm diameter, heated by connection to the two ends (i.e. single, not double).

Solution

(a) $$\text{energy} = (700)\,[0.25 \times (\pi/4)\,(100/1000)^2]\,(5 \times 10^6) = 6.9 \quad \text{MJ}$$

This is substantially the same as Example 1.1(a). Next,

$$\text{resistance} = (0.5/10^6)\,(0.25)/(\pi/4)\,(100/1000)^2 = 15.9 \quad \mu\Omega$$
$$I^2 = (6.9 \times 10^6)/(15.9 \times 10^6)\,(60)$$
$$I = 85\,000 \quad \text{A}$$
$$V = 1.35 \quad \text{V}$$
$$P = 115 \quad \text{kW}$$

(b) This is left as an exercise.

The energy and power are much the same in the two examples, but the short fat billet needs a very substantial current at a low voltage. If there were any voltage drop in the leads, or in the connections to the billet, this would represent a drop in efficiency which would also present cooling problems in the apparatus.

Example 1.3
Consider the rods of Example 1.1, except that one rod is tapered from end to end in such a way that its overall resistance is still the same, but its smallest diameter is 11.5 mm. Calculate the heating rate at this point.

Solution
The cross-section for diameter 11.5 mm is $(11.5/12)^2$ of the original area, so the current density J is increased by $(12/11.5)^2 \simeq 1.09$.

The power density P_v is proportional to J^2, i.e. 1.186. The heating rate will be increased by this amount. Previously it was (700/60) = 11.67 K/s, so now it will be 13.83 K/s, and θ_m at the narrow section will be 830°C rise instead of 700°C.

This example has been included to emphasise how important uniformity of section is in DRH. The specified 4% change in diameter has resulted in an 18.5% change in heating rate. If the dimension changes by $\pm x\%$, the heating rate becomes $\simeq 1 \mp 4x + 10x^2$.

It is interesting to note that eqns 1.8 and 1.5 combine to give

$$\theta_m/t = [\rho/(c\gamma)]J^2 \tag{1.9}$$

i.e. the rate-of-rise of temperature is proportional to the current density squared and the physical constants of the material. It is independent of the length of the workpiece: the length only controls the applied voltage needed to produce J. Also, if convenient, the rate-of-rise can be expressed in terms of E:

$$\theta_m/t = E^2/\rho(c\gamma) \tag{1.10}$$

1.4 Practical modifications

If the theory of DRH were as simple as given in the last two sections, there would be no problems. In practice, the 'constants' ρ and $(c\gamma)$ vary with temperature. Appendix A.1 gives the values for steel, stainless steel, copper and aluminium. It is well-known that ρ varies linearly with temperature for many metals (see Chapter 22), such as copper, brass and aluminium (to mention only the main engineering materials). Iron and steel do not obey this law above the Curie temperature (760°C), but are reasonably linear up to that temperature.

Appendix A.1 has a set of curves showing the variation of ρ with temperature for various metals.

It is not as well-known that specific heat $(c\gamma)$ is also a reasonably linear function of temperature. This can be deduced from examination of published curves for heat content of metals at various temperatures (kW h/tonne) (see Chapter 22), but it can also be obtained by plotting the data in the tables of Appendix A.1. This has also been done in Appendix A.1; again, the behaviour of steel is anomalous.

From this, it will be seen that the simple assumptions of constant ρ and $(c\gamma)$ in the examples have to be modified in practice. Over the temperature range used, ρ varies from 0.16 $\mu\Omega$ m to 0.92 $\mu\Omega$ m and $(c\gamma)$ from 3.65 MJ/m^{3}K to 6.66 MJ/m^{3}K. The principles used in the equations are still correct, but it is necessary to repeat the calculations iteratively, with the values corresponding to the instantaneous temperature. The method used to do this depends on what is kept constant. In static heating applications, as in the examples, the voltage across the load will stay constant and the current will vary, falling off as the metal heats and the resistance increases. The temperature in the load varies through the heating cycle. The quantity staying constant is E, the electric field intensity, expressed in V/m.

If, however, DRH is being used to heat moving strip, where the circuit is constantly being replaced, then the total circuit conditions are constant, E is variable along the length of the strip, and the quantities that remain constant (for a fixed voltage between the contacts) are J, the current density in A/m^2, and the current I (A).

Example 1.4
The rods of the first example have a constant voltage of 100 V DC applied to the unshorted ends. Calculate the time taken to raise the temperature through 700°C, as before, using the values of ρ and $(c\gamma)$ from Appendix A.1.

Solution – iterative method
The controlling equation is eqn 1.10, where the rate-of-rise is expressed in terms of E, since this is being kept constant. To do this in steps, we rewrite it as

$$\Delta\theta_m/\Delta t = E^2/[\rho(c\gamma)]_\theta$$

$$\Delta t = \Delta\theta_m[\rho(c\gamma)]_\theta/E^2 \qquad (1.11)$$

where E is constant $(100/16 = 6.25$ V/m) and $[\rho(c\gamma)]_\theta$ is a mean value over the temperature step Δt being considered.

First, a plot must be made of $\rho(c\gamma)$ against θ from the data of Appendix A.1; this is shown in Fig. 1.2. For simplicity, we shall take the mean values over each 100°C range. If more precision were needed, the range could be shortened. We have

$$\Delta t = [100/(6.25)^2]\,[\rho(c\gamma)] = 2.56\,\rho(c\gamma)$$

Range (°C)	$\rho(c\gamma)$ ($\Omega J/m^2K$)	Δt(s)
0–100	0.7	1.8
100–200	1.0	2.6
200–300	1.4	3.6
300–400	1.95	5.0
400–500	2.72	7.0
500–600	3.7	9.4
600–700	5.0	12.8
	Heating time	42 s

The values have been read from mean temperatures, e.g. 0–100 at 50°C. If more accuracy were needed, take a shorter range. If the individual Δt are not needed,

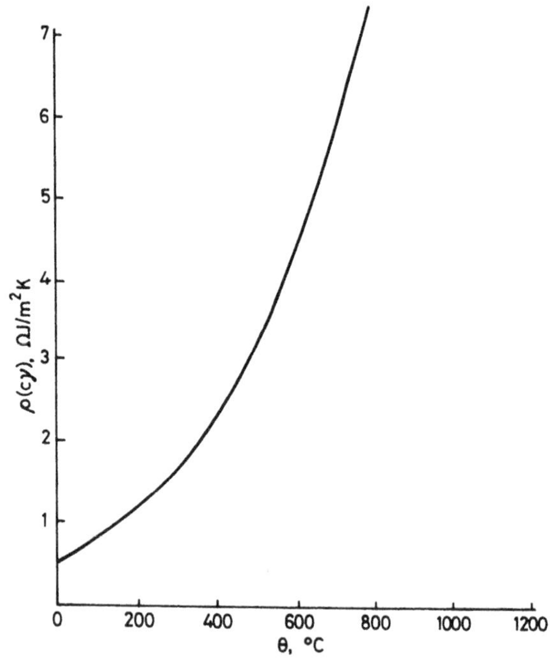

Fig. 1.2 *Variation of $\rho(c\gamma)$ with temperature for 0.23% C mild steel*

it is simpler to add $\rho(c\gamma)$ and multiply by 2.56.

The heating time is shorter because E has increased from 5.4 to 6.25 V/m. If the time is reduced proportionately to $1/E^2$, we obtain 60 $(5.4/6.25)^2$ = 44.8 s, so the iterative method does not greatly change the total heating time. What is significant are the heating times associated with each temperature range. The assumption of constant ρ and $(c\gamma)$ gives a constant time for each temperature range, whereas the more accurate calculation shows very rapid heating at first, falling to 1/7 of the initial value at 700°C. This is because ρ increases 5.8 times and $(c\gamma)$ by 1.82 over the range, i.e. the heating rate at 0°C is $(5.8 \times 1.82) = 10.5$ times that at 700°C. It is important to remember this, as radiation losses increase with temperature, so the longer times at the higher temperatures result in lower efficiency, unless the material is well insulated.

It is interesting to see the change in power input as the workpiece heats, which can be obtained from E^2/ρ W/m^3, where $E = 6.25$ V/m, volume $= 2 \times 8 \times (\pi/4)(12/1000)^2 = 0.001\,81$, so that $P = 0.0707/\rho$.

Range (°C)	ρ ($\mu\Omega$ m)	P(kW)
0–100	0.18	393
100–200	0.25	283
200–300	0.33	214
300–400	0.43	164
400–500	0.54	131
500–600	0.67	106
600–700	0.82	86

Note that, if this table is to be compared with the results of Example 1.1, the value of E has to be changed from 6.25 to 5.4 V/m. The table suggests that a practical heater would have to be designed with variable voltage to give more uniform heating over the temperature range.

All the examples so far have been for mild steel, which has a relatively high resistivity. It is instructive to repeat Example 1.2, using a copper workpiece.

Example 1.5
As Example 1.2, except in copper. Use $(c\gamma) = 3.66 \times 10^6$ J/m^{3}K and $\rho = 0.042\,\mu\Omega$ m.

Solution

$$\text{energy} = (6.9)(3.66 \times 10^6)/(5 \times 10^6) = 5.05 \quad \text{MJ}$$

$$\text{resistance} = (15.9 \times 10^{-6})(0.042/0.5) = 1.33 \times 10^{-6} \quad \Omega$$

$$I^2 = (5.05 \times 10^6)/(1.33 \times 10^{-6})(60)$$

$$I = 252\,000 \quad \text{A}$$

$$V = 0.33 \quad \text{V}$$

$$P = 84.2 \quad \text{kW}$$

These figures show that it is very difficult to heat copper billets *of this shape* by DRH. The energy needed is of the same order as for steel, since the ratio of the specific heats is only 1:0.73, but the lower resistivity makes the voltage impractically low and the current intolerably high. This does not mean that copper and aluminium cannot be heated by DRH, but that it is more difficult, so that the shape of the billet is even more important.

1.5 Contacts

In this introduction, it is not proposed to discuss practical contacts in detail (see Chapter 26); it is sufficient to use the numbers derived so far as the basis for a cautionary statement.

In Example 1.1, the applied voltage was 86 V and the current 1220 A. If the contact volt-drop were approximately 1 V, this would only reduce the efficiency by about 1%. In Examples 1.2 and 1.5, however, a volt-drop of this magnitude would reduce the efficiency to unacceptable proportions – more heat being produced in the contacts than in the workpiece!

There are two major mechanical problems with the contacts:

- The contacts are pressing on a hot object. They receive heat from the workpiece, in addition to that from their own losses. Copper, or its alloys, is the most attractive material for the contacts, because of its low resistivity, but its melting point is only 1064°C which is lower than the forging temperature of steel. So the contacts have to be water-cooled. This has the detrimental effect of cooling the workpiece under the contacts.
- The contacts have to connect to workpieces that are rough-cropped, with a scaly irregular surface. Ingenious designs are available to cope with this, but it must be admitted that the complication detracts from the apparent simplicity of DRH.

Chapter 2

Alternating currents in conductors: the semi-infinite slab

2.1 The problem

The simple solutions of Chapter 1 assumed that direct current was flowing in the workpiece and that the electrodes (contacts) were arranged to allow uniform current flow throughout the cross-section.

It would be rare and fortuitous if the voltage needed for the heating duty coincided with mains voltage (see examples in Chapter 1), so a transformation is necessary and it is often more convenient to use the current in its alternating form than to rectify it to DC. Consequently it is very important to understand how AC flows in a conductor and the effect that this has on heating. These matters will be described in the chapters to follow.

Since there is no conceptual difference between current used for heating and current just flowing in a conductor, the same results are more general than DRH and apply to bus-bars, wires, etc., that is universally[109].

When AC flows in a conductor, the problem of inductance is superimposed on simple resistive flow. Its own alternating magnetic field links with the rest of the conductor to concentrate the current in the surface layers. This is the phenomenon of *skin-effect*, which has to be carefully considered when using AC. There is no way of dealing with the subject non-mathematically, so it has to be expressed in mathematical language, but we shall try to give a physical interpretation of the theory wherever possible. Since we are dealing with what is happening at every point in the conductor and skin-effect makes the electromagnetic quantities different at every depth, circuit concepts are inadequate, which is why the apparently superfluous field ideas were introduced in the first chapter. These field quantities are converted to the more-familiar circuit values at the end.

Many textbooks start with round conductors, since most conductors are round, but this immediately gets involved with Bessel functions, which are not universally understood. It is better to lay a firm foundation of understanding with simpler configurations first so that, when the Bessel functions are needed, the basic ideas only need to be extended. We are going to start with the 'impractical' mathematical abstraction of the *semi-infinite slab*, because it gives simple understandable results that can easily be extended to practical problems.

2.2 Current flow in a semi-infinite slab

What is a semi-infinite slab? It is a block that occupies the whole of space behind a plane which itself stretches to infinity in both directions, i.e. on one side of the plane there is air and on the other side there is conductor. In other words, it is a big thick flat-faced conducting sheet.

Let us bring that abstraction down to something we can handle. In cartesian coordinates, let there be current in the slab (without asking how it got there, where it comes from or where it goes to) *in only one direction*; we shall choose this arbitrarily as the z direction, parallel with the plane. If we then, equally arbitrarily, define the y direction as perpendicular to the plane, it follows that we have an x–z plane at $y = 0$. This has tagged the directions, but has still left us with infinity in all three directions, which is difficult to draw.

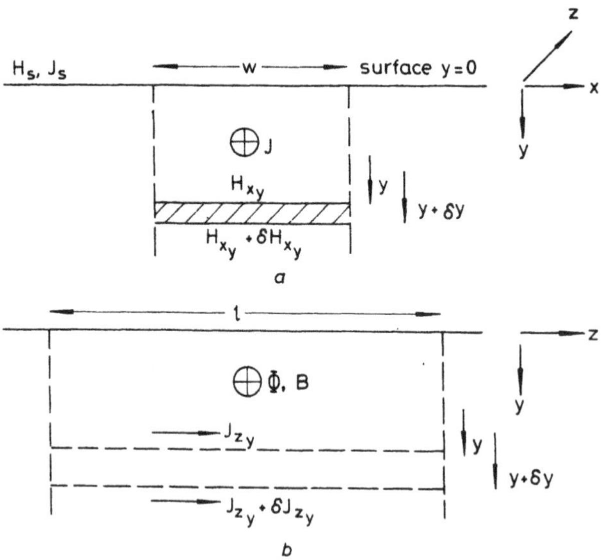

Fig. 2.1 *Current flow in semi-infinite slab, defining symbols*

From symmetry, the same thing must be happening at *any* value of x, i.e. if we take any slice Δx at *any* z, we must have the same current density distribution in the y direction. There is variation with y, but it is the same at every point in any x–z plane. We can therefore simplify the problem by choosing a width w in the x direction in the sure knowledge that what applies there will apply everywhere else.

Finally, the 'volt-drop' in the z direction must be constant, so it is legitimate to look at a finite length l. This has simplified the problem to an element w wide and l long in the x–z plane and very deep in the y direction. We shall see that w and l cancel in the equations, as might be expected, so all we have really done is reduce the problem to one that we can represent on paper (Fig. 2.1).

Current is flowing in the z direction only. Magnetic intensity H is in the x direction. Using Ampère's law $\oint H\mathrm{d}l = I$ around the shaded area (Fig. 2.1a),

$$(H_{xy})\, w + (-)\,(H_{xy} + \delta H_{xy})\, w = J_{zy}(w\,\delta y)$$

where H_{xy} means H in the x direction as a function of y. There are no components of H in the y direction, so (omitting the second subscript)

$$\frac{\mathrm{d}H_x}{\mathrm{d}y} = -J_z \tag{2.1}$$

Next, using Faraday's law $\oint E\mathrm{d}l = -(\mathrm{d}\Phi/\mathrm{d}t)$ around the dashed area (Fig. 2.1b), and remembering that $\boldsymbol{E} = \rho \boldsymbol{J}$ and $\delta\Phi = Bl\delta y$ (where Φ is magnetic flux and B is magnetic flux density),

$$\oint \rho J \mathrm{d}l = -(Bl\delta y)/\mathrm{d}t$$

By definition, we are dealing with sinusoidally-alternating quantities, at an angular frequency ω. Thus

$$B = B_x \cos \omega t = \mathrm{Re}[B_x \exp \mathrm{j}\omega t]$$

where Re means 'the real part'. Then $\mathrm{d}B/\mathrm{d}t = \mathrm{j}\omega B_x \exp \mathrm{j}\omega t$. Hence

$$-\rho(J_{zy})l + \rho(J_{zy} + \delta J_{zy})l = -\mathrm{j}\omega B_{xy} l\,\delta y$$

as the $\exp \mathrm{j}\omega t$ cancel. Thus, again omitting the second subscript,

$$\rho\frac{\mathrm{d}J_z}{\mathrm{d}y} = \mathrm{j}\omega B_x = \mathrm{j}\omega\mu H_x \tag{2.2}$$

since $B_x = \mu H_x$ (where $\mu = \mu_r\mu_0$: see Section 2.2.1).

Eliminating H from eqns 2.1 and 2.2,

$$\rho\frac{\mathrm{d}^2 J_z}{\mathrm{d}y^2} = -\mathrm{j}\omega\mu\frac{\mathrm{d}H_x}{\mathrm{d}y} = -\mathrm{j}\omega\mu(-J_z)$$

$$\frac{\mathrm{d}^2 J_z}{\mathrm{d}y^2} - \frac{\mathrm{j}\omega\mu}{\rho} J_z = 0 \tag{2.3}$$

Similarly, eliminating J,

$$\frac{\mathrm{d}^2 H_x}{\mathrm{d}y^2} = -\frac{\mathrm{d}J_z}{\mathrm{d}y} = \frac{\mathrm{j}\omega\mu}{\rho} H_x$$

Therefore

$$\frac{d^2 H_x}{dy^2} - \frac{j\omega\mu}{\rho} H_x = 0 \tag{2.4}$$

Alternatively, as $B_x = \mu H_x$,

$$\frac{d^2 B_x}{dy^2} - \frac{j\omega\mu}{\rho} B_x = 0 \tag{2.5}$$

We could have started with the diffusion equation in cartesian coordinates:

$$\nabla^2 J = \frac{\mu}{\rho}\frac{\partial J}{\partial t} = \frac{\partial^2 J}{\partial x^2} + \frac{\partial^2 J}{\partial y^2} + \frac{\partial^2 J}{\partial z^2}$$

But, as we have defined the problem, there are no J variations in x and z; so

$$\frac{\partial^2 J_z}{\partial y^2} = \frac{\mu}{\rho}\frac{\partial J}{\partial t}$$

If $J = \mathrm{Re}[J_z \exp j\omega t]$ then, as in eqn 2.3,

$$\frac{d^2 J_z}{dy^2} = \frac{j\omega\mu}{\rho} J_z$$

Equations 2.3, 2.4 and 2.5 have the same form, so the solutions are similar. Proceeding from eqn 2.3, let $\alpha^2 = \mu\omega/2\rho$. Thus

$$\frac{d^2 J_z}{dy^2} - 2j\alpha^2 J_z = 0$$

This is a straightforward differential equation of the form

$$\frac{d^2 J_z}{dy^2} - k^2 J_z = 0$$

where $k^2 = 2j\alpha^2$, and the solution is

$$J_z = A_1 \exp(ky) + A_2 \exp(-ky)$$

To find A_1 and A_2, we use the known boundary conditions. When $y = \infty$, J_z must be finite, i.e. not infinite, so $A_1 = 0$. When $y = 0$, $J_z = J_s$, so $A_2 = J_s$. Hence

$$J_z = J_s \exp[-\alpha y\sqrt{(2j)}]$$

But $\sqrt{(2j)} = 1 + j$, giving

$$J_z = J_s \exp[-(1 + j)\alpha y]$$

Thus

$$J = \mathrm{Re}[J_z \exp j\omega t] = J_s \exp(-\alpha y)\cos(\omega t - \alpha y) \tag{2.6}$$

This is a very important solution in current flow. J_s is the RMS value of the surface current-density, sinusoidally varying.

2.2.1 Skin depth

From eqn 2.6, we can make some important deductions:

- At the surface, where $y = 0$, $J_z = J_s$.
- At every other depth, the *magnitude* of the current density falls off exponentially.
- The variation is everywhere at the same frequency, given by ω.
- There is a phase-lag between the current density at the surface and at depth y, given by αy.

When $y = 1/\alpha$, the magnitude falls to $\exp(-1)$ times its surface value and the

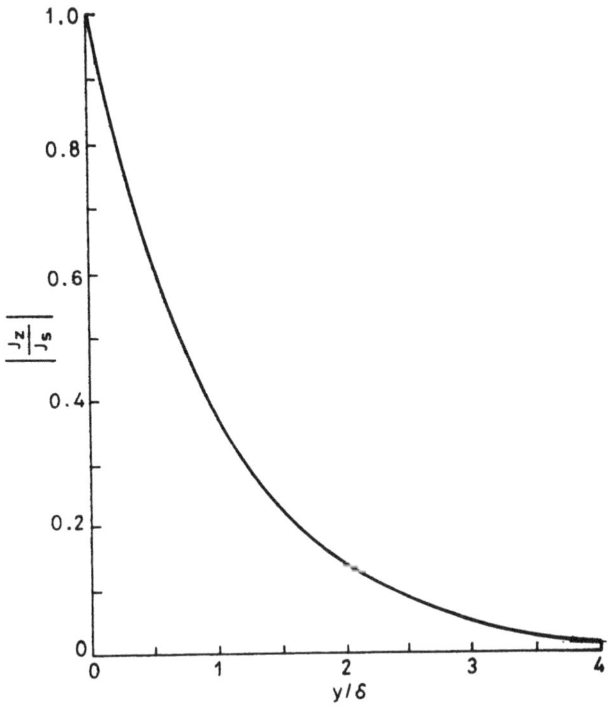

Fig. 2.2 *Variation of current density with depth, semi-infinite slab*

phase has changed by one radian (57.3°). This depth, where the magnitude is $1/e$ (= 0.368) of its value at the surface, is given the special names of *skin depth* or *depth of penetration* and the symbol δ.

It is important not to be misled by these names; they do *not* mean that beyond δ there is no more current density – for example, at twice the skin depth,

$y = 2/\alpha$, the magnitude is $J_s \exp(-2) = 0.135\ J_s$ – but only that the current density is falling to a very small value. Figure 2.2 shows the variation of J with depth. It is often useful to employ the ratio y/δ in this work.

Now α^2 was defined as $\mu\omega/2\rho$. Hence

$$\text{depth of penetration } \delta = \sqrt{(2\rho/\mu\omega)} \tag{2.7}$$

where $\mu = \mu_r \mu_0$.

Depth of penetration is proportional to $\sqrt{(\text{resistivity})}$, to $1/\sqrt{(\text{permeability})}$, and to $1/\sqrt{f}$. A useful figure to remember is that δ for warm copper at power frequencies is 10 mm, i.e. $\mu_r = 1, \rho \simeq 0.02\ \mu\Omega$ m at 50°C, $f = 50$ Hz. Similarly, for silicon iron at 50 Hz, the skin depth is 1.5 mm. A useful approximation is $\delta \simeq 500\sqrt{(\rho/\mu_r f)}$ metres to better than 1% accuracy.

This straightforward theory is satisfactory for ferromagnetic materials below the Curie point *if they are unsaturated*, but even then there is a problem about the value of μ_r to take. The currents used to heat steel are likely to be large, having substantial magnetic fields associated with them. These fields can be found from Ampère's law, $\oint H\,dl = I$. For example, using the results of Example 1.1, the surface H is $1220/\pi(0.012) = 32\,362$ A/m, which takes us a good way up the saturation curve.

This topic is dealt with in Chapter 12. Suffice it here to say that the value of μ_r is much reduced. If the value of H is known, the nomogram on Fig. 7.3 can be used to get μ_r. If H is not known at that stage in the calculation, a temporary estimated value of 20 is recommended, which can be modified later. For 0.23% carbon mild steel at 20°C, with $\rho = 0.16\ \mu\Omega$ m, this gives $\delta = 6.4$ mm at 50 Hz.

A corollary of this is that if a very shallow depth of penetration is wanted for heating thin material, then suitably low H must be used, implying slow heating.

Above the Curie point, steel is non-magnetic ($\mu_r = 1$).

2.2.2 Current

Obviously, if we add up all the current in the semi-infinite slab, the answer is infinite – a futile exercise. If, however, we take the current I flowing in a finite width w, then this is finite. So

$$
\begin{aligned}
I &= \int_0^\infty J_z\, w\, dy = \int_0^\infty J_s\, w \exp(-\alpha y - j\alpha y) dy \\
&= J_s\, w \left[\frac{\exp - \alpha y(1 + j)}{-\alpha(1 + j)} \right]_0^\infty \\
&= \frac{J_s\, w}{\alpha(1 + j)} \\
&= \frac{J_s\, w}{\alpha} \frac{1 - j}{2} \\
&= \frac{J_s\, w}{\alpha\sqrt{2}} \exp(-j\pi/4)
\end{aligned} \tag{2.8}
$$

since $\exp - j\pi/4 = \cos\pi/4 - j\sin\pi/4 = (1 - j)/\sqrt{2}$. Hence the total current lags the surface current by 45°.

2.2.3 Power loss per unit surface area

This is a very important quantity which allows us to compare heating effects. We shall see later that it is a key quantity in the equations of heat flow. The loss per square metre of surface $\mathscr{P}$ in any conductor is obtained by integrating ρJ^2 throughout the volume, over a number of cycles:

$$\mathscr{P} = \int_0^\infty \rho|J_z|^2\,dy(\times\ 1\ \times\ 1)$$

$$= \rho J_s^2 \int_0^\infty \exp(-2\alpha y)dy \qquad (2.9)$$

$$= \frac{\rho\delta}{2} J_s^2 \qquad (2.10)$$

where J_s is the RMS value of surface current-density. This loss is very unevenly distributed through the depth of the slab. The per-unit loss up to a given depth is easily obtained by taking a different upper limit in eqn 2.10, i.e.

$$\frac{\mathscr{P}_y}{\mathscr{P}} = \frac{\text{loss up to a depth } y}{\text{total loss}} = \frac{[\exp(-2\alpha y)]_0^y}{[\exp(-2\alpha y)]_0^\infty}$$

$$= \frac{\exp(-2\alpha y) - 1}{(0 - 1)} = 1 - \exp(-2\alpha y) \qquad (2.11)$$

At $y = (1/\alpha) = \delta$, $\mathscr{P}_y/\mathscr{P} = 1 - \exp(-2) = 1 - 0.135 = 0.865$.
At $y = (2/\alpha) = 2\delta$, $\mathscr{P}_y/\mathscr{P} = 1 - \exp(-4) = 1 - 0.018 = 0.982$.

This shows that most of the loss occurs within one skin depth δ and that only a minute amount occurs beyond two skin depths 2δ.

Equation 2.11 is shown graphically in Fig. 2.3. It illustrates that there is little error in saying 'nearly all of the loss occurs in about one skin depth'; this is a useful thing to remember. The obvious corollary is that all the rest of the slab is heated not directly but by conduction from this surface layer of heated material. The effect is disadvantageous when we wish to heat thick slabs, as we have to wait for the centre to heat.

Note that the symbol $\mathscr{P}$ has been used for power density throughout this book. BSS have allocated this symbol to the Poynting vector, but since loss density is of the same kind, no ambiguity is involved.

2.2.4 Effective resistance

We have

$$\text{loss/surface area} = (\text{RMS current/width})^2 \bar{R}$$

where $\bar{R}$ is the effective resistance per unit length, i.e. Ω/m. Thus

$$\frac{\rho\delta}{2} J_s^2 = \left(\frac{J_s}{\alpha\sqrt{2}}\right)^2 \bar{R}$$

$$\bar{R} = \frac{\rho\delta}{\delta^2} = \frac{\rho}{\delta} \quad \Omega/\text{m} \tag{2.12}$$

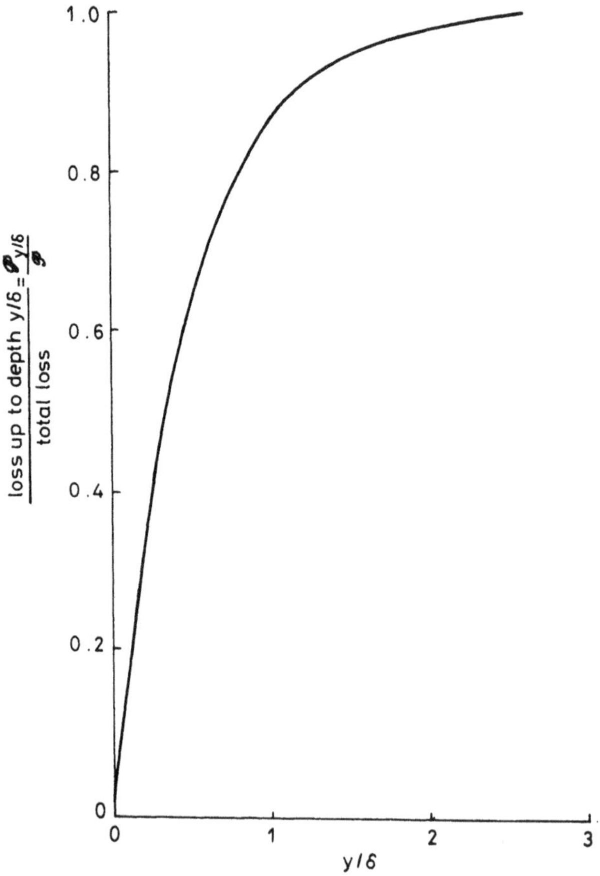

Fig. 2.3 *Loss variation with depth, semi-infinite slab*

Compare this with $\rho l/A$, where $l = 1$ and $A = (1 \times \delta)$. Therefore the energy loss is the same as if the total current were uniformly distributed in one skin depth.

2.2.5 Electric field intensity

Since $E = \rho J$, we can relate the power density $\mathscr{P}$ to RMS surface E from eqn 2.10:

$$\mathscr{P} = \frac{\rho\delta}{2} J_s^2 = \frac{\delta}{2} J_s E_s = \frac{\delta}{2} \frac{E_s^2}{\rho} \qquad (2.13)$$

This equation is much more important than eqn 2.10, since in conduction heating it is usual to work with a certain applied voltage rather than a current density. Note that the loss is now inversely proportional to ρ.

2.2.6 Surface magnetic field intensity

Using $\oint H dl = I$, we can find the surface H simply by following a path of width w at the surface, then in the y direction to infinity, back w at infinity in the x direction, and down to the starting point in the y direction. By definition, there are no components of H in the y direction and at infinity H has fallen to zero, so the only contribution to $\oint H dl$ is $H_s w$.

Equating this to the RMS current, from eqn 2.8,

$$H_s w = w \frac{J_s}{\alpha\sqrt{2}} \qquad (2.14)$$

2.2.7 Total flux

Equation 2.14 can be used to find the total flux per metre width by integration of μH. If $H_s = J_s/\alpha\sqrt{2}$ and it falls off as $\exp(-\alpha y)\cos(\omega t - \alpha y)$, then by analogy with Section 2.2.2 the total flux in width w is given by

$$\Phi = \mu \int_0^\infty H_x w \, dy = \mathrm{Re}\, \mu\, H_s w \exp(-\alpha y - j\alpha y) \exp j\omega t \, dy$$

$$= \mathrm{Re}\, \mu H_s w \exp(j\omega t) \left[\frac{\exp - \alpha y(1 + j)}{-\alpha(1 + j)} \right]_0^\infty$$

$$= \mathrm{Re} \frac{\mu H_s w \exp(j\omega t)}{\alpha(1 + j)}$$

$$= \mathrm{Re} \frac{\mu H_s w}{\alpha} \exp(j\omega t) \left(\frac{1 - j}{2} \right)$$

$$= \mathrm{Re} \frac{\mu H_s w}{\alpha\sqrt{2}} \exp(j\omega t - \pi/4)$$

since $\exp - j\pi/4 = \cos \pi/4 - j\sin \pi/4 = (1 - j)/\sqrt{2}$. Taking the real part,

$$\Phi = \frac{\mu w H_s}{\alpha\sqrt{2}} \cos\left(\omega t - \frac{\pi}{4}\right)$$

But $H_s = J_s/\alpha\sqrt{2} = E_s/\alpha\rho\sqrt{2}$. So

$$\frac{\Phi}{w} = \frac{\mu E_s}{2\rho\alpha^2} \cos\left(\omega t - \frac{\pi}{4}\right)$$

$$= \frac{E_s}{\omega} \cos\left(\omega t - \frac{\pi}{4}\right)$$

Hence the total flux lags the surface electric field intensity $\boldsymbol{E}$ by 45° and, more importantly, is completely independent of μ and ρ, the properties of the metal.

2.3 Physical interpretation

Summarising the mathematics:

- The current density $J = J_s \exp(-\alpha y) \cos(\omega t - \alpha y)$ (eqn 2.6).
- The depth of penetration $\delta = 1/\alpha = \surd(2\rho/\mu\omega)$ (eqn 2.7).
- 86.5% of the total loss occurs in one skin depth δ from the surface.
- 98.2% occurs in two skin depths (2δ) from the surface.
- The energy loss is the same as if the total current were uniformly distributed in one skin depth δ.

It is worth thinking for a moment about eqn 2.6. The cosine term shows that the phase is constantly changing relative to the surface current density and is doing so very rapidly. At $y = \delta$, the phase angle is one radian (57.3°) and, when $y = \pi\delta$, the current is in anti-phase to the surface J, i.e. it is actually flowing backwards! If heating is the only concern then this does not matter, because the phase angle has no effect on the heating, but it certainly does affect the magnitude of the total current. Fortunately the effect is small, since the current density has fallen dramatically by the time that it is in anti-phase: at $y = \pi\delta$, $\exp(-\pi) = 0.043$.

We are not used to thinking of a conductor carrying alternating current having currents flowing in the opposite direction to the voltage, so these are important concepts.

Chapter 3

Alternating currents in conductors: wide rectangular slab

3.1 Introduction

Equations 2.3, 2.4 and 2.5 were derived from first principles and are universally true in any system of cartesian coordinates. If we now move from the semi-infinite slab, with its mathematically simple solutions but abstract realisation, to a slab of finite thickness, where the current can flow in both sides at the same time, then we get slightly more complex solutions but more practical conditions. The flat bus-bar which is much wider than its thickness is a good approximation to this; although small surface currents will flow in the two narrow edges, these can be ignored in the first instance. As in Chapter 2, we need infinite width for a perfect solution, but it is often valid to consider finite widths.

In Section 2.2.1, we saw that the depth of penetration at power frequencies

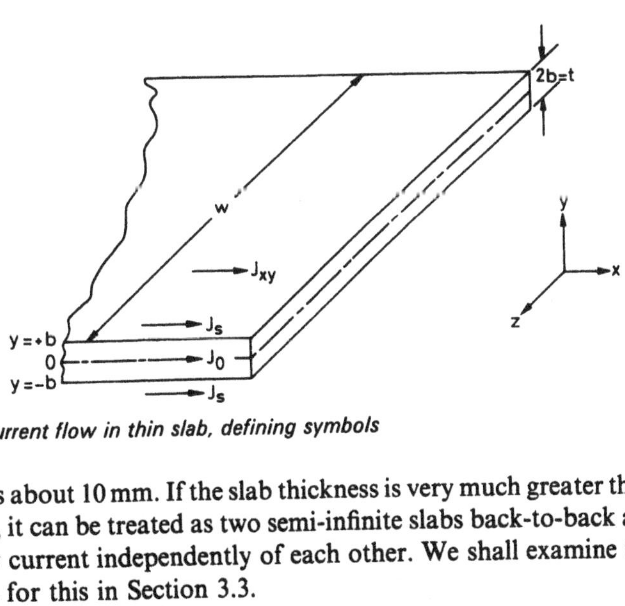

Fig. 3.1 *Current flow in thin slab, defining symbols*

in copper is about 10 mm. If the slab thickness is very much greater than this (the *thick* slab), it can be treated as two semi-infinite slabs back-to-back and the two faces carry current independently of each other. We shall examine the limiting conditions for this in Section 3.3.

Define the slab thickness t as $2b$. We are interested in the general solution of eqn 2.3 for all b. Physically, we have a superposition of two of the solutions of Chapter 2, one from each side, and would expect that the two different exponential terms will combine to give hyperbolic functions. Figure 3.1 shows the configuration, assuming finite width for simplicity of drawing, but remembering that this implies neglect of end-effects. Define the current as flowing in the x direction only. There are no components in the y and z directions. There is variation in magnitude and phase in the y direction, as before; at any y, the same current density exists for all values of z.

It is convenient to define the y axis at the centre plane of the slab, since the current density is clearly symmetrical on either side of the centre-line. The thickness t has been defined as $2b$ to avoid tedious repetition of $b/2$ in the equations.

By symmetry, the current density at the two surfaces will be the same in both magnitude and direction. Define the RMS value of the surface current-density as J_s. The x-direction tag is unnecessary, since all the current flows in the x direction.

3.2 Current-density distribution

The differential equation for J is (see Section 2.2)

$$\frac{\partial^2 J}{\partial y^2} = \frac{\mu}{\rho}\frac{\partial J}{\partial t}$$

and the general solution is of the same shape as in Section 2.2, except for the changed direction of J:

$$J_x = A_1 \exp(ky) + A_2 \exp(-ky)$$

The boundary conditions are that J is the same at $\pm b$:

$$J_{+b} = A_1 \exp(kb) + A_2 \exp(-kb) = J_s$$

$$J_{-b} = A_1 \exp(-kb) + A_2 \exp(+kb) = J_s$$

Only $A_1 = A_2 = A$ can make these equal, i.e. $J_s = 2A \cosh kb$. At $y = 0$, $J_0 = A \exp(0) + A \exp(0) = 2A$, so that $J_s = J_0 \cosh kb$, and, more generally, $J = J_0 \cosh ky$. Hence

$$\frac{J}{J_s} = \frac{\cosh ky}{\cosh kb} = \frac{\cosh (1 + \mathrm{j})\alpha y}{\cosh (1 + \mathrm{j})\alpha b} \tag{3.1}$$

It is interesting to compare the magnitude (modulus) of the current density at various depths with the surface current-density. Since they are complex, the $\cosh (1 + \mathrm{j})$ terms have to be expanded:

$$\left|\frac{J}{J_s}\right| = \left|\frac{\cosh(1+j)\alpha y}{\cosh(1+j)\alpha b}\right| = \left|\frac{\cosh \alpha y \cos \alpha y + j \sinh \alpha y \sin \alpha y}{\cosh \alpha b \cos \alpha b + j \sinh \alpha b \sin \alpha b}\right|$$

$$= \left|\frac{A + jB}{C + jD}\right| = \left|\frac{(A + jB)(C - jD)}{C^2 + D^2}\right|$$

$$= \left|\frac{(AC + BD) + j(BC - AD)}{C^2 + D^2}\right|$$

$$= \frac{\sqrt{(A^2C^2 + B^2D^2 + 2ABCD + B^2C^2 + A^2D^2 - 2ABCD)}}{C^2 + D^2}$$

$$= \sqrt{\left(\frac{A^2 + B^2}{C^2 + D^2}\right)}$$

$$= \sqrt{\left(\frac{\cosh^2\alpha y \cos^2\alpha y + \sinh^2\alpha y \sin^2\alpha y}{\cosh^2\alpha b \cos^2\alpha b + \sinh^2\alpha b \sin^2\alpha b}\right)}$$

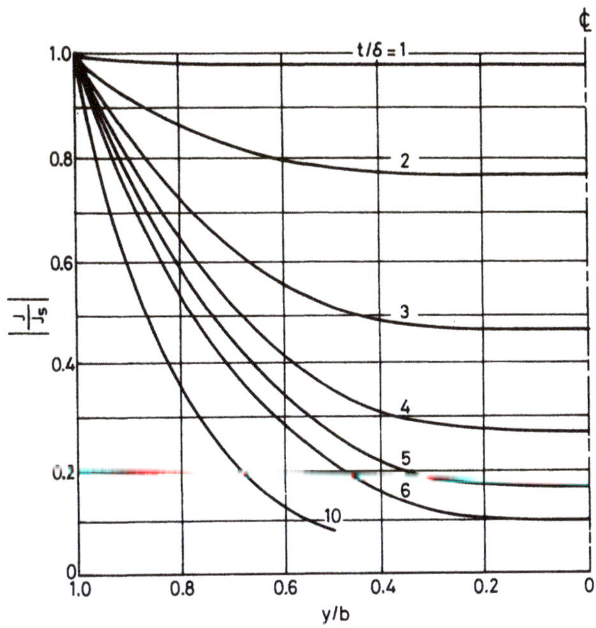

Fig. 3.2 *Fall-off of current density with depth in thin slab, for various ratios of thickness to depth of penetration, expressed as the modulus of J_x/J_s*

Using $2\cosh^2\theta = (\cosh 2\theta) + 1$, $2\sinh^2\theta = (\cosh 2\theta) - 1$, $2\cos^2\theta = \cos 2\theta + 1$, $2\sin^2\theta = 1 - \cos 2\theta$, and substituting and simplifying, we obtain

$$\left|\frac{J}{J_s}\right| = \sqrt{\left(\frac{\cosh 2\alpha y + \cos 2\alpha y}{\cosh 2\alpha b + \cos 2\alpha b}\right)} \tag{3.2}$$

$$= \sqrt{\left(\frac{\cosh 2y/\delta + \cos 2y/\delta}{\cosh 2b/\delta + \cos 2b/\delta}\right)}$$

This can be plotted for various ratios of thickness t (= $2b$) to skin depth δ (Fig. 3.2).

3.3 Total current

Let the width of the slab be w (Fig. 3.1). The total current is the integral of ($J \times$ area) in the slab:

$$\text{total current} = \int_{-b}^{b} wJ\,\mathrm{d}y = \int_{-b}^{b} \frac{wJ_s \cosh(1+\mathrm{j})\alpha y}{\cosh(1+\mathrm{j})\alpha b}\,\mathrm{d}y$$

$$= \frac{J_s w}{\cosh(1+\mathrm{j})\alpha b}\left[\frac{\sinh(1+\mathrm{j})\alpha y}{(1+\mathrm{j})\alpha}\right]_{-b}^{b}$$

$$= \frac{2J_s \delta w}{(1+\mathrm{j})} \tanh(1+\mathrm{j})\frac{b}{\delta} \qquad (3.3)$$

Comparing this with an intermediate result of Section 2.2.2, namely $J_s w/\alpha(1+\mathrm{j})$, it is double when $\tanh(1+\mathrm{j})b/\delta \rightarrow 1$, i.e. the slab behaves like two semi-infinite slabs with independent surfaces. But

$$\begin{aligned} \tanh^{-1}x &= 0.99 \quad \text{when} \quad & x &= 2.65 \\ &= 0.98 & &= 2.30 \\ &= 0.97 & &= 2.09 \\ &= 0.96 & &= 1.95 \\ &= 0.95 & &= 1.83 \end{aligned}$$

So the sides can be considered to be independent to about 1% accuracy if $b/\delta > 2.65$, and to 5% if $b/\delta > 1.83$, ignoring the $(1 + \mathrm{j})$ part of the term. For steel conductors thicker than 32 mm and copper slabs thicker than about 50 mm, it is permissible to use the simpler concepts of semi-infinite slab theory at 50 Hz.

3.4 Power density

Using the same ideas as in Section 2.2.3, remembering that the power is entering through two surfaces and that we can employ the RMS value of current density directly, the loss per square metre of surface is

$$\mathscr{P} = \int_0^b \rho|J|^2\,\mathrm{d}y(1 \times 1)$$

Using eqn 3.2,

$$\mathscr{P} = \rho J_s^2 \int_0^b \frac{\cosh 2\alpha y + \cos 2\alpha y}{\cosh 2\alpha b + \cos 2\alpha b}\,\mathrm{d}y$$

$$= \frac{\rho J_s^2}{2\alpha}\left[\frac{\sinh 2\alpha y + \sin 2\alpha y}{\cosh 2\alpha b + \cos 2\alpha b}\right]_0^b$$

$$= \frac{\rho J_s^2}{2\alpha}\left[\frac{\sinh 2\alpha b + \sin 2\alpha b}{\cosh 2\alpha b + \cos 2\alpha b}\right] \qquad (3.4)$$

For large values of $2\alpha b$, i.e. when the slab is much thicker than the depth of penetration, the bracket tends to $\tanh \infty = 1$ (since sin and cos are always negligible compared with sinh and cosh for large angles) and the value of $\mathscr{P}$ goes to its value for the semi-infinite slab (cf. eqn 2.10), as expected.

For thin slabs, eqn 3.4 can be rewritten as

$$\mathscr{P} = \rho b J_s^2 \left\{\frac{1}{2\alpha b}\left[\frac{\sinh 2\alpha b + \sin 2\alpha b}{\cosh 2\alpha b + \cos 2\alpha b}\right]\right\}$$

$$= \rho b J_s^2 \{p_r\} \qquad \text{W/m}^2 \qquad (3.5)$$

This can be arranged in terms of total workpiece power:

$$P_w = \rho J_s^2 (2wbl) p_r \qquad \text{W} \qquad (3.6)$$

Table 3.1 shows how p_r varies with $2b/\delta$, and these values are plotted in Fig. 3.3. By inspection, for $2b/\delta \geqslant 4$, p_r is increasingly accurately represented by $1/(2b/\delta)$. If we substitute this value in eqn 3.5,

$$\mathscr{P} = \rho b J_s^2 \frac{1}{2b/\delta} = \frac{\rho J_s^2 \delta}{2}$$

which agrees with eqn 2.10.

Note that p_r has been used instead of p to dissociate this expression from the similar, but different, loss function used in work on induction heating (see Chapter 8). The suffix r stands for 'resistance'.

3.4.1 *Real and reactive power density using EI*

Instead of integrating ρJ^2, as in the previous section, the power can be obtained by multiplying volts by current. Since these are not in phase, this gives components of real and reactive power density ($\mathscr{P}$ and $\mathscr{Q}$). We have

$$\text{power} = \text{volts} \times \text{total current}$$

Table 3.1

$2b/\delta$	0	0.5	1	1.5	2	2.5	3	4	5	6
p_r	1	0.998	0.968	0.860	0.678	0.499	0.373	0.249	0.197	0.166
q_r	0	0.42	0.160	0.311	0.406	0.409	0.363	0.263	0.202	0.166
$\cos\phi$	1	0.999	0.987	0.940	0.858	0.773	0.717	0.688	0.698	0.707

$$= [\text{(volts/metre)length}][\text{(total current/width)width}]$$

$$= EI(\text{length} \times \text{width})$$

But power enters from both sides, so the area is (2 × length × width). Therefore

$$\text{power density } \mathscr{P} + j\mathscr{Q} = EI/2 \quad \text{and} \quad E = \rho J_s$$

The expression for the total current I in Section 3.3 is in complex form, so the complex conjugate has to be used in the multiplication, namely EI^* rather than EI. Hence, since power enters from both sides,

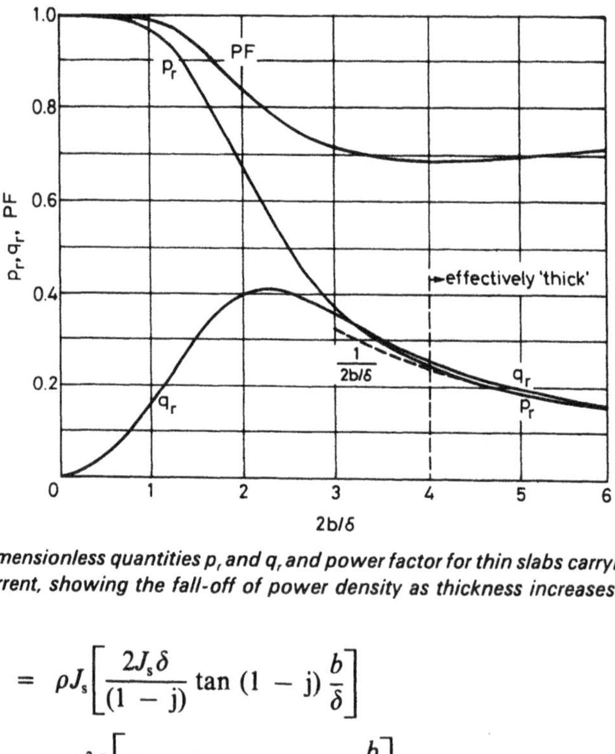

Fig. 3.3 *Dimensionless quantities p_r and q_r and power factor for thin slabs carrying alternating current, showing the fall-off of power density as thickness increases*

$$2(\mathscr{P} + j\mathscr{Q}) = \rho J_s\left[\frac{2J_s\delta}{(1 - j)} \tan (1 - j)\frac{b}{\delta}\right]$$

$$= \rho J_s^2\delta\left[(1 + j) \tanh (1 - j)\frac{b}{\delta}\right]$$

$$= \rho J_s^2\delta\left[(1 + j)\left(\frac{\sinh 2b/\delta - j \sin 2b/\delta}{\cosh 2b/\delta + \cos 2b/\delta}\right)\right] \qquad \text{(Appendix A.2)}$$

$$= \rho J_s^2\delta\left[\frac{\sinh 2b/\delta + \sin 2b/\delta}{\cosh 2b/\delta + \cos 2b/\delta} + j\frac{\sinh 2b/\delta - \sin 2b/\delta}{\cosh 2b/\delta + \cos 2b/\delta}\right]$$

Therefore

$$\mathscr{P} + j\mathscr{Q} = \rho J_s^2 b[p_r + jq_r]$$

Since we have used the RMS surface current-density J_s, this is the mean power density. Note that p_r is the same as in Section 3.4 (cf. eqn 3.5). The new quantity q_r is also given in Table 3.1 and plotted on Fig. 3.3.

3.4.2 Power in circuit terms

Equations 3.4 to 3.6 and Section 3.4 give the power density in terms of J_s. From $E = \rho J$, these expressions can be converted to applied volts by the route V → V/m → $J\,(= E/\rho)$. At the surfaces of the slab there are no flux-produced voltages, so the only component of voltage is the resistive volt-drop, given by $\rho J \times$ length, and the calculations can be done entirely from Section 3.4, with the bonus that we continue the close analogy with the treatment of induction heating, thus increasing our understanding of both (see Chapter 8).

We are more used, however, to thinking about circuit quantities: volts, amperes and ohms. Since we already have the total current from Section 3.3, it is easy to express the solution in these terms. This gives the impedance of the slab, from which we can get the resistance and reactance. Before doing so, it is important to be clear about which reactance is being calculated. In the derivations so far, we have seen that the alternating currents flowing in the slab produce alternating flux which generates voltages in the slab. These voltages are imposed on the applied voltage. The applied voltage is the same, between two y–z planes, everywhere in the thickness, i.e. for all values of y from $+b$ to $-b$, the y–z planes are equipotentials as far as the applied voltage is concerned. The $\mathrm{d}\Phi/\mathrm{d}t$ voltages are then generated and modify the applied voltage at every depth except at the surface, where $\mathrm{d}\Phi/\mathrm{d}t$ is zero. We are about to express these flux-linkages as a reactance; it is therefore an internal reactance. In addition, there will be leakage fluxes in the air outside the slab: these make an external reactance which is not part of our present work, but which must be included if the total impedance of the bar is being calculated.

From eqn 3.3, the total current in width w is

$$I = \frac{2J_s\delta w}{(1 + \mathrm{j})}\tanh(1 + \mathrm{j})b/\delta$$

We also know that the voltage in length l is given by $\rho \times$ surface $J \times$ length:

$$V = \rho J_s l$$

So the impedance of a bar w wide and l long is

$$Z = \frac{V}{I} = \frac{\rho J_s l}{2J_s\delta w}(1 + \mathrm{j})/\tanh(1 + \mathrm{j})b/\delta \qquad \text{ohms} \tag{3.7}$$

The impedance per unit length and unit width is

$$Z' = \frac{\rho(1 + \mathrm{j})}{2\delta}\bigg/\tanh(1 + \mathrm{j})b/\delta \tag{3.8}$$

where the prime reminds us that we are dealing with the property of unit surface area. Note that this is a complex quantity.

By a similar method to the derivation of eqn 3.2, the terms in $\sinh(1 + j)b/\delta$ and $\cosh(1 + j)b/\delta$ can be expanded and associated with the $1 + j$ term to give

$$Z' = R' + jX'$$

$$= \frac{\rho}{2\delta}\left[\frac{\sinh 2b/\delta + \sin 2b/\delta}{\cosh 2b/\delta - \cos 2b/\delta}\right] + j\frac{\rho}{2\delta}\left[\frac{\sinh 2b/\delta - \sin 2b/\delta}{\cosh 2b/\delta - \cos 2b/\delta}\right] \quad (3.9)$$

If the slab were carrying direct current, then $R_{dc} = \rho/2b\ \Omega$ (for 1 m long and 1 m wide). Therefore

$$\frac{R_{ac}}{R_{dc}} = \frac{1}{2}\left(\frac{2b}{\delta}\right)\left[\frac{\sinh 2b/\delta + \sin 2b/\delta}{\cosh 2b/\delta - \cos 2b/\delta}\right] \quad (3.10)$$

$$\frac{X_{ac}}{R_{dc}} = \frac{1}{2}\left(\frac{2b}{\delta}\right)\left[\frac{\sinh 2b/\delta - \sin 2b/\delta}{\cosh 2b/\delta - \cos 2b/\delta}\right] \quad (3.11)$$

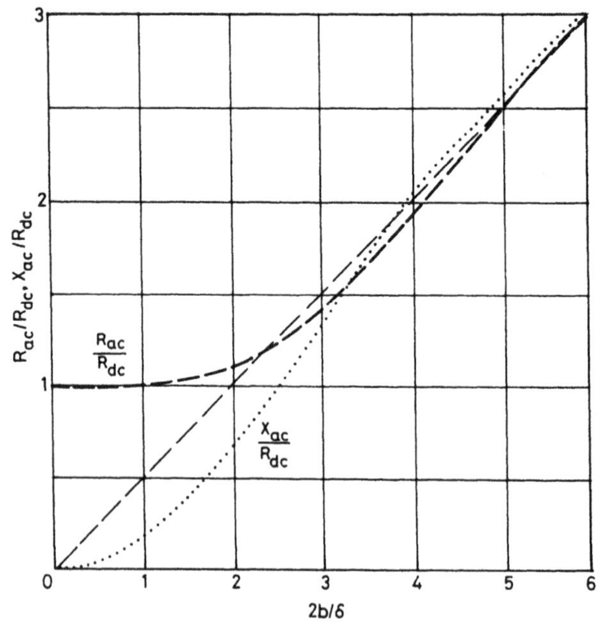

Fig. 3.4 *R_{ac}/R_{dc} and X_{ac}/R_{dc} in thin slabs for various ratios of thickness to skin depth*

When the values of the argument $2b/\delta$ are large, the sin and cos terms become negligible and both expressions tend to b/δ.

Equations 3.10 and 3.11 are plotted on Fig. 3.4. Although instructive, these curves should be used cautiously. At first sight, there is a tendency to say that, because the AC resistance is greater than the DC resistance, the loss is also greater, but it must be remembered that the total current flowing is considerably less for AC than for DC for a given E at the surface, so the loss will be less.

3.4.3 Power under AC and DC conditions

We can compare the AC and DC power, using eqn 3.5, for given applied volts, i.e. the same E and the same J_s for AC and DC:

$$\text{AC power} = \rho b J_s^2 p_r \text{ per side} = 2\rho b J_s^2 p_r$$

$$\text{DC power} = \rho J^2(\text{volume}) = \rho J^2(2b \times 1 \times 1)$$

If we make the surface conditions identical for AC and DC, i.e. $J = J_s$, then

$$\text{AC power/DC power} = p_r \tag{3.12}$$

So Fig. 3.3 shows the power fall-off, making it very useful as a measure of the price that is being paid for using AC.

3.4.4 Power factor

The power factor $\cos\phi$ of the strip carrying AC is given by $\cos[\tan^{-1}(X_{ac}/R_{ac})]$ or by $\cos[\tan^{-1}(q_r/p_r)]$. This is also given in Table 3.1 and plotted on Fig. 3.3. It is 0.9 or better when the thickness is less than about twice the depth of penetration. After that, there is a rapid fall-off towards 0.7 at 3δ; above this thickness, the power factor angle stabilises at 45° ($\cos\phi = 0.707$).

3.4.5 Limits of size – strip

The combination of the fall-off of power density from Fig. 3.3 and power factor from Fig. 3.4 makes the AC heating of strip less attractive for steel thicker than $2 \times 6.4 \approx 13$ mm and copper thicker than 20 mm at 50 Hz, although these thicknesses increase as the material heats, since $\delta \propto \sqrt{\rho}$.

3.5 Loss approximations

Equations 3.10 and 3.11 are the ones usually found in the textbooks, since most of the literature is concerned with the circuit properties of, say, bus-bars and not electroheating effects. Since the concepts of this book are equally useful for both, we shall first obtain the classical approximations for eqn 3.10, before passing to eqn 3.11 for electroheat[2,3,5].

3.5.1 Resistance

In eqn 3.10 we have

$$\frac{R_{ac}}{R_{dc}} = \frac{1}{2}\left(\frac{2b}{\delta}\right)\left[\frac{\sinh 2b/\delta + \sin 2b/\delta}{\cosh 2b/\delta - \cos 2b/\delta}\right]$$

but

$$\sinh x = x/1! + x^3/3! + x^5/5! + \ldots$$
$$\cosh x = 1 + x^2/2! + x^4/4! + \ldots$$
$$\sin x = x/1! - x^3/3! + x^5/5! - \ldots$$
$$\cos x = 1 - x^2/2! + x^4/4! - \ldots$$

Hence

$$\frac{1}{2}x\left[\frac{\sinh x + \sin x}{\cosh x - \cos x}\right] = \frac{x}{2}\left[\frac{x + x^5/5! + x^9/9! + \ldots}{x^2/2! + x^6/6! + x^{10}/10! + \ldots}\right]$$

$$\simeq \frac{1 + x^4/5! + x^8/9!}{1 + 2x^4/6! + 2x^8/10!}$$

If $x = t/\delta$ is small, we can simplify this to

$$\left(1 + \frac{x^4}{5!} + \ldots\right)\left(1 - \frac{2x^4}{6!} + \ldots\right) = 1 + \frac{x^4}{5!}\left(1 - \frac{1}{3}\right) + O(x^8)$$

$$= 1 + \frac{x^4}{180} + O(x^8)$$

Therefore, when t/δ is small,

$$\frac{R_{ac}}{R_{dc}} \simeq 1 + \frac{(t/\delta)^4}{180} + \ldots$$

At high frequencies, δ is small and t/δ becomes big. We have already seen that, for these conditions, sin and cos are small compared with sinh and cosh, so the [ ] expression in eqn 3.10 tends to tanh t/δ, which becomes unity at high t/δ. The resistance per unit area then becomes $R' \simeq \rho/2\delta$, which is independent of thickness, assuming that the thickness is large compared with δ.

3.5.2 Power density

From eqn 3.5, this is controlled by

$$p_r = \frac{1}{2\alpha b}\left[\frac{\sinh 2\alpha b + \sin 2\alpha b}{\cosh 2\alpha b + \cos 2\alpha b}\right]$$

Using the substitutions of Section 3.5.1, we have

$$\frac{1}{x}\left[\frac{\sinh x + \sin x}{\cosh x + \cos x}\right] = \frac{1}{x}\left[\frac{x + x^5/5! + x^9/9! + \ldots}{1 + x^4/4! + x^8/8! + \ldots}\right]$$

$$\simeq \frac{1 + x^4/5! + x^8/9!}{1 + x^4/4! + x^8/8!}$$

When $x = t/\delta$ is small, this becomes

$$\left(1 + \frac{x^4}{5!} + \ldots\right)\left(1 - \frac{x^4}{4!} + \ldots\right) = 1 + x^4\left(\frac{1}{5!} - \frac{1}{4!}\right) + O(x^8)$$

Therefore

$$p_r \rightarrow 1 - x^4/30$$

When t/δ becomes big, the square bracket in the expression for p_r again goes to unity, and $p_r \propto \delta/t$, as already plotted on Fig. 3.3.

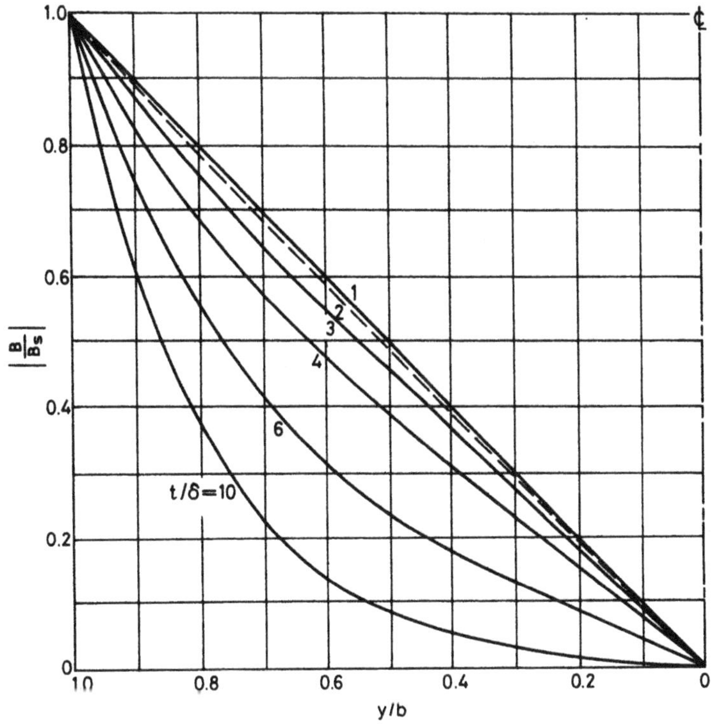

Fig. 3.5 *Flux-density distribution in thin slabs*

3.6 Flux-density distribution

Although we are usually not very interested in the magnetic conditions for conduction heating, these will be useful later. We shall not derive the equations here from first principles: in Section 8.7 the expressions are given as duals of the induction equations, so we shall simply quote them here (the reader may like to set up the appropriate boundary conditions for the diffusion equation solution: see Section 8.2). From Section 8.7:

$$\frac{B_z}{B_s} = \frac{H_z}{H_s} = \frac{\sinh(1 + j)y/\delta}{\sinh(1 + j)b/\delta} \tag{3.13}$$

$$\left|\frac{B_z}{B_s}\right| = \left|\frac{H_z}{H_s}\right| = \frac{\cosh\,(y/b)(t/\delta) - \cos\,(y/b)(t/\delta)}{\cosh\,(t/\delta) - \cos\,(t/\delta)} \tag{3.14}$$

Figure 3.5 shows eqn 3.14 plotted. For $t/\delta = 1$, this is almost a straight line and only approaches an exponential for higher values. This leads us naturally to a discussion of the meaning of 'depth of penetration' δ.

3.7 Depth of penetration

This topic is discussed more fully in Reference 1.

In Section 2.2.1 δ was defined for the semi-infinite slab, and in the present chapter it has been shown that the solutions simplify to the results of Chapter 2 for *higher* values of b/δ. However, for lower b/δ, Figs 3.2 and 3.5 tell a different story. Figure 3.2 shows that per-unit J (and hence E) becomes more uniform as b/δ gets smaller, and certainly has not reached 1/e times its surface value at depth δ. Similarly, per-unit B and H tend to straight lines (Fig. 3.5) and do not then obey the exponential law. When we come to thin tubes, we shall find the current density gets increasingly uniform.

In fact, from here on, the idealised theory of Chapter 2 is not obeyed, *except* when conditions approach those of the semi-infinite slab. It is important to read the rest of the text with these caveats in mind and to treat δ as a characteristic parameter of a conductor subject to alternating phenomena (B, J, H and E), having the dimension of length and being numerically equal to:

$$\left(\frac{2 \times \text{electrical resistivity}}{\text{angular frequency} \times \text{magnetic permeability}}\right)^{1/2}$$

This is the definition proposed by the author in Reference 1.

Chapter 4

Alternating currents in conductors: circular cross-section

This is the problem that is normally solved in the literature,[2-5] both because of its inherent importance and because it is an excellent example of the use of Bessel functions. Most of these solutions are aimed at the circuit viewpoint, i.e. resistance, reactance and impedance, rather than power loss, which is the purpose of DRH. Warren[2] has an excellent section (pp. 243ff.) on current in a circular conductor but the book has long been out-of-print; also, it uses unrationalised CGS concepts, so we shall start again from first principles. For the reader not conversant with Bessel functions, we shall not make a lot of fuss about them here: suffice it to say that they are the equivalent in a cylindrical world of the sinusoids and exponentials of the previous chapters and, like them, are described by infinite series[4], which will be quoted in the appropriate place. We shall find that there is a close similarity between the solutions for the slab and those for the cylinder. Just as, in the thin slab, we found that the two sides interacted to give hyperbolic functions, Bessel functions are needed because every point in the section of a circular conductor is affected by the currents flowing in the rest of the conductor.

4.1 Solid conductor

Current flows in the z direction only, and B and H exist only in the peripheral (θ) direction (Fig. 4.1a). Using $\oint H\mathrm{d}l = I$ on the shaded area:

$$(H_{\theta r})2\pi r + (-)(H_{\theta r} + \delta H_{\theta r})2\pi(r + \delta r) = J_{zr}(2\pi r\,\delta r)$$

Then (omitting the second subscript)

$$-H_\theta\,\delta r - \delta H_\theta(r + \delta r) = J_z \delta r\, r$$

$$-H_\theta\,\delta r - \delta H_\theta r = J_z \delta r\, r$$

since $\delta H_\theta \delta r$ is negligibly small. Thus

$$-H_\theta - r\frac{\mathrm{d}H_\theta}{\mathrm{d}r} = J_z\, r \tag{4.1}$$

As in Section 2.2 we are dealing with sinusoidal quantities, so $J = \mathrm{Re}[J_z \exp \mathrm{j}\omega t]$ and $\mathrm{d}B/\mathrm{d}t = \mathrm{j}\omega B_\theta \exp \mathrm{j}\omega t$. Taking $\oint E\mathrm{d}l = -\mathrm{d}\Phi/\mathrm{d}t$ for the hatched area of Fig. 4.1b, we have

$$-\rho(J_{zr})l + \rho(J_{zr} + \delta J_{zr})l = -\mathrm{j}\omega B_{\theta r} l \delta r$$

Thus, again omitting the second subscript,

$$\rho \frac{\mathrm{d}J_z}{\mathrm{d}r} = -\mathrm{j}\omega\mu H_\theta \tag{4.2}$$

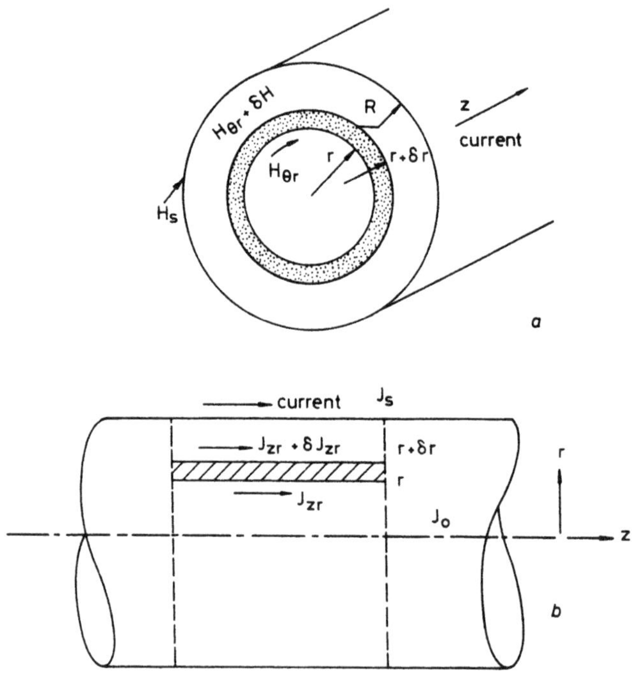

Fig. 4.1 *Current flow in a circular conductor, defining symbols*

From eqns 4.1 and 4.2,

$$\frac{\rho}{\mathrm{j}\omega\mu}\frac{\mathrm{d}J}{\mathrm{d}r} + \frac{\rho}{\mathrm{j}\omega\mu}\frac{\mathrm{d}^2 J}{\mathrm{d}r^2} r = Jr$$

$$\frac{\mathrm{d}^2 J}{\mathrm{d}r^2} + \frac{\mathrm{d}J}{\mathrm{d}r}\frac{1}{r} = 2\mathrm{j}\alpha^2 J \tag{4.3}$$

The direction of J has been omitted from the last two equations as it is now unambiguous. $\alpha^2 = \omega\mu/2\rho$ has again been substituted, as in Section 2.2.

Equation 4.3 is the cylindrical equivalent of eqn. 2.3. However, unlike Section

2.2, the equations for B and H are *not* of the same shape as that in J, because of the cylindrical form. Equation 4.3 is a Bessel equation of order zero.

4.1.1 Solutions of the Bessel equation

The solution of

$$\frac{d^2 y}{dx^2} + \frac{1}{x}\frac{dy}{dx} - k^2 y = 0 \tag{4.4}$$

is

$$y = AI_0(kx) + BK_0(kx)$$

where I_0 and K_0 are Bessel functions defined as follows:

For all x:

$$I_0(x) = 1 + \left(\frac{x}{2}\right)^2 + \frac{(x/2)^4}{2!^2} + \frac{(x/2)^6}{3!^2} + \text{etc.}$$

$$K_0(x) = I_0(x)(\ln 2 - \gamma - \ln x) + \left(\frac{x}{2}\right)^2 + \left(1 + \frac{1}{2}\right)\frac{(x/2)^4}{2!^2} + \left(1 + \frac{1}{2} + \frac{1}{3}\right)\frac{(x/2)^6}{3!^2} + \text{etc.}$$

where γ is Euler's constant, 0.5772157.

For large x:

$$I_0(x) = \frac{e^x}{\sqrt{(2\pi x)}}\left[1 + \frac{1^2}{(8x)} + \frac{1^2 \times 3^2}{2!(8x)^2} + \frac{1^2 \times 3^2 \times 5^2}{3!(8x)^3} + \text{etc.}\right]$$

$$K_0(x) = e^{-x}\sqrt{\left(\frac{\pi}{2x}\right)}\left[1 - \frac{1^2}{(8x)} + \frac{1^2 \times 3^2}{2!(8x)^2} - \frac{1^2 \times 3^2 \times 5^2}{3!(8x)^3} + \text{etc.}\right]$$

4.1.2 Functions ber, bei, ker, kei

If eqns 4.4 and 4.3 are compared, it will be seen that $k^2 = 2j\alpha^2$, i.e. both I_0 and K_0 have $\sqrt{j}$ in their arguments. It is convenient, following Lord Kelvin, to define four new functions to separate the real and imaginary parts of the Bessel functions:

$$\left.\begin{aligned} I_0(x\sqrt{j}) &= \text{ber}\, x + j\, \text{bei}\, x \\ K_0(x\sqrt{j}) &= \text{ker}\, x + j\, \text{kei}\, x \end{aligned}\right\} \tag{4.5}$$

where

ber comes from Bessel real
bei comes from Bessel imaginary
ker comes from Kelvin real
kei comes from Kelvin imaginary.

It is obvious, from the series form of $I_0(x)$ given above, that

$$\text{ber } x = 1 - \frac{(x/2)^4}{2!^2} + \frac{(x/2)^8}{4!^2} - \text{etc.}$$

$$\text{bei } x = \left(\frac{x}{2}\right)^2 - \frac{(x/2)^6}{3!^2} + \frac{(x/2)^{10}}{5!^2} - \text{etc.}$$

Full similar expressions are given by Dwight[6] for ker and kei for small x, and for all four expressions for large x.

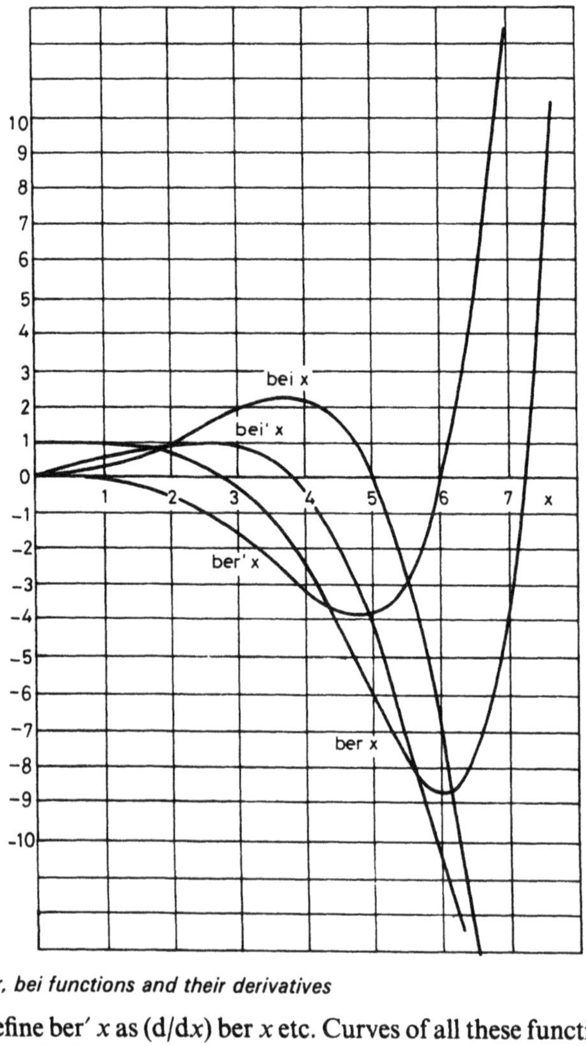

Fig. 4.2a *ber, bei functions and their derivatives*

We also define ber′ x as (d/dx) ber x etc. Curves of all these functions are given in Figs 4.2a and b. By integration of the ber, bei, ker and kei expressions, we get

$$\int x \text{ ber } kx \, dx = \frac{x}{k} \text{ bei}' \, kx$$

$$\int x \text{ bei } kx \, dx = -\frac{x}{k} \text{ber}' \, kx$$

$$\int x \text{ ker } kx \, dx = \frac{x}{k} \text{kei}' \, kx \qquad (4.6)$$

$$\int x \text{ kei } kx \, dx = -\frac{x}{k} \text{ker}' \, kx$$

These identities will be useful later.

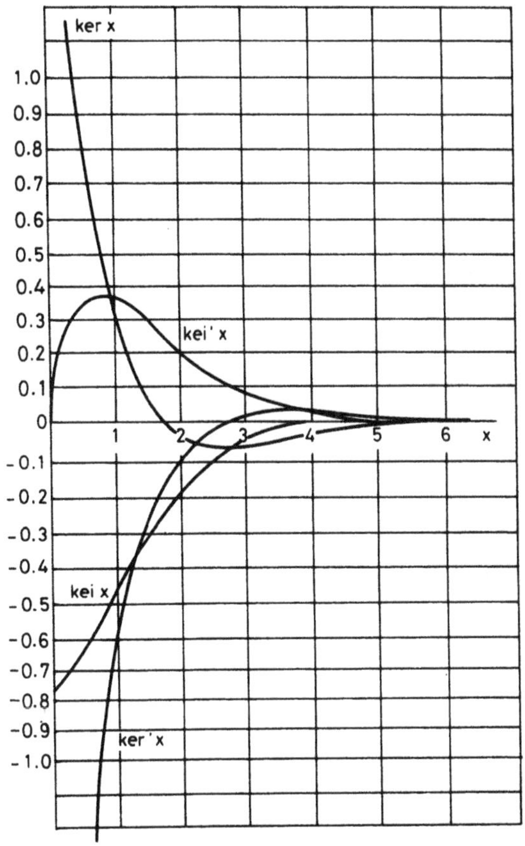

Fig. 4.2*b* *ker, kei functions and their derivatives*

4.1.3 Polar forms

Having introduced ber etc. to show that $I_0(x\sqrt{j})$ has real and imaginary parts, which electrical engineers are used to manipulating, we can combine (ber + j bei) and (ker + j kei) into polar forms. We know that $A + jB$ forms are more convenient for adding and subtracting, whereas $R \angle \theta$ forms are better to multiply and divide, since the result comes simply from $RS \angle \theta + \phi$ or $R/S \angle \theta - \phi$. In the derivations that follow, the polar form is usually simpler, but we can convert to $A + \mathrm{j}B$ forms at any time.

For ber, bei etc. there is a standard notation[4] which we shall follow:

$$\begin{aligned}
\mathrm{ber}\, x &= M_0(x) \cos \theta_0(x) \\
\mathrm{bei}\, x &= M_0(x) \sin \theta_0(x) \\
(\mathrm{ber}\, x + \mathrm{j}\, \mathrm{bei}\, x) &= M_0(x)\, [\cos \theta_0(x) + \mathrm{j} \sin \theta_0(x)] \\
&= M_0(x) \exp \mathrm{j}\, \theta_0(x)
\end{aligned}$$

Similarly

$$\begin{aligned}
\mathrm{ker}\, x &= N_0(x) \cos \phi_0(x) \text{ etc.} \\
(\mathrm{ker}\, x + \mathrm{j}\, \mathrm{kei}\, x) &= N_0(x) \exp \mathrm{j}\, \phi_0(x)
\end{aligned}$$

It follows that

$$\begin{aligned}
M_0(x) &= (\mathrm{ber}^2\, x + \mathrm{bei}^2\, x)^{1/2} \\
N_0(x) &= (\mathrm{ker}^2\, x + \mathrm{kei}^2\, x)^{1/2} \\
\theta_0 &= \tan^{-1} \frac{\mathrm{bei}\, x}{\mathrm{ber}\, x} \\
\phi_0 &= \tan^{-1} \frac{\mathrm{kei}\, x}{\mathrm{ker}\, x}
\end{aligned}$$

Curves of M_0 and N_0 are shown in Figs 4.3a and b; tables of exact values are to be found in McLachlan.[4]

For the derivative functions, the notation is as follows:

$$\begin{aligned}
\mathrm{ber}'\, x &= M_1(x) \cos \theta_1(x) \\
\mathrm{bei}'\, x &= M_1(x) \sin \theta_1(x) \\
(\mathrm{ber}'\, x + \mathrm{bei}'\, x) &= M_1(x) \exp \mathrm{j}\, \theta_1(x) \\
\mathrm{ker}'\, x &= N_1(x) \cos \phi_1(x) \text{ etc.} \\
(\mathrm{ker}'\, x + \mathrm{j}\, \mathrm{kei}'\, x) &= N_1(x) \exp \mathrm{j}\, \phi_1(x) \\
M_1(x) &= (\mathrm{ber}'^2\, x + \mathrm{bei}'^2\, x)^{1/2} \\
N_1(x) &= (\mathrm{ker}'^2\, x + \mathrm{kei}'^2\, x)^{1/2} \\
\theta_1 &= \tan^{-1} \frac{\mathrm{bei}'\, x}{\mathrm{ber}'\, x} \\
\phi_1 &= \tan^{-1} \frac{\mathrm{kei}'\, x}{\mathrm{ker}'\, x}
\end{aligned}$$

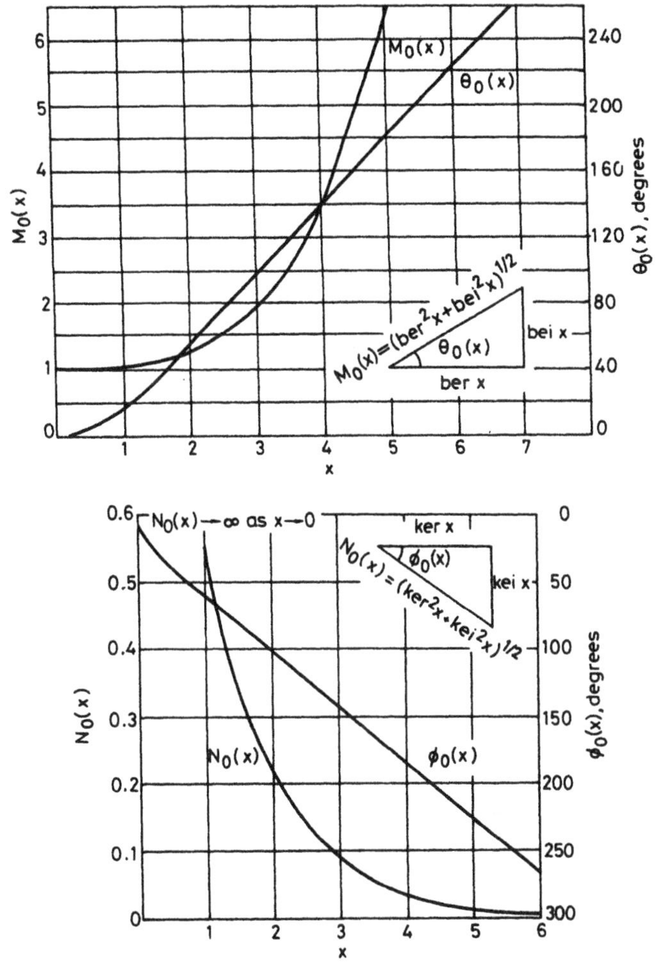

Fig. 4.3 *Function curves and vector representations for (a) $M_0(x)$ and $\theta_0(x)$ (b) $N_0(x)$ and $\phi_0(x)$ (after McLachlan[4])*

4.2 Current-density distribution and total current in a solid conductor

From eqns 4.3 and 4.4, the current-density distribution is given by

$$J = AI_0[\sqrt{(2j)}\alpha r] + BK_0[\sqrt{(2j)}\alpha r] \tag{4.7}$$

The current density must be finite at $r = 0$; from Section 4.1.1, for small x, $K_0(0) = \infty$ since $\ln 0 = \infty$, so B must be zero. Thus

$$\begin{aligned} J &= AI_0[\sqrt{(2j)}\alpha r] \\ &= A[\text{ber}\,\sqrt{(2)}\alpha r + \text{j bei}\,\sqrt{(2)}\alpha r] \\ &= AM_0(mr)\exp \text{j}\theta_0(mr) \end{aligned}$$

writing $m = \sqrt{(2)}\alpha$ for simplicity.

At $r = 0$, let $J = J_0$. From Sections 4.1.2 and 4.1.3, $M_0(0) = 1$. Therefore $J_0 = A$ and so

$$J = J_0 M_0 (mr) \exp j\theta_0 (mr)$$

Let the surface current-density be $J_s = J_0 M_0 (mR) \exp j\theta_0(mR)$(see Fig. 4.1). Then

$$\frac{J}{J_s} = \frac{M_0 (mr) \exp j\theta_0 (mr)}{M_0 (mR) \exp j\theta_0 (mR)} \tag{4.8}$$

At the surface, the current density is in phase with the applied voltage and has a known value, given by E/ρ, so the denominator is completely defined; hence the magnitude and phase at every other radius is known.

Having formally defined our Bessel functions in their polar form, it will make writing and understanding easier if we change to the simpler notation of

$$\frac{J}{J_s} = \frac{M_0 (mr)}{M_0 (mR)} \frac{\angle \theta_0 (mr)}{\angle \theta_0 (mR)} \tag{4.9}$$

which engineers are used to handling. From eqn 4.9 we can take the modulus of the ratio:

$$\left|\frac{J}{J_s}\right| = \frac{M_0 (mr)}{M_0 (mR)} \tag{4.10}$$

This is plotted on Fig. 4.4. It is similar in form to Fig. 3.2. Tables of Bessel functions ($x\sqrt{2}$) are given in Appendix A.3 for convenience of calculation, together with values of M_0, M_1, N_0 and N_1.

The phase angle relative to the surface is given by $\angle \theta_0 (mr) - \angle \theta_0 (mR)$. Fig. 4.3a shows that $\theta_0 (x)$ is a very good straight line, except very near the origin. From this, it follows that the phase change with radius, measured from the surface, is linear except at the centre of the wire.

The total current I flowing in the wire is obtained by integrating $J\,\mathrm{d}A$ from $r = 0$ to $r = R$:

$$I = \int_0^R 2\pi r J \,\mathrm{d}r$$

$$= 2\pi J_s \int_0^R r \frac{(\mathrm{ber}\, mr + \mathrm{bei}\, mr)}{(\mathrm{ber}\, mR + \mathrm{j\, bei}\, mR)} \mathrm{d}r$$

$$= 2\pi J_s \frac{\delta R}{\sqrt{2}} \frac{(\mathrm{bei}'\, mR - \mathrm{j\, ber}'\, mR)}{(\mathrm{ber}\, mR + \mathrm{j\, bei}\, mR)} \tag{4.11}$$

When mR is large, i.e. $R \gg \delta$, due either to large R or to high frequency producing small δ, then

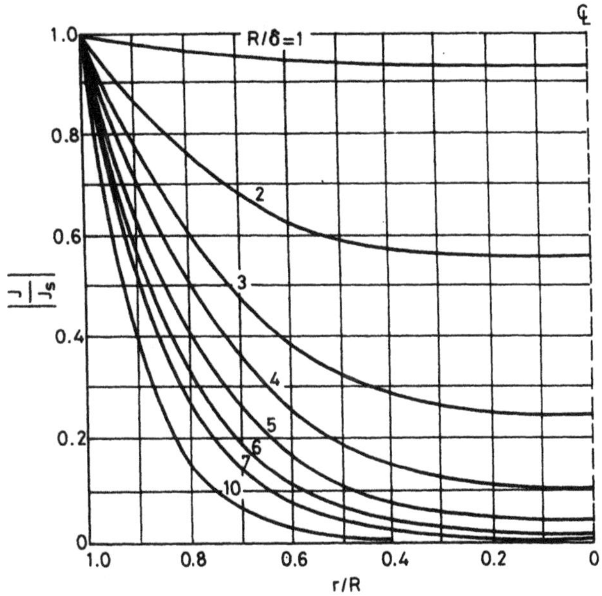

Fig. 4.4 *Fall-off of current density with radius in solid cylinder, for various ratios of radius to depth of penetration, expressed as the modulus of J/J_s (given by $M_0(mr)/M_0(mr)$)*

$$\text{ber}'\, mR \rightarrow \frac{\exp mR/\sqrt{2}}{\sqrt{(2\pi mR)}} \cos\left(\frac{mR}{\sqrt{2}} + \frac{\pi}{8}\right)$$

$$\text{bei}'\, mR \rightarrow \frac{\exp mR/\sqrt{2}}{\sqrt{(2\pi mR)}} \sin\left(\frac{mR}{\sqrt{2}} + \frac{\pi}{8}\right)$$

$$\text{ber}\, mR \rightarrow \frac{\exp mR/\sqrt{2}}{\sqrt{(2\pi mR)}} \cos\left(\frac{mR}{\sqrt{2}} - \frac{\pi}{8}\right)$$

$$\text{bei}\, mR \rightarrow \frac{\exp mR/\sqrt{2}}{\sqrt{(2\pi mR)}} \sin\left(\frac{mR}{\sqrt{2}} - \frac{\pi}{8}\right)$$

Equation 4.11 then reduces to

$$I = 2\pi R\delta \frac{J_s}{\sqrt{2}} \exp(-j\pi/4)$$

If we compare this with the results of Section 2.2.2, and substitute the periphery $2\pi R$ for w, the result is identical, i.e. the circular problem reduces to the same answer as the semi-infinite block.

4.2.1 Phase angle and power factor

From eqn 4.9 we saw that the phase angle between the current at any radius r and the surface current is $\angle\theta_0(mr) - \angle\theta_0(mR)$. The phase angle between the total current and the surface current (which is obviously in phase with the surface current-density J_s) is the angle of I/J_s:

$$\angle I/J_s = \angle -j\left[\frac{\text{ber}'\, mR + j\text{bei}'\, mR}{\text{ber}\, mR + j\text{bei}\, mR}\right] \tag{4.12}$$

$$= \angle -j[\angle\theta_1(mR) - \angle\theta_0(mR)] = \theta_T$$

So the power factor (PF) angle is

$$\cos\theta_T = \cos[\theta_1(mR) - \theta_0(mR) - 90°]$$

the 90° coming from $-j$. For large mR, $(\theta_1 - \theta_0) \to 45°$, so the PF $\to 0.707$. Other power factors are obtained from Section 4.3.3 for all values of mR.

4.3 Power

In Section 3.4, we obtained the power by two methods: we integrated ρJ^2 through the volume, giving the real power; and we multiplied voltage and current, which gave both real and reactive power. Here we shall use the latter, since it gives more information, and we shall leave the other as an exercise.

The electric field $E = \rho J_s$ V/m, and the total current I is given by eqn 4.11. As indicated in Section 3.4.1, the complex conjugate I^* must be used. Hence the power per metre is given by

$$EI^* = (\rho J_s)\left[2\pi J_s \frac{R\delta}{\sqrt{2}} \frac{(\text{bei}'\, mR + j\,\text{ber}'\, mR)}{(\text{ber}\, mR - j\,\text{bei}\, mR)}\right]$$

$$= \left(2\pi\rho J_s^2 \frac{R\delta}{\sqrt{2}}\right)\left[\frac{(\text{bei}'\, mR + j\,\text{ber}'\, mR)(\text{ber}\, mR + j\,\text{bei}\, mR)}{\text{ber}^2\, mR + \text{bei}^2\, mR}\right]$$

$$= \left(2\pi\rho J_s^2 \frac{R\delta}{\sqrt{2}}\right)\left[\frac{(\text{ber}\, mR\,\text{bei}'\, mR - \text{bei}\, mR\,\text{ber}'\, mR) + j(\text{ber}\, mR\,\text{ber}'\, mR + \text{bei}\, mR\,\text{bei}'\, mR)}{\text{ber}^2\, mR + \text{bei}^2\, mR}\right] \tag{4.13}$$

The power has real and imaginary parts, since the total current is not in phase with the voltage, so the expression can be split into two separate parts – the real power and the reactive power. Equation 4.13 gives correct values for both, if RMS current density is used. The ratio of the two gives the tangent of the phase angle.

4.3.1 Real power

To distinguish real power from $\mathcal{P}$ and P_v we shall call it P_w, where w stands for workpiece. Since eqn 4.13 is for unit length, we must incorporate the length l to get the real power:

$$P_w = (2\pi\rho J_s^2 l)\left(\frac{R\delta}{\sqrt{2}}\right)\left[\frac{\text{ber } mR \text{ bei}' mR - \text{bei } mR \text{ ber}' mR}{\text{ber}^2 mR + \text{bei}^2 mR}\right] \quad (4.14)$$

$$= (\pi\rho J_s^2 \, l \, R^2)\left\{\frac{2}{mR}[\quad]\right\}$$

$$= (\pi\rho J_s^2 \, l \, R^2)\,\{p_r\} \quad \text{W} \quad (4.15)$$

Function p_r is plotted on Fig. 4.5 in terms of the ratio of workpiece diameter $d = 2R$ to depth of penetration.

As in Section 3.4.3, we can compare AC and DC power for given applied volts:

$$\text{AC power} = \pi\rho J_s^2 \, l \, R^2 \, p_r$$

$$\text{DC power} = \rho J^2 \text{ (volume)} = \rho J^2 \, l \, \pi R^2$$

If $J = J_s$ as before (Section 3.4.3),

$$\text{AC power/DC power} = p_r$$

So Fig. 4.5 shows the fall-off in power due to using AC.

Example 4.1
Use ρJ^2, analogously to Section 3.4, to obtain the real power.

Solution
We have

$$P_w = l \int_0^R \rho |J|^2 \, 2\pi r \, dr$$

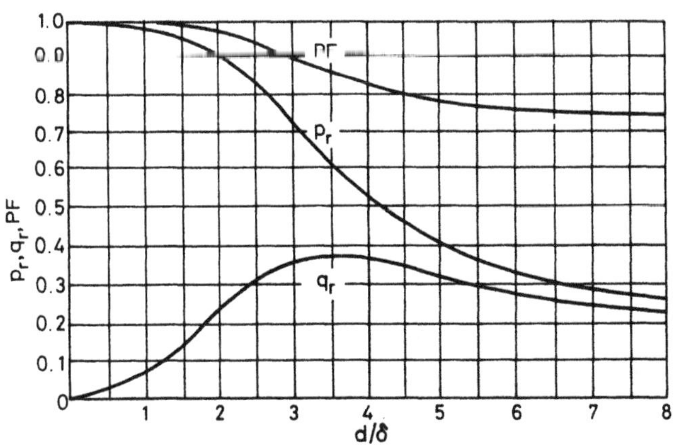

Fig. 4.5 *p_r, q_r and power factor for solid cylinder*

But from eqn 4.8,

$$\left|\frac{J}{J_s}\right|^2 = \frac{\text{ber}^2\, mr + \text{bei}^2\, mr}{\text{ber}^2\, mR + \text{bei}^2\, mR}$$

Therefore

$$P_w = \frac{2l\rho\pi J_s^2}{\text{ber}^2\, mR + \text{bei}^2\, mR}\int_0^R r\,(\text{ber}^2\, mr + \text{bei}^2\, mr)\text{d}r$$

But

$$\int x\,(\text{ber}^2\, x + \text{bei}^2\, x)\,\text{d}x = x(\text{ber}\, x\ \text{bei}'\, x - \text{bei}\, x\ \text{ber}'\, x)$$

So

$$\int r\,(\text{ber}^2\, mr + \text{bei}^2\, mr)\,\text{d}r = \frac{r}{m}(\text{ber}\, mr\ \text{bei}'\, mr - \text{bei}\, mr\ \text{ber}'\, mr)$$

Therefore

$$P_w = 2l\,\rho\pi J_s^2\,\frac{R}{m}\left[\frac{\text{ber}\, mR\ \text{bei}'\, mR - \text{bei}\, mR\ \text{ber}'\, mR}{\text{ber}^2\, mR + \text{bei}^2\, mR}\right]$$

$$= \pi l\,\rho\, J_s^2\, R^2\left\{\frac{2}{mR}[\quad]\right\}$$

$$= \pi\rho J_s^2\, R^2\, l\,\{p_r\} \qquad \text{W}$$

4.3.2 Reactive power

By similar arguments,

$$Q_w = (\pi\rho J_s^2\, l\, R^2)\left\{\frac{2}{mR}\left[\frac{\text{ber}\, mR\ \text{ber}'\, mR + \text{bei}\, mR\ \text{bei}'\, mR}{\text{ber}^2\, mR + \text{bei}^2\, mR}\right]\right\} \qquad (4.16)$$

$$= (\pi\rho J_s^2\, l\, R^2)\,\{q_r\} \qquad \text{var} \qquad (4.17)$$

The function q_r is plotted on Fig. 4.5.

4.3.3 Power factor, using p_r and q_r

Power factor was not plotted earlier since, having obtained p_r and q_r, it is now simpler to take $\cos\phi = \cos[\tan^{-1}(q_r/p_r)]$. This is also plotted on Fig. 4.5. The PF is 0.9 or better for $d/\delta < 3$, after which it falls off to 0.76 at $d/\delta = 6$. Thereafter, it falls slowly towards its ultimate value of 0.707. From the approximation given, for $d/\delta > 8$,

$$\cos\phi = \cos\left[\tan^{-1}\left(\frac{d/\delta}{1.23 + d/\delta}\right)\right]$$

This formula gives the following values for higher d/δ:

d/δ	10	20	30	100	1000
$\cos\phi$	0.747	0.728	0.721	0.711	0.7075

4.4 Circuit quantities

We follow the method of Section 3.4.2. First, from eqn 4.11,

$$I = 2\pi J_s \frac{R\delta}{\sqrt{2}}\left[\frac{\text{bei}'\, mR - \text{j ber}'\, mR}{\text{ber}\, mR + \text{j bei}\, mR}\right]$$

We also have $V = \rho J_s l$. Therefore the impedance is given by

$$Z = \frac{V}{I} = \frac{\rho J_s\, l}{2\pi J_s}\frac{\sqrt{2}}{R\delta}\left[\frac{\text{ber}\, mR + \text{j bei}\, mR}{\text{bei}'\, mR - \text{j ber}'\, mR}\right] \quad \text{ohms} \tag{4.18}$$

But $R_{dc} = \rho l/\pi R^2$. Therefore, remembering $m = \sqrt{(2)}/\delta$,

$$\frac{Z}{R_{dc}} = \frac{mR}{2}\left[\frac{\text{ber}\, mR + \text{j bei}\, mR}{\text{bei}'\, mR - \text{j ber}'\, mR}\right] \tag{4.19}$$

As before, we must separate [ ] into real and imaginary parts, by rationalisation:

$$\frac{Z}{R_{dc}} = \frac{mR}{2}\left[\frac{(\text{ber}\, mR\, \text{bei}'\, mR - \text{bei}\, mR\, \text{ber}'\, mR) + \text{j}(\text{bei}\, mR\, \text{bei}'\, mR + \text{ber}\, mR\, \text{ber}'\, mR)}{\text{bei}'^2\, mR + \text{ber}'^2\, mR}\right] \tag{4.20}$$

$$\frac{R_{ac}}{R_{dc}} = \frac{mR}{2}\left[\frac{\text{ber}\, mR\, \text{bei}'\, mR - \text{bei}\, mR\, \text{ber}'\, mR}{\text{bei}'^2\, mR + \text{ber}'^2\, mR}\right] \tag{4.21}$$

$$\frac{X_{ac}}{R_{dc}} = \frac{mR}{2}\left[\frac{\text{ber}\, mR\, \text{ber}'\, mR + \text{bei}\, mR\, \text{bei}'\, mR}{\text{bei}'^2\, mR + \text{ber}'^2\, mR}\right] \tag{4.22}$$

Equations 4.21 and 4.22 are plotted on Fig. 4.6. If we compare this figure with the similar data on Fig. 3.4, we see that both X_{ac}/R_{dc} curves tend quickly to linear values, given by b/δ on Fig. 3.4 and $R/2\delta$ on Fig. 4.6. However, whereas the R_{ac}/R_{dc} ratio also goes quickly to b/δ in Fig. 3.4, it goes to $(R/2\delta) + (1/4)$ in the cylinder. This explains why the power factor approaches its asymptotic value of 0.707 so slowly in Section 4.3.3. $(R/2\delta) + (1/4)$ is only a good approximation, the actual value of the constant being about 0.27 for these values of R/δ.

For large values of R/δ, a close approximation can be made by taking the AC resistance as that of a tube δ thick and of radius R. Therefore

$$\frac{R_{ac}}{R_{dc}} = \frac{\rho l/2\pi R\delta}{\rho l/\pi R^2} = \frac{R}{2\delta}$$

which is the actual value when $R/2\delta \gg 1/4$.

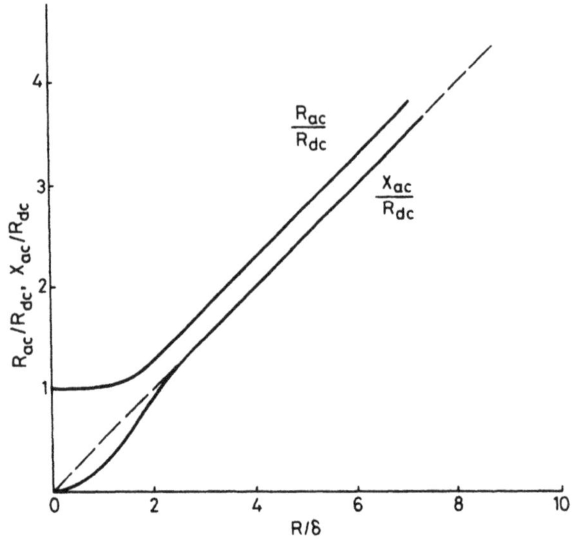

Fig. 4.6 *R_{ac}/R_{dc} and X_{ac}/R_{dc} in solid cylinder for various ratios of radius to skin depth*

4.5 Loss approximations

4.5.1 Resistance

From eqn 4.21,

$$\frac{R_{ac}}{R_{dc}} = \frac{mR}{2}\left[\frac{\text{ber}\, mR\ \text{bei}'\, mR - \text{bei}\, mR\ \text{ber}'\, mR}{\text{bei}'^2\, mR + \text{ber}'^2\, mR}\right]$$

However, for small x, ignoring $(x/2)^9$ and greater,

$$\text{ber}\, x = 1 - \frac{(x/2)^4}{(2!)^2} + \frac{(x/2)^8}{(4!)^2} \qquad \text{ber}'\, x = -\frac{2(x/2)^3}{(2!)^2} + \frac{4(x/2)^7}{(4!)^2}$$

$$\text{bei}\, x = (x/2)^2 - \frac{(x/2)^6}{(3!)^2} \qquad \text{bei}'\, x = \frac{x}{2} - \frac{3(x/2)^5}{(3!)^2}$$

Therefore, using $y = mR/2$,

$$\frac{R_{ac}}{R_{dc}} = y\left[\frac{(1 - y^4/4)\,(y - y^5/12) - (y^2 - y^6/36)\,(-\,y^3/2 + y^7/144)}{(y - y^5/12)^2 + (y^3/2 - y^7/144)^2}\right]$$

$$= \left[\frac{(1 - y^4/4)\,(1 - y^4/12) + (y - y^5/36)\,(y^3/2 - y^7/144)}{(1 - y^4/12)^2 + (y^2/2 - y^6/144)^2}\right]$$

If we ignore anything greater than y^4,

$$\frac{R_{ac}}{R_{dc}} \simeq \frac{1 - y^4\,(1/4 + 1/12) + (y^4/2)}{1 - y^4/6 + y^4/4} \simeq \frac{1 + y^4/6}{1 + y^4/12}$$

$$\simeq \left(1 + \frac{y^4}{6}\right)\left(1 - \frac{y^4}{12}\right) \simeq 1 + \frac{y^4}{12}$$

But $y = mR/2 = \surd(2)\alpha R/2$, and so $y^4 = \alpha^4 R^4/4$. Therefore

$$\frac{R_{ac}}{R_{dc}} \simeq 1 + \frac{1}{48}\left(\frac{R}{\delta}\right)^4 \tag{4.23}$$

This is the well-known approximation for thin conductors when mR is small.

For copper conductors at 20°C, using $\delta = \surd(2\rho/\mu\omega)$,

$$\delta = \surd[2(0.017 \times 10^{-6})/(4\pi \times 10^{-7})2\pi f]$$

$$= 0.0656/\surd f \quad \text{m}$$

Therefore

$$\frac{R_{ac}}{R_{dc}} \simeq 1 + 1124\, R^4 f^2 \ldots$$

For $f = 50$ Hz and $R = 10$ mm, $R_{ac}/R_{dc} = 1.028$. That is, for a conductor with radius approximately equal to the depth of penetration, there is only a small increase in resistance.

4.5.2 Resistance at high frequencies

At high freque icies, δ is small, and α and mR are large (assuming that R is not small). For large x, using Dwight,[6] it can be shown that eqn 4.21 becomes

$$\frac{R_{ac}}{R_{dc}} = \frac{R}{2\delta} + \frac{1}{4} + \frac{3}{32(R/\delta)} \ldots$$

$$= \frac{R\surd f}{0.1312} + 0.25 + \frac{0.00615}{R\surd f} \tag{4.24}$$

For example, for $f = 1$ MHz and $R = 1$ mm, $R\surd f = 1$. Therefore

$$\frac{R_{ac}}{R_{dc}} = \frac{1}{0.1312} + 0.25 + \frac{0.006\,15}{1} = 7.88$$

That is, the resistance of this small-diameter wire is greatly increased at 1 MHz.

4.5.3 Power

From eqns 4.14 and 4.15, for the cylinder,

$$p_r = \frac{2}{x}\left[\frac{\text{ber } x \text{ bei}' x - \text{bei } x \text{ ber}' x}{\text{ber}^2 x + \text{bei}^2 x}\right]$$

where $x = mR$. Using the approximations of the previous sections and $y = x/2$, as before, for small x we have

$$p_r = \frac{1}{y}\left[\frac{\text{ber } 2y \text{ bei}' \, 2y - \text{bei } 2y \text{ ber}' \, 2y}{(1 - y^4/4)^2 + (y^2 + y^6/36)^2}\right]$$

$$\simeq \left[\frac{(1 - y^4/4)\,(1 - y^4/12) + (y^3/3)}{1 - y^4/2 + y^4}\right] \qquad \text{ignoring} > y^4$$

$$\simeq \frac{1 + y^4\,(1/3 - 1/4 - 1/12)}{1 + y^4/2} \simeq 1 - \frac{y^4}{2}$$

Therefore

$$p_r \simeq 1 - \frac{1}{8}\left(\frac{R}{\delta}\right)^4$$

for small values of R/δ. This is not a very useful approximation as, over the range where it is valid, p_r is so close to unity that little error is caused by using 1.

Chapter 5

Hollow conductors: tubes

5.1 Introduction

It is obviously easier to heat round tubes (or rectangular tubes) by DRH than the equivalent solid section. If the tube wall is thin enough, regardless of diameter, the current will occupy the cross-section substantially uniformly and the problem becomes simple. In this chapter we shall examine the equations already derived, but with new boundary conditions, to try to arrive at the rules for 'thick' and 'thin' tubes[4,98–100].

5.2 Equations for tubes

We can start with eqn 4.3:

$$\frac{d^2J}{dr^2} + \frac{dJ}{dr}\frac{1}{r} = 2j\alpha^2 J$$

As before, the solution is eqn 4.7:

$$J = AI_0\,[\sqrt{(2j)}\alpha r] + BK_0\,[\sqrt{(2j)}\alpha r]$$

However, the constant B will no longer disappear at $r = 0$ and the solution will contain both ber and ker, bei and kei functions, i.e.

$$J = A(\text{ber}\,mr + \text{j bei}\,mr) + B(\text{ker}\,mr + \text{j kei}\,mr) \tag{5.1}$$

where $m = \sqrt{(2)}\alpha$, as before. The surface current density J_s is known from ρ and the applied E, so we can get one relation between the two unknowns, A and B:

$$J_s = A(\text{ber}\,mR + \text{j bei}\,mR) + B(\text{ker}\,mR + j\,\text{kei}\,mR) \tag{5.2}$$

To find another relation between A and B, we use the known fact that, at the inner radius b, a path around the inner surface encloses no current. Thus $\oint H dl = 0$, and so $H = 0$, $B = 0$, curl E = curl $\rho J = 0$ and curl $J = 0$. However in polar form,

$$\text{curl } J = \begin{vmatrix} a_r/r & a_\psi & a_z/r \\ \partial/\partial r & \partial/\partial\psi & \partial/\partial z \\ J_r & rJ_\psi & J_z \end{vmatrix}$$

For our problem, there are no components of J_r and J_ψ; nor are there rates-of-change in the ψ and z directions. Hence

$$\text{curl } J = \begin{vmatrix} a_r/r & a_\psi & a_z/r \\ \partial/\partial r & 0 & 0 \\ 0 & 0 & J_z \end{vmatrix}$$

So curl $J = \partial J_z/\partial r = 0$ at $r = b$ is our second condition. From eqn 5.1,

$$\frac{\mathrm{d}J}{\mathrm{d}r} = mA(\text{ber}' \, mr + \text{j bei}' \, mr) + mB \, (\text{ker}' \, mr + \text{j kei}' mr) \tag{5.3}$$

Equating to zero when $r = b$ gives

$$\frac{B}{A} = -\frac{\text{ber}' mb + \text{j bei}' \, mb}{\text{ker}' \, mb + \text{j kei}' \, mb} \tag{5.4}$$

This relationship occurs frequently in this chapter: it is a constant for a given value of b.

5.3 Current-density distribution

Combining eqns 5.1 and 5.2, we have

$$\frac{J}{J_s} = \frac{(A/B) \, (\text{ber} \, mr + \text{j bei} \, mr) + (\text{ker} \, mr + \text{j kei} \, mr)}{(A/B) \, (\text{ber} \, mR + \text{j bei} \, mR) + (\text{ker} \, mR + \text{j kei} \, mR)}$$

This can be rearranged so that the first term is J/J_s for the solid:

$$\frac{J}{J_s} = \frac{(\text{ber} \, mr + \text{j bei} \, mr)}{(\text{ber} \, mR + \text{j bei} \, mR)} \times \left[\frac{1 + (B/A)(\text{ker} \, mr + \text{j kei} \, mr)/(\text{ber} \, mr + \text{j bei} \, mr)}{1 + (B/A)(\text{ker} \, mR + \text{j kei} \, mR)/(\text{ber} \, mR + \text{j bei} \, mR)} \right] \tag{5.5}$$

The [ ] is the tube effect. After manipulation, eqn 5.5 has the form $(C + \text{j}D)/(E + \text{j}F)$, the modulus of which is $\sqrt{\{(C^2 + D^2)/(E^2 + F^2)\}}$. This modulus is the ratio of the absolute value of current density at any radius to its value at the surface. It is plotted on Figs 5.1a to h. Only the active section is shown (there

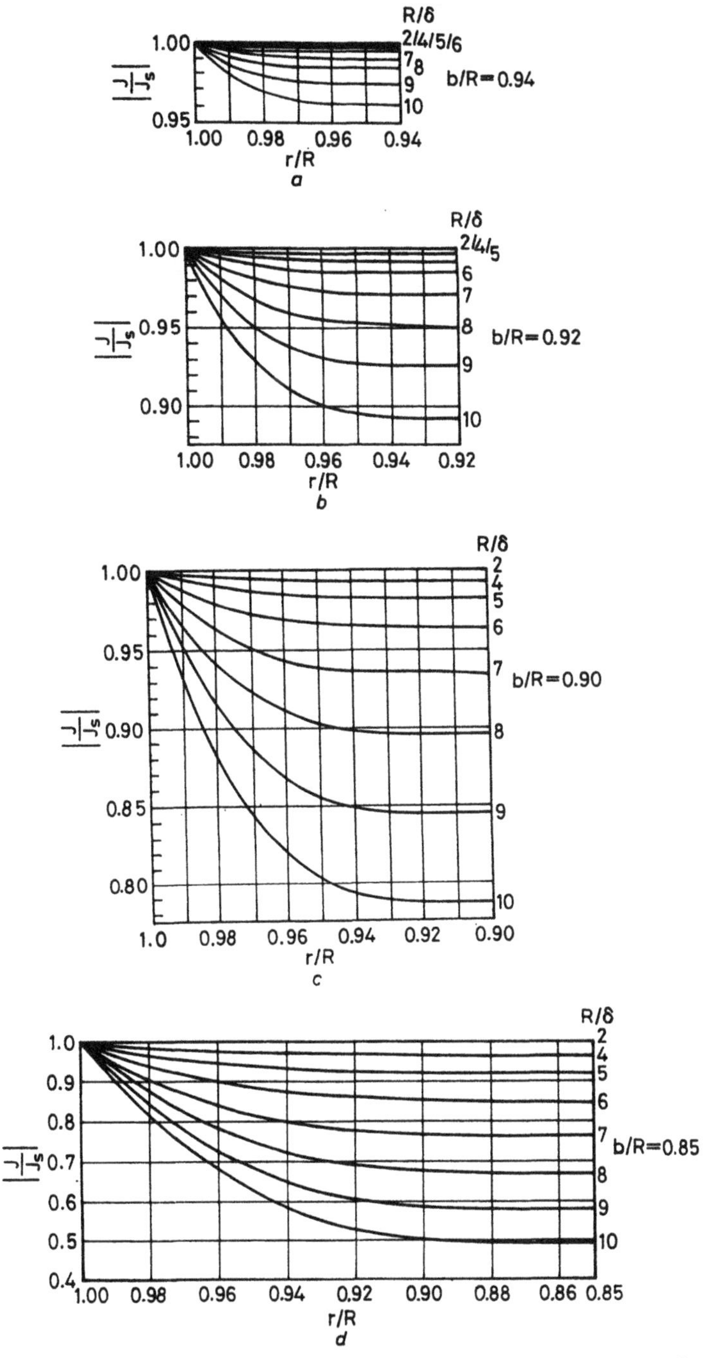

Fig. 5.1 *Fall-off of current density for tubes, expressed as the modulus of J/J_s, for various ratios of radius to depth of penetration, and for selected ratios of internal radius b to external radius R (Cont. over page)*

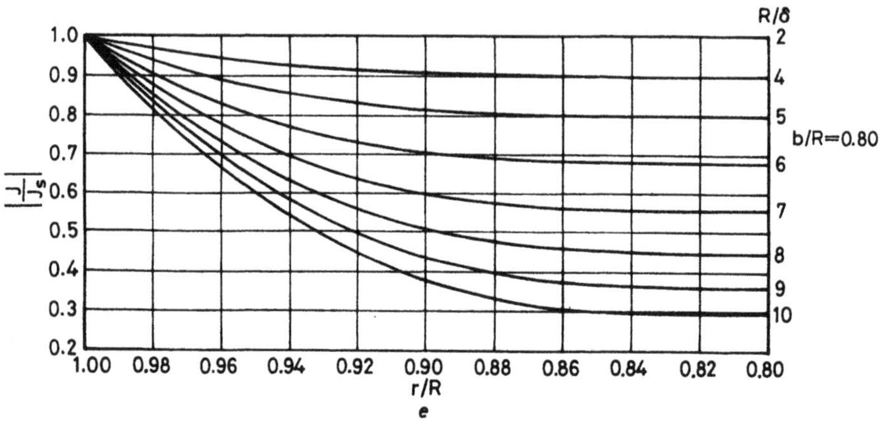
R/δ
1.0
0.9
0.8
0.7
0.6
0.5
0.4
0.3
0.2
|J/Js|
2
4
5
6
7
8
9
10
b/R=0.80
1.00 0.98 0.96 0.94 0.92 0.90 0.88 0.86 0.84 0.82 0.80
r/R
e

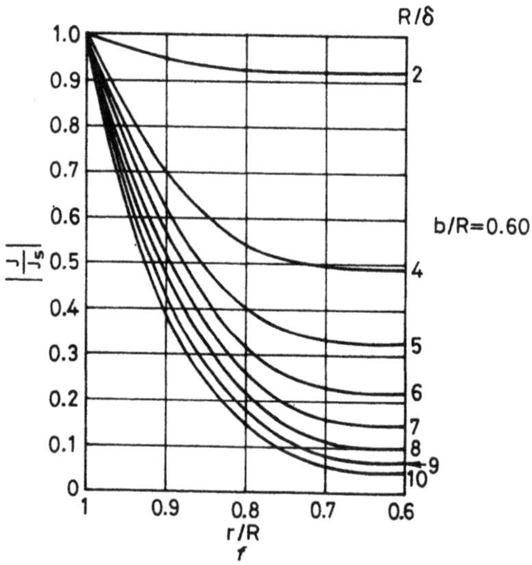
R/δ
1.0
0.9
0.8
0.7
0.6
0.5
0.4
0.3
0.2
0.1
0
|J/Js|
2
b/R=0.60
4
5
6
7
8
9
10
1 0.9 0.8 0.7 0.6
r/R
f

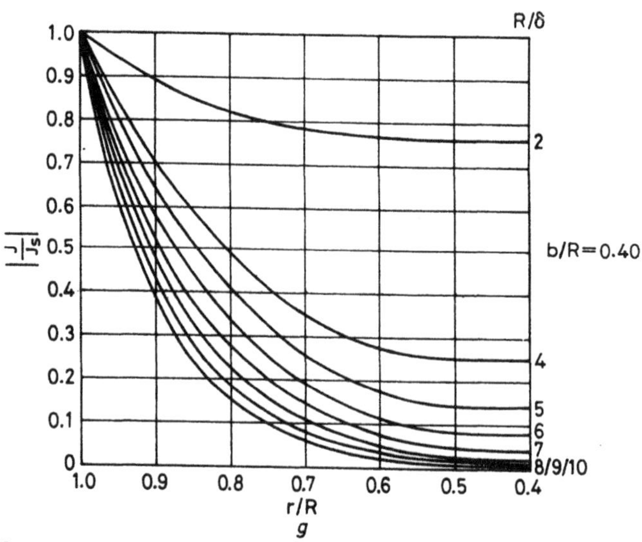
R/δ
1.0
0.9
0.8
0.7
0.6
0.5
0.4
0.3
0.2
0.1
0
|J/Js|
2
b/R=0.40
4
5
6
7
8/9/10
1.0 0.9 0.8 0.7 0.6 0.5 0.4
r/R
g

Fig. 5.1 *(Cont.)*

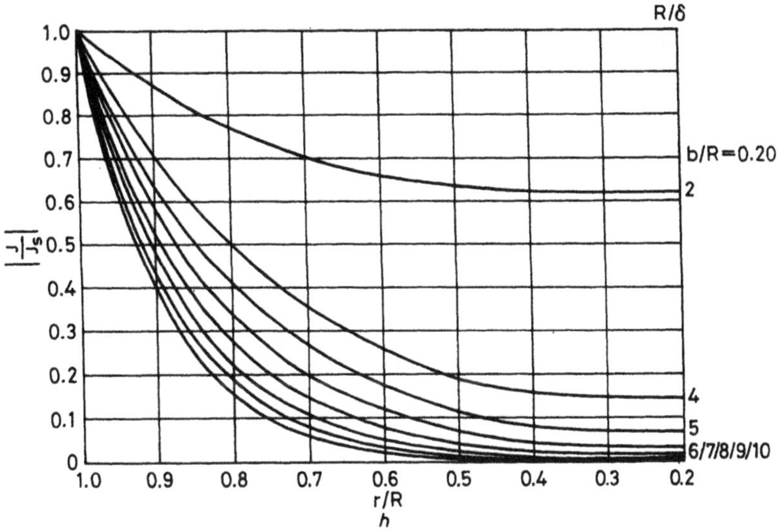

Fig. 5.1 *(Cont.) Fall-off of current density for tubes, expressed as the modulus of J/J_s, for various ratios of radius to depth of penetration, and for selected ratios of internal radius b to external radius R*

is no current in the hole) and the other side is obviously the mirror image. All are plotted for $R/\delta = 2, 4, 5, 6, 7, 8, 9$ and 10. The b/R ratios are 0.94, 0.92, 0.9, 0.85, 0.8, 0.6, 0.4 and 0.2, equivalent to t/D ratios of 3, 4, 5, 7.5, 10, 20, 30 and 40%, respectively. For $t/D < 3\%$, the curves are all practically indistinguishable from unity, which also applies to b/R down to 0.8 for $R/\delta = 2$. As the tube gets thicker, the current distribution approaches that of the solid.

5.4 Loss density

The loss distribution relative to the surface is given by the square of the modulus used in Section 5.3.

5.5 Total current

We can integrate the current density eqn 5.5 over the area of the conductor to give the total current I in terms of the surface current density J_s (using the known integrals given in eqn 4.6):

$$I = \int_b^R 2\pi r J \,\mathrm{d}r$$

$$= 2\pi J_s \int_b^R r \left[\frac{A(\mathrm{ber}\, mr + \mathrm{j}\ \mathrm{bei}\, mr) + B(\mathrm{ker}\, mr + \mathrm{j}\ \mathrm{kei}\, mr)}{A(\mathrm{ber}\, mR + \mathrm{j}\,\mathrm{bei}\, mR) + B(\mathrm{ker}\, mR + \mathrm{j}\ \mathrm{kei}\, mR)} \right] \mathrm{d}r$$

$$= \frac{2\pi J_s}{m}\left[\frac{r\{A(\text{bei}'\,mr - \text{j}\,\text{ber}'\,mr) + B(\text{kei}'\,mr - \text{j}\,\text{ker}'\,mr)\}}{A(\text{ber}\,mR + \text{j}\,\text{bei}\,mR) + B(\text{ker}\,mR + \text{j}\,\text{kei}\,mR)}\right]_b^R$$

$$= \frac{2\pi J_s}{m}\left[\frac{\begin{array}{c}R\{A(\text{bei}'\,mR - \text{j}\,\text{ber}'mR) + B(\text{kei}'\,mR - \text{j}\,\text{ker}'\,mR)\}\\ -b\{A(\text{bei}'\,mb - \text{j}\,\text{ber}'\,mb) + B(\text{kei}'\,mb - \text{j}\,\text{ker}'\,mb)\}\end{array}}{A(\text{ber}\,mR + \text{j}\,\text{bei}\,mR) + B(\text{ker}\,mR + \text{j}\,\text{kei}\,mR)}\right] \quad (5.6)$$

If we divide eqn 5.3 by mj, remembering that $1/\text{j} = -\text{j}$,

$$\frac{1}{m\text{j}}\frac{\text{d}J}{\text{d}r} = A(\text{bei}'\,mr - \text{j}\,\text{ber}'\,mr) + B(\text{kei}'\,mr - \text{j}\,\text{ber}'\,mr)$$

When $r = b$, $\text{d}J/\text{d}r = 0$ (see Section 5.2), so

$$A(\text{bei}'\,mb - \text{j}\,\text{ber}'\,mb) + B(\text{kei}'\,mb - \text{j}\,\text{ker}'\,mb) = 0$$

Substituting this into eqn 5.6, we obtain

$$I = \frac{2\pi J_s R}{m}\left[\frac{A(\text{bei}'\,mR - \text{j}\,\text{ber}'\,mR) + B(\text{kei}'\,mR - \text{j}\,\text{ker}'\,mR)}{A(\text{ber}\,mR + \text{j}\,\text{bei}\,mR) + B(\text{ker}\,mR + \text{j}\,\text{kei}\,mR)}\right] \quad (5.7)$$

$$= \frac{2\pi J_s R}{m}\left[\frac{(\text{bei}'\,mR - \text{j}\,\text{ber}'\,mR)}{(\text{ber}\,mR + \text{j}\,\text{bei}\,mR)}\right]$$

$$\times \left[\frac{1 + (B/A)\,(\text{ker}'\,mR + \text{j}\,\text{kei}'\,mR)/(\text{ber}'\,mR + \text{j}\,\text{bei}'\,mR)}{1 + (B/A)(\text{ker}\,mR + \text{j}\,\text{kei}\,mR)/(\text{ber}\,mR + \text{j}\,\text{bei}\,mR)}\right] \quad (5.8)$$

where the first [ ] is the total current for the solid cylinder and the second [ ] modifies it for tubes. The effects of b are all in the B/A term.

5.6 Power

As in Section 4.3, we can use the total current and the known electric field to obtain the real and imaginary parts of power per metre:

$$\text{power/metre} = EI^* = (\rho J_s) \times (\text{eqn 5.7})^* \qquad \text{W/m}$$

This gives mean power if the RMS value of J is used:

$$\text{power} = \frac{\rho 2\pi J_s^2 R^2 l}{mR}\left[\frac{(\text{bei}'\,mR + \text{j}\,\text{ber}'\,mR) + (B/A)^*(\text{kei}'\,mR + \text{j}\,\text{ker}'\,mR)}{(\text{ber}\,mR - \text{j}\,\text{bei}\,mR) + (B/A)^*\,(\text{ker}\,mR - \text{j}\,\text{kei}\,mR)}\right] \quad (5.9)$$

The conditions at the central hole are implicit in $(B/A)^*$. When $b \to 0$, both ber′ and bei′ $\to 0$; the power is then identical with eqn 4.13, as expected.

In Sections 3.4.3 and 4.3.1, we compared AC and DC power for the condition $J_{dc} = J_s$, i.e. the DC current-density throughout the material equal to the AC surface RMS current-density. For the tube, we shall do a similar comparison to give the expression $p_{rtube} + jq_{rtube}$.

For the tube,

$$\text{DC power} = \rho J_{dc}^2 l \pi (R^2 - b^2) \tag{5.10}$$

Hence

$$\frac{\text{AC power}}{\text{DC power}}$$

$$= \frac{1}{1 - (b/R)^2} \frac{2}{mR} \left[\frac{(\text{bei}' mR + \text{j ber}' mR) + (B/A)^* (\text{kei}' mR + \text{j ker}' mR)}{(\text{ber}\, mR - \text{j bei}\, mR) + (B/A)^* (\text{ker}\, mR - \text{j kei}\, mR)} \right]$$

$$= \left\{ \frac{2}{mR} \left(\frac{\text{bei}' mR + \text{j ber}' mR}{\text{ber}\, mR - \text{j bei}\, mR} \right) \right\}$$

$$\times \left[\frac{1 + (B/A)^* (\text{kei}' mR + \text{j ker}' mR)/(\text{bei}' mR + \text{j ber}' mR)}{1 + (B/A)^* (\text{ker}\, mR - \text{j kei}\, mR)/(\text{ber}\, mR - \text{j bei}\, mR)} \right] \frac{1}{1 - (b/R)^2} \tag{5.11}$$

where

$$(B/A)^* = -(\text{bei}' mb + \text{j ber}' mb)/(\text{kei}' mb + \text{j ker}' mb)$$

The terms in { } in eqn 5.11 are the same as those in eqn 4.13 that led to p_r and q_r for the solid cylinder. The other terms represent the effect of the tube. The combination of all the terms in eqn 5.11 gives $p_{rtube} + jq_{rtube}$.

Note how little simplification or rationalisation has been done in this derivation. The aim has been to follow the same thought processes as in Chapters 3 and 4, with the added complication of ker and kei functions. The modern calculator or computer copes easily and accurately with rationalisation, using the logic:

$$\frac{s + \text{j}t}{u + \text{j}v} \to \frac{R_1 \angle \theta_1}{R_2 \angle \theta_2} \to \frac{R_1}{R_2} \angle (\theta_1 - \theta_2) \to e + \text{j}f$$

or

$$(s + \text{j}t)(u + \text{j}v) \to R_1 R_2 \angle (\theta_1 + \theta_2) \to e + \text{j}f$$

as appropriate.

With assumed values of mR and b/R, using tables of ber, bei etc. (see Appendix A.3), with some tedium, we get curves of p_{rtube} and q_{rtube} (Fig. 5.2). An example of the calculation is given in Section 5.8. These curves allow us to answer the question posed at the start of the chapter about 'thin' and 'thick' tubes. For small values of the thickness/diameter ratio t/d, the AC power is almost the same as the DC power and the PF is good. The real power is high, even for $t/d = 0.25$, up to $d/\delta \simeq 3$, but the PF is getting increasingly bad.

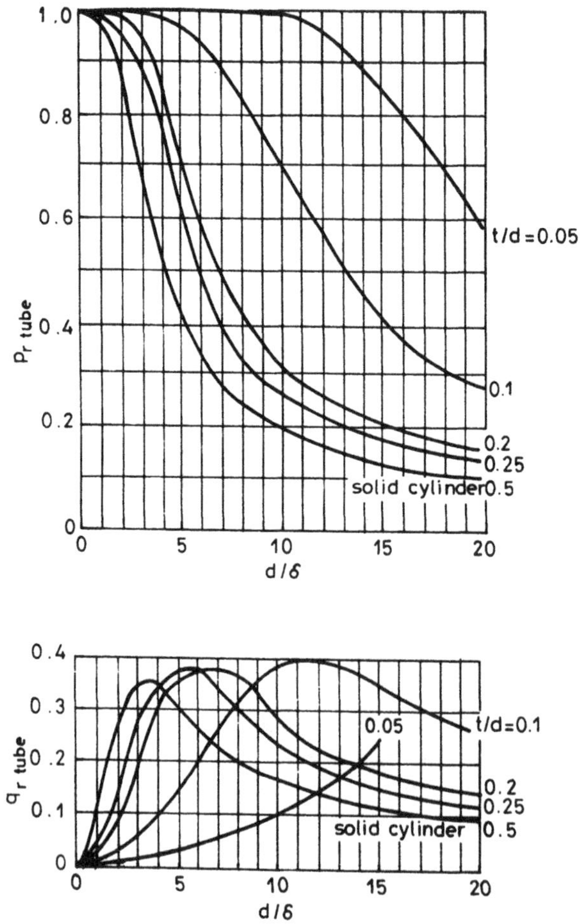

Fig. 5.2 *p_r and q_r for tubes, for various ratios of diameter to depth of penetration and various t/d ratios*

5.7 Circuit quantities

In a manner similar to Section 4.4, we have $Z = V/I$, where $V = \rho J_s l$ and I is given by eqn 5.7:

$$I = \frac{2\pi J_s R}{m}\left[\frac{A(\text{bei}' mR - \text{j ber}' mR) + B(\text{kei}' mR - \text{j ker}' mR)}{A(\text{ber}\, mR + \text{j bei}\, mR) + B(\text{ker}\, mR + \text{j kei}\, mR)}\right]$$

But, for a tube, $R_{dc} = \rho l/\pi R^2[1 - (b/R)^2]$. So

$$\frac{Z}{R_{dc}} = \frac{mR}{2}\left[\frac{(\text{ber}\, mR + \text{j bei}\, mR) + (B/A)(\text{ker}\, mR + \text{j kei}\, mR)}{(\text{bei}' mR - \text{j ber}' mR) + (B/A)(\text{kei}' mR - \text{j ker}' mR)}\right] \times \left[1 - \left(\frac{b}{R}\right)^2\right] \qquad (5.12)$$

This reduces to eqn 4.19 when $b \to 0$.

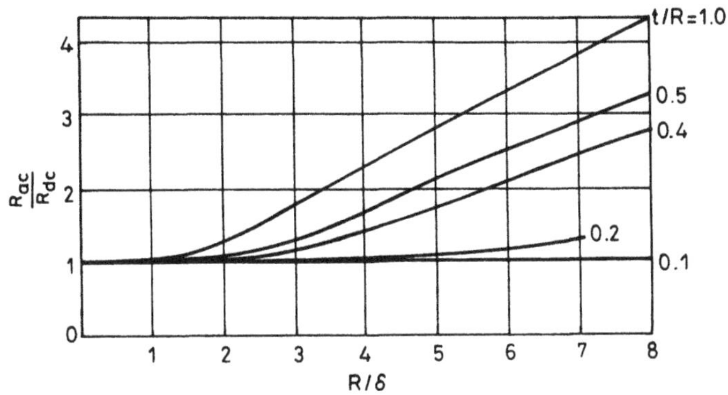

Fig. 5.3 *R_{ac}/R_{dc} for various ratios of radius to skin depth for tubes of various thicknesses. Asymptotic lines are given by $(t/R)(2 - t/R)$ times the asymptotic value for $t/R = 1$: see text*

If we compare eqn 5.12 with the first form of eqn 5.11, we see that each is the inverse of the complex conjugate of the other. In the process of calculating Fig. 5.2, we can get $p + jq$, change it to $p - jq$ and invert to give R_{ac}/R_{dc} and X_{ac}/R_{dc}, similar to Fig. 4.6. These ratios are shown in similar form on Figs 5.3 and 5.4. For values of $mR > 4$, i.e. $R/\delta > 4/\sqrt{2}$, the ker terms are negligible relative to the ber terms and the curves tend to the full cylinder values multiplied by $1 - (b/R)^2$ or, in the symbols of Figs 5.3 and 5.4, by $(t/R)[2 - (t/R)]$ since $b = R - t$:

t/R	1	0.5	0.4	0.2	0.1
$(t/R)[2 - (t/R)]$	1	0.75	0.64	0.36	0.19

This is true for large t/R, but must be used with caution for smaller values of t/R, since the terms in ker′ and kei′ in the denominator of B/A are also small and leave terms in ber′mb and bei′mb, which have to be calculated. As mR increases, the effect of the ker and kei terms gets increasingly negligible, so the above multipliers are true, in the limit.

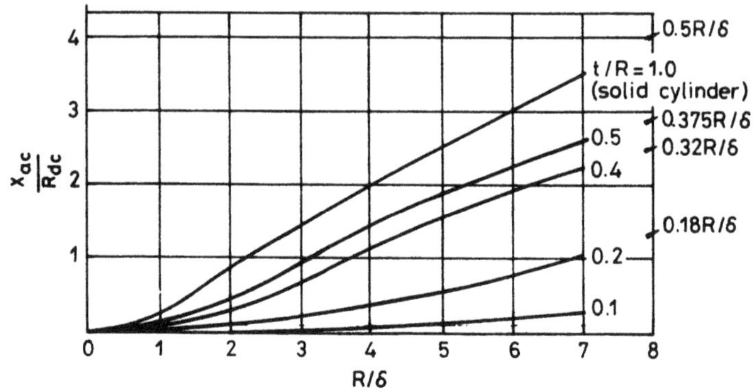

Fig. 5.4 *X_{ac}/R_{dc} for various ratios of radius to skin depth for tubes of various thicknesses. Asymptotic lines are shown: see text*

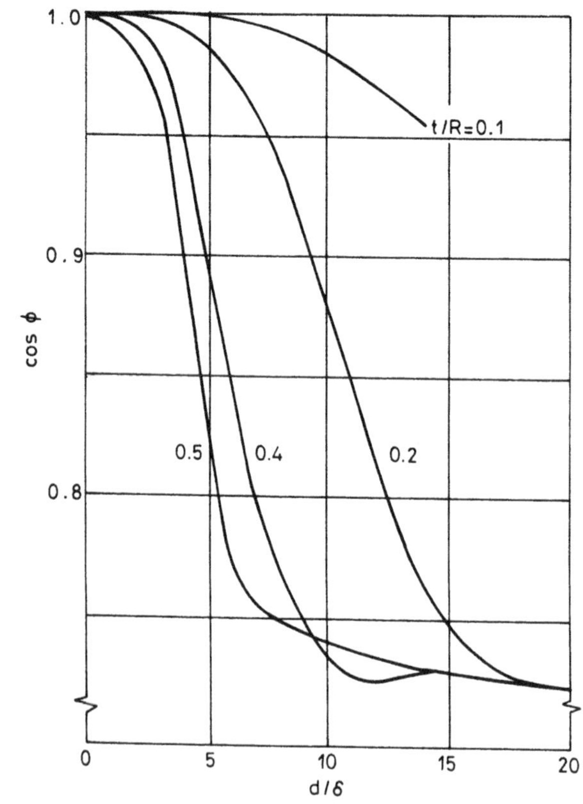

Fig. 5.5 *Power factor cos φ as a function of the ratio of diameter to skin depth for tubes of various thicknesses*

Asymptote lines have been drawn on Figs 5.3 and 5.4.

5.7.1 Power factor

As before, the power factor $\cos\phi = \cos[\tan^{-1}(X/R)] = \cos[\tan^{-1}(q_{tube}/p_{tube})]$. This has been plotted on Fig. 5.5. For thin tubes, e.g. $t/R < 0.1$, the power factor is good (> 0.95) over the whole range of d/δ shown. Obviously, if $\delta \ll d$, because of high frequency or otherwise, the power factor is bound to fall off, since the tube can be considered as a slab with a thin active layer on the outside.

From Fig. 5.5, the power factor falls off rapidly towards an asymptotic value of

$$\cos\phi = \cos\left[\tan^{-1}\left(\frac{1}{1 + 0.54\,\delta/R}\right)\right]$$

i.e. it is independent of t/R.

Example 5.1

Prove the last statement, using the information from the rest of this chapter.

Solution

We have, for $\delta/R \to 0$, $\cos\phi \to 0.707$, i.e. $\delta \ll R$. For $\delta/R = 0.1$, i.e. $d/\delta = 20$, $\cos\phi = \cos[\tan^{-1}(1/1.054)] = 0.725$. This value, greater than 0.707, is due to the asymptotic value of R_{ac}/R_{dc} being $(R/2\delta + 0.27)$ instead of $R/2\delta$.

The curves show clearly that, except for small ratios of d/δ or for $t/R \to 0.1$, the use of AC to heat tubes produces rapidly-falling power factors. Unless the tube falls in the region with high p_r and low q_r, the use of DC is preferable. Nowadays, when rectification is readily available, DC is nearly always preferable.

5.8 Example calculation: power loss constants and impedance ratios for tube

We can use a tabular form. First, the point being calculated is defined. Next, $(B/A)^*$ is calculated from the equation below eqn 5.11. The top line of eqn 5.11 is calculated and divided by the denominator. Multiplying this by $2/[mR(1-(b/R)^2)]$, numeric, gives $p_{rtube} + jq_{rtube}$. This is then converted to $(R_{ac}/R_{dc})_{tube} + j(X_{ac}/R_{dc})_{tube}$ by taking the reciprocal of the complex conjugate, $p_{rtube} - jq_{rtube}$. The power factor can then be filled in.

Tables 5.1a and b show the blank and completed calculation forms. Note that the calculations in the table were done with more than three significant figures from Dwight's tables:[6] the values given may not exactly agree with recalculations.

Table 5.1

(a) *Blank calculation form*

$mR =$ $\qquad$ $mb =$

$b/R = mb/mR =$ $\qquad$ $t/R = 1 - (b/R) =$ $\qquad$ $R/\delta = \alpha R = mR/\sqrt{2} =$

$t/d =$ $\qquad$ $d/\delta =$

$2/[mR(1 - (b/R)^2)] =$

$$\left(\frac{B}{A}\right)^* = (-)\frac{\text{ber}'mb - \text{j bei}'mb}{\text{ker}'mb - \text{j kei}'mb} = (-)\frac{\qquad}{\qquad} = \qquad \text{j}$$

Numerator:

$[2/mR\,(1 - (b/R)^2)]\,[(\text{bei}'\,mR + \text{j ber}'\,mR) + \quad (B/A)^* \quad (\text{kei}'\,mR + \text{j ker}'\,mR)]$

$[\qquad\qquad]\,[(\qquad \text{j} \qquad) + (\qquad \text{j} \qquad)(\qquad \text{j} \qquad)]$

$= \qquad \text{j}$

Denominator:

$[(\text{ber}\,mR - \text{j bei}\,mR) + \quad (B/A)^* \quad (\text{ker}\,mR - \text{j kei}\,mR)]$

$[(\qquad \text{j} \qquad) + (\qquad \text{j} \qquad)(\qquad \text{j} \qquad)]$

$= \qquad \text{j}$

$$\frac{\text{Numerator}}{\text{Denominator}} = p_{\text{rtube}} + \text{j}\ q_{\text{rtube}}$$

$$\frac{1}{p_{\text{rtube}} - jq_{\text{rtube}}} = \frac{1}{\qquad - \text{j} \qquad} = \qquad + j$$

Power factor $= \cos\,[\tan^{-1}((X_{ac}/R_{dc})/\,(R_{ac}/R_{dc}))]$
$= \cos\,[\tan^{-1}(\qquad / \qquad)] =$

(b) *Completed calculation form*

Chosen data: $mR = 3 \quad mb = 1.5$

Derived data:

$b/R = mb/mR = 1.5/3 = 0.5 \quad t/R = 1 - (b/R) = 0.5 \quad R/\delta = \alpha R = mR/\sqrt{2} = 2.12$

$t/d = 0.25 \quad d/\delta = 4.24$

$2/[mR(1 - (b/R)^2)] = 2/3(0.75) = 0.889$

$$\left(\frac{B}{A}\right)^* = (-)\frac{\text{ber}'\,mb - \text{j}\,\text{bei}'\,mb}{\text{ker}'\,mb - \text{j}\,\text{kei}'\,mb} = (-)\frac{-0.210 - \text{j}0.730}{-0.294 - \text{j}\,0.296} = -\,1.596 - \text{j}\,0.876$$

Numerator:

$[2/mR\,(1 - (b/R)^2)]\,[(\text{bei}'\,mR + \text{j}\,\text{ber}'\,mR) + (B/A)^*\,(\text{kei}'\,mR + \text{j}\,\text{ker}'\,mR)]$

$[0.889]\,[(0.880 - \text{j}\,1.57) + (-1.596 - \text{j}0.876)\,(0.092 + \text{j}0.021)]$

$= 0.668 - \text{j}\,1.497$

Denominator:

$[(\text{ber}\,mR - \text{j}\,\text{bei}\,mR) + (B/A)^*\,(\text{ker}\,mR - \text{j}\,\text{kei}\,mR)]$

$[(-0.221 - \text{j}\,1.938) + (-1.596 - \text{j}\,0.876)\,(-0.067 + \text{j}\,0.051)]$

$= -0.069 - \text{j}\,1.962$

$$\frac{\text{Numerator}}{\text{Denominator}} = \underset{p_{\text{rtube}}}{0.751} + \text{j}\,\underset{q_{\text{rtube}}}{0.367}$$

$$\frac{1}{p_{\text{rtube}} - \text{j}q_{\text{rtube}}} = \frac{1}{0.751 - \text{j}\,0.367} = \underset{(R_{ac}/R_{dc})_{\text{tube}}}{1.075} + \underset{(X_{ac}/R_{dc})_{\text{tube}}}{\text{j}\,0.526}$$

Power factor $= \cos\,[\tan^{-1}((X_{ac}/R_{dc})/\,(R_{ac}/R_{dc}))]$

$= \cos\,[\tan^{-1}(0.526/1.075)] = 0.898$

Exercise

Choose values from the range ber etc. given in Dwight's tables[6] or Appendix A.3, i.e. 0–10, and repeat the calculations. Check the results against Figs 5.3, 5.4 and 5.5. (If the exercise is carried out by a class, choose various values to plot a curve.)

Chapter 6

Wire and strip heating

6.1 Introduction

Under stationary conditions, the conditions along the length of the workpiece are uniform at any instant. The applied voltage is the same throughout the heating cycle, so the electric field strength E is the same at any point along the length. As the workpiece heats, the resistivity rises, so that the surface current-density falls. The specific heat $(c\gamma)$ also changes with temperature (see Appendix A.1).

In many industrial metallurgical processes, the material needs to be heated as part of a moving system. Examples are rolling, intermediate annealing, tempering, etc., where the material starts off on one reel and, after going through various processes, including DRH, ends up on another reel. We are not here concerned with the process in general, only the part of the run where the work is heated. For present purposes, since the material is usually flexible – wire, small rod or thin narrow strip – we can ignore the problems of current distribution across the section (x/δ small, where x is the appropriate linear dimension) and assume uniform current density and heating.

As the material passes between the contacts (e.g. rollers) at constant velocity, the same current must flow axially everywhere in the work, independently of temperature, but the volt-drop/metre E is different because of changes in ρ and $(c\gamma)$. We are dealing with conditions of constant total resistance between the contacts, but the resistance/metre changes along the length as the temperature rises. Before introducing these complications, examine the new conditions with ρ and $(c\gamma)$ constant. Note that the analysis is for steady velocity conditions; variable velocity will be discussed later (Section 6.4).

6.2 Constant ρ and $(c\gamma)$

Figure 6.1 illustrates wire and strip heating. Let the velocity be v m/s, the cross-section be A m^2 and the temperature rise be $\Delta\theta$ K. The power input is calculated from the heat content of the material. The volume heated through $\Delta\theta$ per second is given by velocity (m/s) × area (m^2), i.e. vA. Hence

$$\text{power } P = vA\,(c\gamma)\,(\Delta\theta) \tag{6.1}$$

$$\mathrm{W} = \frac{\mathrm{m}^3}{\mathrm{s}}\frac{\mathrm{J}}{\mathrm{m}^3\mathrm{K}}\mathrm{K} = \frac{\mathrm{J}}{\mathrm{s}} = \mathrm{W}$$

Note that this is independent of ρ and contact spacing. If $(c\gamma)$ is variable, the

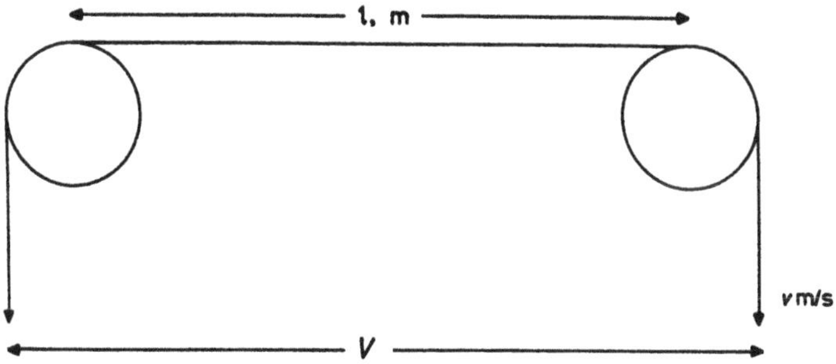

Fig. 6.1 *Wire and strip heating*

heat content at the start and finish have to be subtracted; intermediate values only influence potential distribution. For constant ρ, the overall resistance R is given by

$$R = \frac{\rho l}{A} \quad \Omega$$

From $P = V^2/R$,

$$V^2 = RP$$

$$= \frac{\rho l}{A} vA(c\gamma)(\Delta\theta)$$

$$V = \sqrt{[\rho l v(c\gamma)(\Delta\theta)]} \tag{6.2}$$

Thus V is independent of cross-section. From $I = P/V$,

$$I = A\sqrt{\left[\frac{v(c\gamma)(\Delta\theta)}{\rho l}\right]} \tag{6.3}$$

If the work is subdivided – for example 10 wires side-by-side over the same contacts – it is clear that the subdivision does not affect the volts, but that power and the total current are both proportional to the total area.

In practice, it is desirable to use higher voltages and lower currents to reduce the currents carried by the contacts and the proportion of voltage (and hence power) lost at the contact volt-drop. For a given installation ρ, v, $(c\gamma)$ and $(\Delta\theta)$

are fixed, so that the designer has only the spacing of contacts to vary and V is only affected as $\sqrt{l}$. Distance l should obviously be as big as is practical.

Example 6.1

Steel wire, 1.2 mm diameter, is to be heated through 850°C at a speed of 0.75 m/s. Calculate the electrical conditions for constant spacings between contacts of 1 and 10 metres, assuming average values of $\rho = 0.5\,\mu\Omega\,\text{m}$ and $(c\gamma) = 5\,\text{MJ/m}^3\,\text{K}$.

Solution

We have

$$P = vA(c\gamma)(\Delta\theta)$$

$$= (0.75)\frac{\pi}{4}\left(\frac{1.2}{10^3}\right)^2(5 \times 10^6)(850) = 3605 \quad \text{W}$$

This is independent of length between contacts. Next,

$$V = \sqrt{[\rho l v(c\gamma)(\Delta\theta)]}$$

$$= \sqrt{[(0.5/10^6)0.75(5 \times 10^6)850]}\sqrt{l} = 39.92\sqrt{l} \quad \text{V}$$

Finally, $I = P/V$. Thus we have:

Spacing	V	I
1 m	39.92 V	90.3 A
10 m	126.2 V	28.6 A

The 10 m spacing gives more desirable V and I, but may present practical problems of support for the thin, hot wire.

The speed of 0.75 m/s was chosen as a practical example. The transit times are 1.33 s and 13.3 s, respectively. If the process were speeded up 10 times to 7.5 m/s, the power would increase 10 times to 36 kW. The powers needed for DRH are high, because the heat is being put into the material very quickly, i.e. from 0°C → 850°C in 1.33 s or 13.3 s in this example, or in 0.13 s or 1.3 s for 7.5 m/s.

Example 6.2

Steel strip 10 mm wide and 1.2 mm thick is to be heated through 850 K at 0.75 m/s. The conditions are the same as those in Example 6.1.

Solution

Here nothing has changed but the area, which has increased from 1.13 mm^2 to 12 mm^2, i.e. a 10.6-fold increase. The power increases to 38.3 kW; the voltage remains at 39.9 V and 126.2 V; and the current increases to 958 A and 303 A.

6.3 Variable ρ and $(c\gamma)$

6.3.1 General:

Given the actual measured and tabulated values of ρ and $(c\gamma)$, it is always possible to use the equations of the following section in an incremental form using a table. This method will not be pursued here, in the light of the analytical solution that follows, but is available for difficult materials.

6.3.2 Analytical solution

Consider an element dx at x (Fig. 6.2). The power in the element is given by

$$P = I^2R = I^2\rho_\theta \mathrm{d}x/A$$

where ρ_θ is the resistivity at a temperature θ. We also have that $(c\gamma)_\theta$ is the

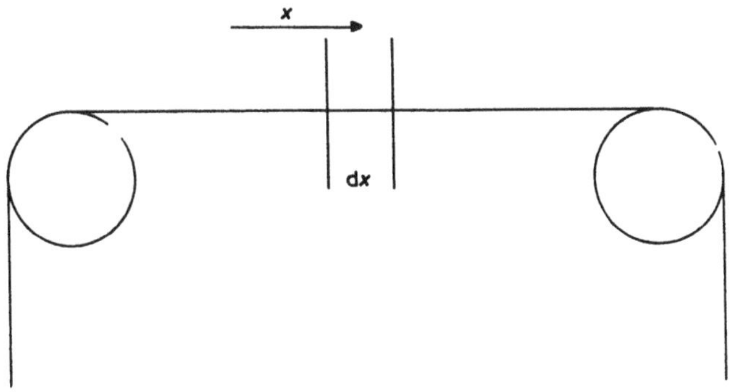

Fig. 6.2 *Element dx at x*

specific heat at a temperature θ. The transit time of the element is $\mathrm{d}t = \mathrm{d}x/v$, where v is the velocity (m/s). Hence the energy into the element is

$$P\,\mathrm{d}t = \rho_\theta I^2 \frac{\mathrm{d}x}{A}\frac{\mathrm{d}x}{v}$$

The rise in the temperature of the element is given by

$$\mathrm{d}\theta = \frac{\text{energy}}{(c\gamma)_\theta \text{ volume}}$$

$$= \left(\frac{\rho}{c\gamma}\right)_\theta \frac{I^2(\mathrm{d}x)^2}{Av(A\,\mathrm{d}x)}$$

$$= \left(\frac{\rho}{c\gamma}\right)_\theta \frac{J^2\,\mathrm{d}x}{v} \qquad (6.4)$$

This is the equation that can always be used in an incremental solution.

If the data from Appendix A.1 are used to plot $(\rho/c\gamma)_\theta$ against θ for mild steel, we get

$$0\text{–}600^\circ\mathrm{C}: \quad \left(\frac{\rho}{c\gamma}\right)_\theta = \frac{1}{10^{16}}(408 + 1.530\theta)$$

(correlation factor = 0.999 77)

$$600°\text{–}700°\text{C:}\quad \left(\frac{\rho}{c\gamma}\right)_\theta = \frac{1}{10^{16}}(1030 + 0.5\theta) \tag{6.5}$$

$$700\text{–}900°\text{C:}\quad \left(\frac{\rho}{c\gamma}\right)_\theta = \frac{1}{10^{16}}(-465 + 2.63\theta)$$

(correlation factor = 0.998 99)

These excellent linear relations can be substituted for $(\rho/c\gamma)_\theta$ in eqn 6.4 to give very accurate answers. Alternatively, an overall line can be found from 20°C to 800°C with a correlation of 0.997, which is accurate enough for our present purposes:

$$\left(\frac{\rho}{c\gamma}\right)_\theta = \frac{1}{10^{16}}(418 + 1.48\theta) = a + b\theta \tag{6.6}$$

Rearranging eqn 6.4 and solving, we obtain

$$\mathrm{d}x = \left(\frac{v}{J^2}\right)\frac{\mathrm{d}\theta}{(a + b\theta)}$$

$$[x]_0^l = \left(\frac{v}{J^2}\right)\left[\frac{\ln(a + b\theta)}{b}\right]_{\theta_1}^{\theta_2} \tag{6.7}$$

Also

$$\mathrm{d}V = \rho_\theta J\,\mathrm{d}x$$

$$= \rho_\theta J\frac{v\,\mathrm{d}\theta}{J^2(\rho/c\gamma)_\theta} = (c\gamma)_\theta \frac{v}{J}\,\mathrm{d}\theta \tag{6.8}$$

From Appendix A.1, it will be seen that $(c\gamma)_\theta$ rises up to 700°C and falls after 800°C. Linear regression gives the following:

20–500°C: $(c\gamma)_\theta = 10^6(3.52 + 0.003\,17\theta)$ (correlation 0.994)

20–800°C: $(c\gamma)_\theta = 10^6(3.32 + 0.004\,17\theta)$ (correlation 0.981)

This is an excellent correlation from 20°C to 500°C, but less good from 20°C to 800°C. For present purposes, use

$$(c\gamma)_\theta = 10^6[3.32 + 0.004\,17\,\theta] = e + f\theta \tag{6.9}$$

Hence

$$[V]_0^V = \left(\frac{v}{J}\right)\left[c\theta + \frac{d\theta^2}{2}\right]_{\theta_1}^{\theta_2} \tag{6.10}$$

Example 6.3
Repeat Example 6.1 using the more precise equations above.

Solution
(a) *Spacing 1 m between contacts*
We have

$$[x]_0^1 = \frac{0.75}{J^2}\left(\frac{10^{16}}{1.48}\right)\left[\ln\frac{418 + 1.48\theta}{10}\right]_0^{850}$$

Thus

$$1 = \frac{0.75}{J^2}\left(\frac{10^{16}}{1.48}\right)\left[\ln\frac{418 + (1.48 \times 850)}{418 + (1.48 \times 0)}\right]$$

$$J^2 = 10^{16}\frac{0.75}{1.48}1.39$$

$$J = 83.9 \times 10^6 \quad \text{A/m}^2$$

Then

$$I = \frac{\pi}{4}\left(\frac{1.2}{10^3}\right)^2 J$$

$$= 94.9 \quad \text{A}$$

$$V = \left(\frac{0.75}{83.9 \times 10^6}\right)10^6\left[3.322(850) + \frac{0.004\,17}{2}(850)^2\right] = 38.7 \quad \text{V}$$

$$P = 38.7 \times 94.9 = 3673 \quad \text{W}$$

These values are comparable with the 90.3 A, 39.9 V and 3605 W obtained in Example 6.1. The approximate solutions are sufficiently accurate for many purposes.

(b) *Spacing 10 m between contacts*
Since everything in eqn 6.7 is unchanged except the limits for x, then J and I are reduced by $(1/\sqrt{10})$; hence $I = 30$ A. Only J changes in eqn 6.10; thus $V = 38.7/\sqrt{10} = 12.24$ V. P is unchanged.

Example 6.4
Aluminium 500 mm wide and 1.2 mm thick is to be heated from 20°C to 500°C at a speed of 0.9 m/s between contacts. Using the data of Appendix A.1, estimate the electrical parameters

(a) approximately, using average values
(b) more precisely, by linear approximations for $(\rho/c\gamma)$ and $(c\gamma)$ against temperature.

If the reader does not have a least-squares approximation to a straight line on his calculator, he can use a graph and draw a best straight line by eye.

Solution

(a) Average values of $\rho = 0.055\,\mu\Omega\text{m}$ and $\gamma = 2.7 \times 10^6\,\text{J/m}^3\text{K}$ will be used (these are obtained by adding the values from 20°C to 500°C, dividing by 6 and rounding). We obtain:

$$P = vA(c\gamma)(\Delta\theta)$$

$$= 0.9\left(\frac{500 \times 1.2}{10^6}\right)(2.7 \times 10^6)480 = 0.7 \quad \text{MW}$$

$$V = \sqrt{[\rho l v(c\gamma)(\Delta\theta)]}$$

$$= \sqrt{[(0.055/10^6)(2)(0.9)(2.7 \times 10^6)480]} = 11.3 \quad \text{V}$$

$$I = 62\,000 \quad \text{A}$$

(b) Using straight-line approximation on calculator:

$$\left(\frac{\rho}{c\gamma}\right)_\theta = \frac{1}{10^{15}}(9.84 + 0.041\theta) \qquad \text{(correlation 0.999)}$$

$$(c\gamma)_\theta = 10(2.52 + 0.000\,655\theta) \qquad \text{(correlation 0.999)}$$

These are very good straight lines. Following Example 6.3,

$$l = \left(\frac{v}{J^2}\right)\left[\frac{\ln(a + b\theta)}{b}\right]_{\theta_1}^{\theta_2}$$

$$2 = \frac{0.9}{J^2}\left(\frac{10^{15}}{0.041}\right)\left[\ln\frac{9.84 + 0.041 \times 500}{9.84 + 0.041 \times 20}\right]$$

$$J = 107 \times 10^6 \quad \text{A/m}^2$$

Then

$$I = \left(\frac{500 \times 1.2}{10^6}\right)J = 64\,300 \quad \text{A}$$

$$V = \left(\frac{v}{J}\right)\left[e\theta + \frac{f\theta^2}{2}\right]_{\theta_1}^{\theta_2} = \left(\frac{0.9 \times 10^6}{107 \times 10^6}\right)$$

$$\times\left[2.52(500) + \frac{0.000\,655}{2}(500^2) - 2.52(20) - \frac{0.000\,655}{2}(20^2)\right]$$

$$= \frac{0.9}{107}1291 = 10.8 \quad \text{V}$$

As before, the precise method is not grossly different from the approximate method.

6.4 Practical conditions

Up to now in this chapter, we have assumed an ideal process line in which all the material is moving through the system at constant velocity. This is substantially true for the treatment of wire and thin strip, where very long lengths can be accommodated on the input reel, so that the start and finish of each reel are negligible compared with the steady-running period in between. It is an accepted fact that there will be scrap at each end.

When the material is of larger cross-section, there will be problems associated with acceleration and deceleration. For example, in one rod-drawing mill the acceleration and deceleration times are each 25% of the total transit time, leaving only 50% at constant rated speed. It is beyond the scope of this book to analyse this, as the variation of ρ and $(c\gamma)$ and their relationship to a variable velocity comes into the equations. It will be obvious from the previous section that, if the applied voltage were kept constant with a slow speed and uniformly cold material, i.e. starting conditions, then the material would rapidly overheat. Modern control devices can be used to reduce the voltage in an orderly way to achieve the correct output temperature during the changes of ρ and velocity.

Part II

Induction heating

Chapter 7

Basic induction heating

7.1 Introduction

We started this book with the conduction heating of metals because this is nearer to our everyday experience. We moved from the simple passage of DC in the metal to the complications that arise from the use of alternating current and, during the derivations, saw that it is normally preferable to use DC, whenever possible. Another reason for starting with conduction was that it is a very effective way to heat metals if the shape is right. It allowed us to develop the theory from first principles in a way that closely matches the way we were taught, starting with Ohm's law and with the current passing from end to end of the material between the electrodes. The understanding that has been developed in those chapters can now be extended into the realm of induction heating.[2,4,7,10,12,13,30,31,33,47,107,108]

In induction heating, we still heat by the passage of currents through the material but now they are *induced* from a separate source. Whatever the configuration, induction heating *always* involves a varying magnetic field. The metal to be heated has become the secondary circuit of a transformer; this is not always obvious in some configurations but it is always true. As with conduction heating, the theory of induction heating is simpler if we progress from the thick slab to the thin slab (Chapter 8) to the cylinder (Chapter 9) but, because we are more used to visualising the transformer in its normal cylindrical form, it is probably easier to start our conceptualising with a cylindrical workpiece.

Figure 7.1 shows a transformer with a multi-turn single-layer primary and a single-turn short-circuited secondary, separated by a small air gap. When alternating current flows in the primary, voltages are induced in the secondary which cause currents to flow in it, and these currents tend to cancel the flux that produces them, according to Lenz's law. From the results of Chapter 2, we would suspect that those currents will effectively flow in about a skin depth δ – we shall expand on this later – given by $\sqrt{(2\rho/\mu\omega)}$, i.e. depending on frequency and the two properties of the material, resistivity and relative permeability.

The primary coil is always constructed of copper, since it is important to keep losses to a minimum so that the heating is done at maximum efficiency. It is normally water-cooled to keep the losses down and to remove both its internal losses and the heat reaching it from the billet.

The gap between the coil and the billet is kept to a minimum, so that they are closely coupled. The gap is controlled by the need to move the work in and out

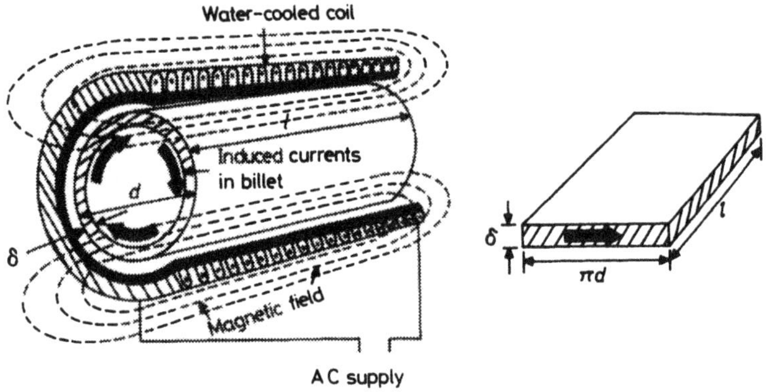

Fig. 7.1 *Principles of billet heating*

of the coil, and by any thermal insulation needed to reduce heat transfer from the hot material to the coil.

The efficiency η is simply expressed by:

$$\eta = \frac{1}{1 + \sqrt{(\rho_c / \rho_w \mu_w)}}$$

where ρ_c is the resistivity of the coil, ρ_w is the resistivity of the workpiece and μ_w is the relative permeability of the workpiece. This idealised equation should be treated with care, but it does give a broad-brush idea of what controls the efficiency. It shows that a material with high resistivity and permeability, such as steel, can have an efficiency approaching 100%, but that copper, where the root term approaches unity, has an efficiency of about 50%. It was formulated for simple coils and is certainly wrong for multi-layer coils (see Chapter 22), where the coil current is not limited to the skin depth.

7.2 The semi-infinite slab

Once again, it is easier to start with the abstraction of a semi-infinite slab, just as for DRH. In Fig. 7.2a, a coil lies close to the face of the slab. There is some difficulty in picturing where the other side of the coil lies and where the currents 'return' in the slab. It may help to think of a coil of infinite radius wound round a cylinder of infinite radius. The length of the slab in the x direction is unimportant and it is convenient to consider one metre, as shown in Fig. 7.2b.

The coil produces a magnetic field H_s at the surface of the slab. Note that it is H_s in A/m that counts, not 'ampere-turns', and that the heating is quite unaffected by the subdivision of that current into individual conductors; the only things affected by subdivision are the resistance of the coil and the voltage needed to drive the current through the coil. The coil is assumed to be so near the slab surface that H_s at the surface is identical to H_s of the coil.

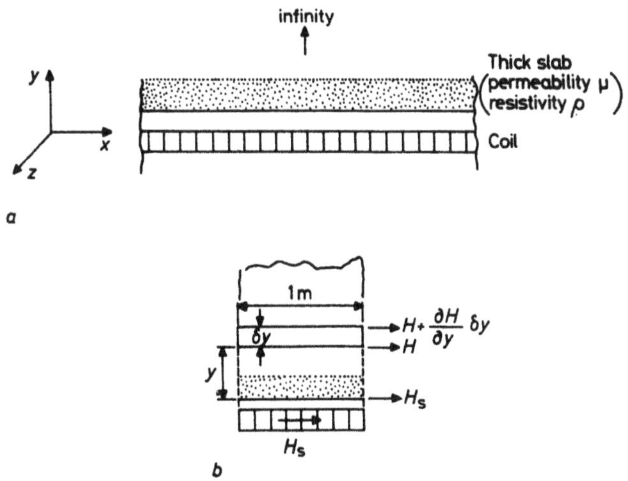

Fig. 7.2 *(a) Coil facing semi-infinite slab. (b) Defining dimensions in the slab*

Unlike DRH, induction heating will not work with DC as no voltage would be induced in the workpiece, so the applied field must be pulsating in magnitude. From physical understanding of the problem, a pulsating H_s in the x direction produces eddy currents in the z direction. Since all the quantities in the slab are vectors, it is important to label them with direction tags.

The boundary conditions are:

$$\begin{aligned} H_x(y=0) &= H_s \\ H_y &= 0 \quad \text{everywhere} \\ H_z &= 0 \quad \text{everywhere} \end{aligned} \tag{7.1}$$

H_s is the RMS value of H_x at the surface ($y = 0$), having angular frequency ω.

Note that the directions of H and I have been chosen to agree with those of the similar derivation in Chapter 2 on DRH. By this choice we save ourselves from the necessity of repeating the theory of Section 2.2; eqns 2.1 to 2.5 are identical. As before, there are no J variations in x and z.

7.3 Duality of DRH and induction heating

This brings us to a very important theme of this book and one reason for

Table 7.1 *Summary of results of Chapter 2*

$J_z(y)$	=	$J_s \exp(-\alpha y) \cos(\omega t - \alpha y)$
$E_z(y)$	=	$E_s \exp(-\alpha y) \cos(\omega t - \alpha y)$
$H_x(y)$	=	$H_s \exp(-\alpha y) \cos(\omega t - \alpha y)$
$B_x(y)$	=	$B_s \exp(-\alpha y) \cos(\omega t - \alpha y)$
Depth of penetration δ	=	$\sqrt{(2\rho/\mu\omega)}$

associating conduction and induction heating in the title. At this stage, it is evident that, by a suitable choice of coordinates, we have made the theory of the two heating methods identical for the semi-infinite slab and all the discussion of depth of penetration, etc. can be carried forward. So there is no difference between a given H inducing a current I in the slab and the same value of current flowing in the slab producing the same H at the surface. The only difference is which is 'cause' and which is 'effect'. *This identity is only true for the idealised case of the semi-infinite slab.* We shall see later that there is a duality relationship between the two heating modes for other configurations. The reader is referred to Chapter 2 for the derivations; the results are summarised in Table 7.1.

Figures 2.2 and 2.3 also apply to induction heating.

7.4 Depth of penetration

The nomogram of Fig. 7.3 allows the rapid calculation of depth of penetration for both conduction and induction. Scales A, B and C give the value of δ for non-magnetic materials, i.e. non-ferrous metals and ferrous metals above the Curie point. For ferrous metals below the Curie point, the value of μ is a variable function of H. This is discussed in more detail in Chapter 13 but, by using scale B as a pivot in conjunction with scale E, a reasonably accurate skin depth can be read from scale D.

7.5 Flux density, current density and electric field

Assuming that the material is magnetically linear,

$$B_x = B_s \exp -\alpha y \exp -j\alpha y = B_s \exp -(1 + j)\alpha y \tag{7.2}$$

The 'low-frequency' Maxwell equation, $\boldsymbol{J} = \text{curl}\, \boldsymbol{H}$, gives $\boldsymbol{J}$ everywhere in the slab.

The determinant for curl $\boldsymbol{H}$ is

$$\begin{vmatrix} \boldsymbol{i} & \boldsymbol{j} & \boldsymbol{k} \\ \dfrac{\partial}{\partial x} & \dfrac{\partial}{\partial y} & \dfrac{\partial}{\partial z} \\ H_x & H_y & H_z \end{vmatrix}$$

But $H_y = H_z = 0$ and there are no H variations in the x and z directions, so

$$\text{curl}\,\boldsymbol{H} = \begin{vmatrix} \boldsymbol{i} & \boldsymbol{j} & \boldsymbol{k} \\ 0 & \dfrac{\partial}{\partial y} & 0 \\ H_x & 0 & 0 \end{vmatrix}$$

$$J_z = -\boldsymbol{k}\,\partial H_x/\partial y = -\boldsymbol{k}\,\partial/\partial y\,[H_s \exp -(1 + \mathrm{j})\alpha y]$$

$$= \boldsymbol{k}\alpha(1 + \mathrm{j})H_s \exp -(1 + \mathrm{j})\alpha y$$

But $\sqrt{(2\mathrm{j})} = 1 + \mathrm{j} = (\sqrt{2}) \exp \mathrm{j}\pi/4$, and so

$$J_z = \boldsymbol{k}(\sqrt{2})\alpha H_x \exp \mathrm{j}\pi/4 \qquad (7.3)$$

This shows that $\boldsymbol{J}$ is orthogonal to $\boldsymbol{H}$, as expected, and is everywhere 45° out of phase with $\boldsymbol{H}$.

When $y = 0$,

$$J_z(0) = \boldsymbol{k}(\sqrt{2})\alpha H_x(0) \exp \mathrm{j}\pi/4$$

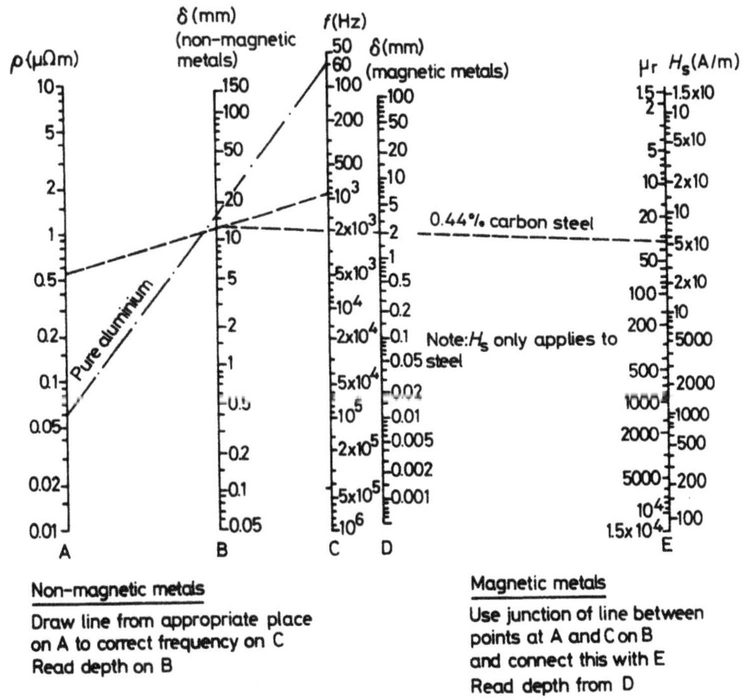

Fig. 7.3 *Depth of penetration nomogram. The scale for μ_r is accurate, in accordance with $\delta = \sqrt{(2\rho/\mu_0\mu_r\omega)}$. The scale for H_s depends on the relationship between μ_r and H_s: for this figure, eqn 13.4 and Fig. 13.1 have been used, giving $\mu_r = 734\,000/H_s^{0.92}$*

So

$$J_z(y)/J_z(0) = H_x(y)/H_x(0) = \exp - (1 + \mathrm{j})\alpha y$$

$$J_z(y) = J_z(0) \exp - (1 + \mathrm{j})\alpha y \tag{7.4}$$

This shows that the variation of J_z is the same as the variation of H and B, relative to its own surface value.

We also know that $\boldsymbol{E} = \rho \boldsymbol{J}$, so that

$$E_z(y) = E_s \exp - (1 + \mathrm{j})\alpha y \tag{7.5}$$

As before, for conduction, all the electromagnetic quantities obey the same 'skin-effect' rules.

7.6 Total current

The total current in one metre width of the slab can be found by integrating J through the slab depth. This was done in Section 2.2.2, giving

$$\text{current/metre width} = [J_s \exp - \mathrm{j}\pi/4]/\alpha\sqrt{2} \tag{7.6}$$

Again, the total current per unit width lags the surface current density by 45°. But, from above eqn 7.4,

$$J_s = (\sqrt{2})\alpha H_s \exp \mathrm{j}\pi/4$$

so

$$\text{current/metre width} = [(\sqrt{2})\alpha H_s \exp \mathrm{j}\pi/4][\exp - \mathrm{j}\pi/4]/\alpha\sqrt{2}$$

$$= H_s \tag{7.7}$$

That is, the current per metre width is numerically equal to the applied H at the surface. This result can also be obtained by taking the line integral $\oint H\mathrm{d}l$ around the total current. Since there are no contributions to H in the y direction or at infinity, it follows that $(H_s \times 1) = I$, so that $I = H_s$, as above, and that the two quantities are in phase.

7.7 Loss

The loss per area $\mathscr{P}$ is found, as for conduction (eqn 2.10), to be

$$\mathscr{P} = \frac{\rho\delta}{2} J_s^2 \tag{7.8}$$

But we need this result in terms of H, using $|J_s| = \sqrt{(2|H_s|/\delta)}$, so

$$\mathscr{P} = \rho H_s^2/\delta \tag{7.9}$$

The loss-density distribution is the same as that described in Section 2.2.3 and Fig. 2.3.

Example 7.1
Find the power loss per unit surface area, given the following data:

Material	0.23% carbon mild steel
Resistivity ρ	0.16 $\mu\Omega$ m at 20°C
Applied field H_s	200 000 A/m RMS
Frequency	50 Hz

Solution
From Fig. 7.3, relative permeability $\mu_r = 10$. Then
$\delta = \surd[(2 \times 0.16 \times 10^{-6})/(10 \times 4\pi \times 10^{-7})(100\pi)] = 9.00\,\text{mm}$
(or use nomogram). From eqn 7.9,

$$\mathscr{P} = \rho H_s^2/\delta = \frac{(0.16 \times 10^{-6})(4 \times 10^{10})}{(9 \times 10^{-3})} = 0.71 \times 10^6 \quad \text{W/m}^2$$

$$= 0.71 \quad \text{MW/m}^2$$

So power densities of about 1 MW/m^2 are reasonable. Note that, as the temperature rises towards the Curie point, the resistivity increases to about 0.9 $\mu\Omega$ m and the power density increases to 1.7 MW/m^2. At the Curie point, the steel becomes non-magnetic and the loss density suddenly decreases, as the next example shows.

Example 7.2
Find the power loss per unit surface area, given the following data:

Material	0.23% carbon mild steel
Resistivity ρ	1.1 $\mu\Omega$ m at 800°C
Relative permeability μ_r	unity at Curie point
Applied field H_s	200 000 A/m RMS
Frequency	50 Hz

Solution
The depth of penetration has *increased* by $\surd(20 \times 1.1/0.16) = 11.7$, to about 105 mm. From eqn 7.9,

$$\mathscr{P} = \rho H_s^2/\delta = \frac{(1.1 \times 10^{-6})(4 \times 10^{10})}{(0.105)} = 4.2 \times 10^5 \quad \text{W/m}^2$$

The elevenfold increase in the depth of penetration is greater than the sevenfold increase in resistivity, and the power density drops.

Example 7.3
Find the power loss per unit surface area, given the following data:

Material	copper
Resistivity ρ	$0.017\,\mu\Omega$ m at 20°C
Relative permeability μ_r	unity
Applied field H_s	200 000 A/m RMS
Frequency	50 Hz

Solution
From nomogram Fig. 7.3, $\delta = 10$ mm. From eqn 7.9,

$$\mathscr{P} = \rho H_s^2/\delta = \frac{(0.017 \times 10^{-6})(4 \times 10^{10})}{(0.01)} = 6.8 \times 10^4\,\mathrm{W/m^2}$$

This shows that the heating of copper at 50 Hz is about ten times worse than steel at the same temperature. At 800°C, the copper resistivity rises to $0.07\,\mu\Omega$ m and the loss density rises to $1.38 \times 10^5\,\mathrm{W/m^2}$.

7.8 Total flux and power factor

When dealing with conduction, we were not interested in the magnitude of the flux. The flux becomes important in induction heating; this is not necessarily so for thick slabs, which are not practical, but the topic is included here for the principles it contains.

The total flux Φ per metre width is obtained by integrating $B_x\,\mathrm{d}y$ throughout the slab (see eqn 7.2):

$$\Phi = \mu H_s \int_0^\infty \exp\left[-(1 + \mathrm{j})\alpha y\right](\mathrm{d}y \times 1)$$

$$= \frac{\mu H_s}{\alpha\sqrt{2}} \exp\left(-\mathrm{j}\pi/4\right) \tag{7.10}$$

This is proportional to the imposed field H_s and is 45° behind it in phase. In eqn 7.7, the total current was shown to be in phase with H_s. Since the surface flux must be in phase with H_s, the total flux lags the surface flux by 45°.

The EMF is given by the time differential of Φ:

$$E = N\frac{\mathrm{d}\Phi}{\mathrm{d}t} = \frac{-\mu H_s N\omega}{\alpha\sqrt{2}} \sin\left(\omega t - \frac{\pi}{4}\right) = \frac{\mu H_s N\omega}{\alpha\sqrt{2}} \cos\left(\omega t + \frac{\pi}{4}\right) \tag{7.11}$$

The power factor of the slab comes from the relationship between E and I, which is shown on the phasor diagram of Fig. 7.4a. The current lags the EMF by 45°, so the power factor is 0.707 lagging, i.e. the resistance of the load is equal to the reactance. Under idealised conditions, this would also be the power factor of the existing coil but, in practice, the coil will itself have leakage reactance X_l, which worsens the overall power factor. In addition, the coil has a resistive volt-drop. Figure 7.4b shows the modified ideal phasor diagram.

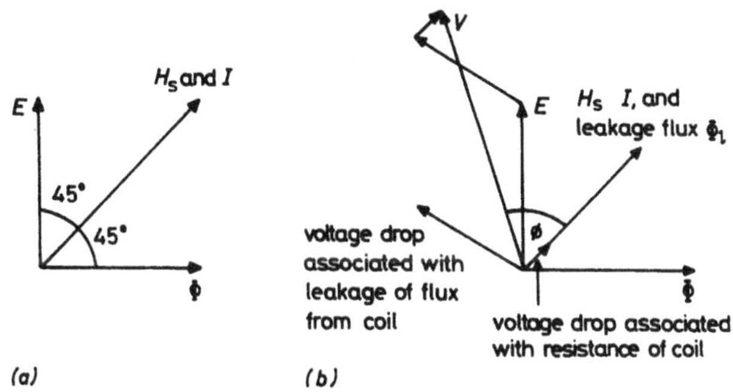

Fig. 7.4 *(a) Ideal phasor diagram for slab. (b) Phasor diagram, including coil resistance and leakage reactance*

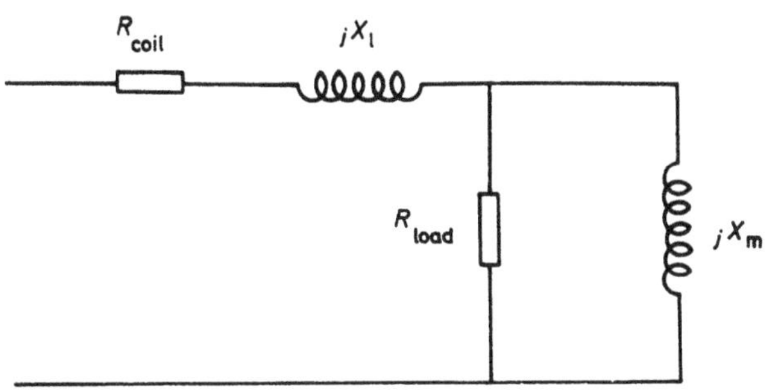

Fig. 7.5 *Equivalent circuit for slab*

The equivalent circuit model for the slab (Fig. 7.5) is best thought of as an iron-cored reactor, in which the load has the special property of having equal resistance and reactance. The flux associated with this reactance X_m has the nature of magnetising flux.

7.9 Power, using the Poynting theorem

In Section 7.7, the loss was obtained by integrating ρJ^2 throughout the volume. The active power through a surface (in W/m^2) is also given by the Poynting theorem[106]:

$$\mathscr{P} = \mathrm{Re}[\boldsymbol{E} \times \boldsymbol{H}^*] \tag{7.12}$$

From eqn 7.3,

$$E_s = \rho J_s = (\sqrt{2})\rho H_s \alpha \exp j\pi/4$$

so that

$$\mathscr{P} = \text{Re}\left[(\sqrt{2})\rho H_s^2 \alpha \frac{1 + j}{\sqrt{2}}\right]$$

$$= \rho H_s^2 \alpha$$

Since H is a sinusoidally-varying quantity, this gives the mean loss per m^2, using the RMS value of H, as in eqn 7.9.

Similarly, the reactive power (in var/m^2) is given by the imaginary part:

$$\mathscr{Q} = \text{Im}[E \times H^*] \qquad (7.13)$$

$$= \text{Im}\left[(\sqrt{2})\rho H_s^2 \alpha \frac{1 + j}{\sqrt{2}}\right]$$

$$= \rho H_s^2 \alpha$$

i.e. the same value as $\mathscr{P}$, which was to be expected for a power factor of 0.707, assuming constant permeability.

Chapter 8

Induction heating of thin slabs

8.1 Introduction

The analysis of Chapter 3 can be used, with different boundary conditions, for the induction heating of a slab surrounded by a coil. The width w is assumed to be large compared with the thickness $2b$. This assumption of $w \gg 2b$ implies

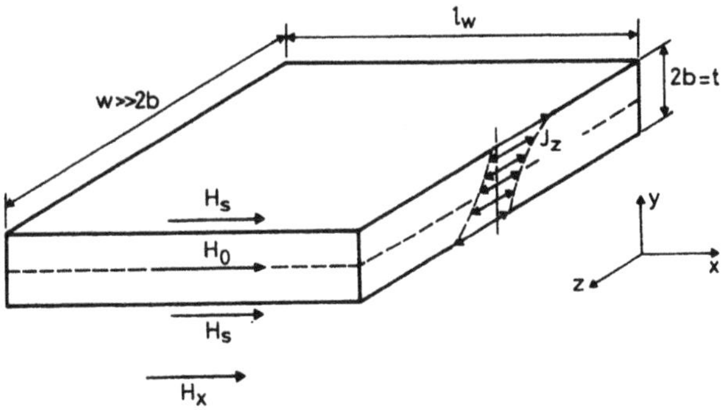

Fig. 8.1 *Thin slab, defining symbols and showing magnitudes of the current density*

negligible end-effects at the extremities of w. If $2b$ is greater than a certain thickness, to be discussed later, the two surfaces will be independent and the solution will be twice the value given in Chapter 7.

If the slab is thin, i.e. b is small, the current distribution will be somewhat like that shown in Fig. 8.1, where the H on each side of the slab induces some current in the other half, tending to oppose the stronger current induced by the nearer magnetic field. Note that the full exponential fall-off is not completed for either current. Figure 8.1 also defines the direction of the exciting field H_x, which implies that the current density is in the z direction. It is convenient to make the problem symmetrical by moving the y axis to the centre-line.

8.2 Magnetic field distribution

For induction heating it is the magnetic field that is applied, so the controlling equation must be in H (for conduction we started with J). Thus

$$\frac{\partial^2 H_x}{\partial y^2} = \frac{\mu}{\rho}\frac{\partial H_x}{\partial t} \tag{8.1}$$

for which the solution is

$$H_x = A_1 \exp(ky) + A_2 \exp(-ky)$$

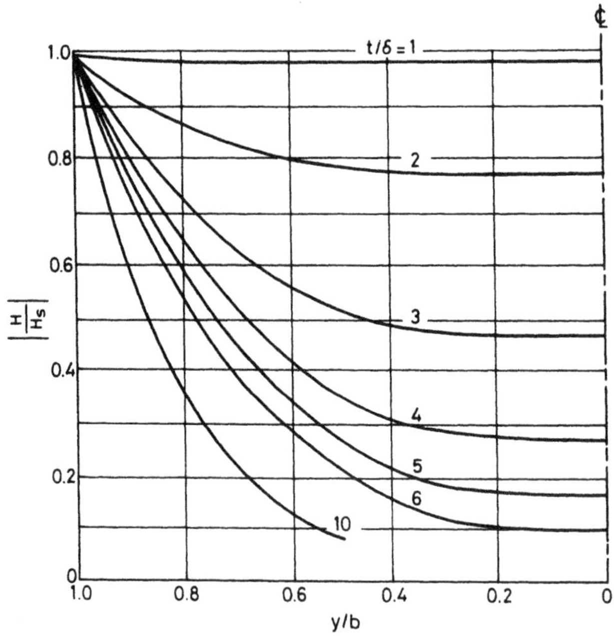

Fig. 8.2 *Fall-off of magnetic field with depth in thin slab, for various ratios of thickness to depth of penetration, expressed as the modulus of the H/H_s ratio*

Defining the surface value of H_x as $H_s = \mathrm{Re}[H_{sm} \exp \mathrm{j}\omega t]$ and noting that H is the same at $\pm b$:

$$H_b = A_1 \exp kb + A_2 \exp -kb = H_s$$

$$H_{-b} = A_1 \exp k(-b) + A_2 \exp kb = H_s$$

Only $A_1 = A_2 = A$ can make $H_b = H_{-b}$. At $y = 0$, $H_0 = 2A$, so that $H_s = H_0 \cosh kb$: more generally, $H = H_0 \cosh ky$. Hence

$$\frac{H_x}{H_s} = \frac{\cosh ky}{\cosh kb} = \frac{\cosh(1 + \mathrm{j})\alpha y}{\cosh(1 + \mathrm{j})\alpha b} \tag{8.2}$$

Since this equation has the same shape as eqn 3.1, the same analysis for the modulus and Fig. 3.2 apply. This has been relabelled as Fig. 8.2.

8.3 Current-density distribution

To obtain the J distribution, curl H could be used. However, it is easier to apply the obvious boundary condition of $J_b = -J_{-b}$ in $J = A_1' \exp ky + A_2' \exp - ky$, giving $A_1' = -A_2' = A'$ and $J_0 = 0$, as expected:

$$J_s = A'[\exp(kb) - \exp(-kb)] = 2A' \sinh kb$$

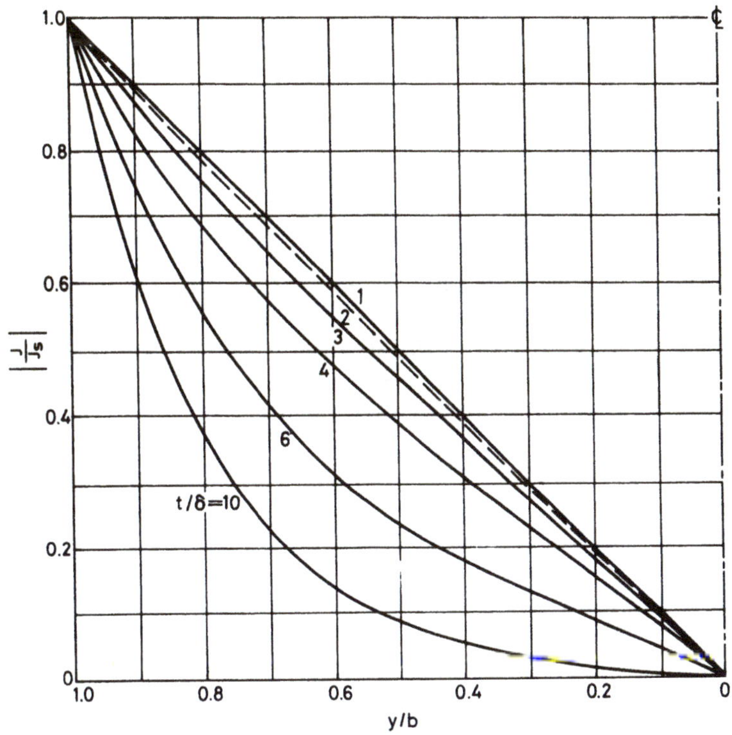

Fig. 8.3 *Fall-off of current density with depth in thin slab, for various ratios of thickness to depth of penetration, expressed as the modulus of the J/J_s ratio*

So

$$\frac{J_z}{J_s} = \frac{\sinh ky}{\sinh kb} = \frac{\sinh(1 + j)\alpha y}{\sinh(1 + j)\alpha b} \tag{8.3}$$

By a rationalisation method similar to that used in Section 3.2, we get

$$\left|\frac{J_z}{J_s}\right| = \left(\frac{\cosh 2\alpha y - \cos 2\alpha y}{\cosh 2\alpha b - \cos 2\alpha b}\right)^{1/2} \tag{8.4}$$

Plotting eqn 8.4 gives Fig. 8.3. The current density must be zero at the centre-line, from symmetry. The fall-off is linear, or nearly so, up to $t = 3\delta$. This is very different from the normal concept of 'depth of penetration', with its exponentials. We examined this more closely in Section 3.7. Whilst it was not needed in the theory of conducting slabs, it is clear that, by analogy, we could have started with the H equation there and used the condition that H must be zero at the centre-line to derive an expression identical in form to eqn 8.4 in H. It is therefore valid to label the y axis of Fig. 8.3 with $|H_x/H_s|_{\text{cond}}$, just as Fig. 8.2 can have dual labels, one for induction in H and one for conduction in J.

From $\boldsymbol{J} = \text{curl}\,\boldsymbol{H}$,

$$J = -k\frac{\partial H_x}{\partial y}$$

$$= -kH_s\frac{\partial}{\partial y}\left[\frac{\cosh(1+\mathrm{j})\alpha y}{\cosh(1+\mathrm{j})\alpha b}\right]$$

$$= -k\alpha H_s(1+\mathrm{j})\left[\frac{\sinh(1+\mathrm{j})\alpha y}{\cosh(1+\mathrm{j})\alpha b}\right] \tag{8.5}$$

At $y = b$, $|J_s| = (\sqrt{2})\alpha H_s \tanh(1+\mathrm{j})\alpha b$ and, for values where the slab is thick compared with the depth of penetration, i.e. large αb, $|J_s| \rightarrow (\sqrt{2})\alpha H_s$. The slab can then be considered as two independent half slabs, each behaving as a semi-infinite slab. This is true to 1% accuracy if $2b > 5.3\delta$; so, at 50 Hz, it is permissible to use the simpler concepts of semi-infinite slab theory for steel thicker than about 32 mm and for copper > 50 mm.

8.4 Flux distribution

The total flux Φ (webers) per metre width of the slab is obtained by integrating $B_x \mathrm{d}y$ across the slab:

$$\Phi = \mu H_s \int_{-b}^{b} \frac{\cosh(1+\mathrm{j})\alpha y}{\cosh(1+\mathrm{j})\alpha b}\,\mathrm{d}y(\times 1)$$

$$= \frac{2\mu H_s}{(1+\mathrm{j})\alpha}\tanh(1+\mathrm{j})\alpha b$$

Replace time and use the identity $(1+\mathrm{j}) = (\sqrt{2})\exp \mathrm{j}45°$:

$$\Phi = \frac{2\mu H_s}{\alpha\sqrt{2}}\tanh(1+\mathrm{j})\alpha b \angle -\pi/4 \tag{8.6}$$

This expression is effectively twice (two sides) that for the semi-infinite slab (eqn 7.10) for $2b > 5.3\delta$. For thicknesses less than this, the flux is reduced by $\tanh(1+\mathrm{j})(\alpha b)$. The modulus of $\tanh(1+\mathrm{j})(\alpha b)$ can be obtained, using various identities, as

$$|\tanh(1 + j)\alpha b| = \sqrt{\left(\frac{\cosh 2\alpha b - \cos 2\alpha b}{\cosh 2\alpha b + \cos 2\alpha b}\right)}$$

8.5 Power

We could get the real power by integrating ρJ^2 throughout the volume, but Poynting gives us both real and reactive power at the same time. This will prove specially useful here. Poynting gives, remembering that power is entering from both sides,

$$\mathscr{P} + j\mathscr{Q} = \boldsymbol{E} \times \boldsymbol{H}^* \qquad \text{VA/m}^2 \tag{8.7}$$

$$= \rho\alpha H_s(1 + j)\left[\frac{\sinh(1 + j)\alpha b}{\cosh(1 + j)\alpha b}\right]H_s$$

$$= \rho\alpha H_s^2(1 + j)[\tanh(1 + j)\alpha b]$$

$$= \rho\alpha H_s^2(1 + j)\left[\frac{\sinh 2\alpha b + j\sin 2\alpha b}{\cosh 2\alpha b + \cos 2\alpha b}\right]$$

$$= \rho\alpha H_s^2\left[\frac{\sinh 2\alpha b - \sin 2\alpha b}{\cosh 2\alpha b + \cos 2\alpha b} + j\,\frac{\sinh 2\alpha b + \sin 2\alpha b}{\cosh 2\alpha b + \cos 2\alpha b}\right] \tag{8.8}$$

Compare this expression with that given in Section 3.4.1. It will be seen that both have the same form, except that the real and imaginary parts inside [ ] are

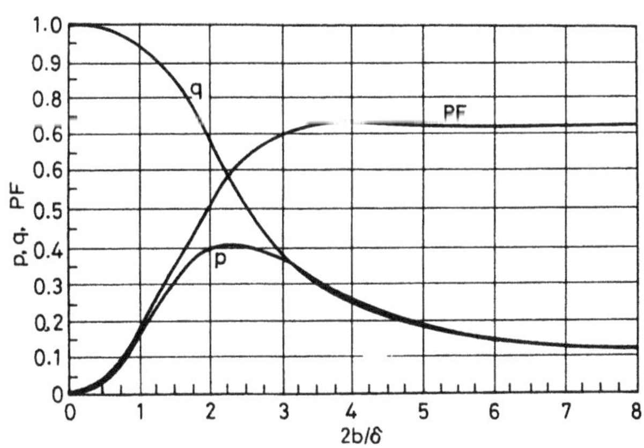

Fig. 8.4 *p and q functions for a thin slab, together with power factor. When $2b/\delta = t/\delta > 8$, $p = q = 1/(t/\delta)$*

in reverse order. We can rearrange this to give:

$$\mathcal{P} + j\mathcal{Q} = 2b\rho\alpha^2 H_s^2[\quad]/2\alpha b$$
$$= 2b\rho\alpha^2 H_s^2[p + jq] \qquad (8.9)$$

The p and q terms correspond to those defined in Reference 7. They have the same values as q_r and p_r for the resistive case, so we do not need to recalculate them but simply relabel the curves of Fig. 3.3; this has been done in Fig. 8.4. Using the RMS of H gives the mean value of $\mathcal{P} + j\mathcal{Q}$.

The power into the slab, P_w, is (both sides)

$$P_w = (2lw)\mathcal{P}$$
$$= (2lbw)2\rho\alpha^2 H_s^2 p$$
$$= (\text{slab volume})(\mu\pi f)H_s^2 p \qquad (8.10)$$

Equations 8.9 and 8.10 and Fig. 8.4 are very important in induction heating:

- For a fixed frequency, the power/volume ratio (i.e. the quantity that defines the rate of heating) is proportional to p, so the optimum comes where $p = 0.417$ at a work thickness/δ ratio of 2.25. The curve is flat-topped, so only 1% is lost between 2.1 and 2.5.
- For a given workpiece and free choice of frequency, the power/volume ratio is proportional to fp. For (thickness/δ) > 8, p is proportional to δ and hence to $1/\sqrt{f}$, so that power/volume and $\mathcal{P}$ are proportional to $\sqrt{f}$. It is always advantageous to increase frequency, other things being equal.

Example 8.1

Given the following data, calculate the RMS magnetising field:

Metal	steel
Required power P_w	50 kW
Frequency	3000 Hz
Resistivity ρ	1 μΩ m
Width w	125 mm
Thickness t	6 mm
Length l_w	250 mm

Solution

Since we do not yet know H, assume that $\mu_r = 20$, as recommended in Chapter 2. The skin depth is given by

$$\delta = \sqrt{(2\rho/\mu\omega)} = \sqrt{\left[\frac{2(10^{-6})}{(20)(4\pi \times 10^{-7})(2\pi \times 3000)}\right]}$$
$$= 2.05 \times 10^{-3} \quad \text{m}$$

and so $t/\delta = 2.92$ and $p = 0.38$ from Fig. 8.4. Using eqn 8.10, P_w = volume × $\mu\pi f \times H_s^2 p$, gives:

$$\begin{aligned} H_s^2 &= 50\,000/(0.25)(0.125)(0.006)(20 \times 4\pi \times 10^{-7})(\pi \times 3000)(0.38) \\ &= 2.96 \times 10^9 \\ H_s &= 54\,430 \quad \text{A/m} \end{aligned}$$

This value of H does not correspond to the assumed μ_r. It is necessary to iterate, giving:

μ_r	δ	t/δ	p	H
33	1.5	4	0.25	50 141
34.8	1.55	3.86	0.26	45 719
40	1.45	4.14	0.24	43 090
40	same for practical purposes			

So H_s = 43 100 A/m.

For this calculation, it is convenient to use the analytical relationship $\mu_r = 734\,000/H^{0.92}$ linking μ_r and H.

8.6 Power factor

As in Section 4.3.3, we derive the power factor from $\cos[\tan^{-1}(q/p)]$, where q/p is given by

$$(\sinh t/\delta + \sin t/\delta)/(\sinh t/\delta - \sin t/\delta)$$

These values are plotted on Fig. 8.4. Very thin slabs give bad power factors. At maximum p, the PF has reached 0.6, after which it slightly overshoots cos 45° before settling down to 0.707. The PF curve, taken in conjunction with that of p, emphasises the importance of working above the maximum p. Even at its best the PF is not as good as it is for conduction heating; this is an inherent feature of induction heating and has to be dealt with by PF correction.

8.7 Duality

The duality of the conduction and induction solutions is beginning to emerge. For slabs, we have:

Induction		*Conduction*		*Value*
$\dfrac{B_x}{B_s}$	$\dfrac{H_x}{H_s}$	$\dfrac{J_x}{J_s}$	$\dfrac{E_x}{E_s}$	$\dfrac{\cosh(1 + j)y/\delta}{\cosh(1 + j)b/\delta}$
$\dfrac{J_z}{J_s}$	$\dfrac{E_z}{E_s}$	$\dfrac{B_z}{B_s}$	$\dfrac{H_z}{H_s}$	$\dfrac{\sinh(1 + j)y/\delta}{\sinh(1 + j)b/\delta}$

Induction		*Conduction*		*Value*
$\left\lvert\frac{B_x}{B_s}\right\rvert$	$\left\lvert\frac{H_x}{H_s}\right\rvert$	$\left\lvert\frac{J_x}{J_s}\right\rvert$	$\left\lvert\frac{E_x}{E_s}\right\rvert$	$\frac{\cosh(y/b)(t/\delta) + \cos(y/b)(t/\delta)}{\cosh(t/\delta) + \cos(t/\delta)}$
$\left\lvert\frac{J_z}{J_s}\right\rvert$	$\left\lvert\frac{E_z}{E_s}\right\rvert$	$\left\lvert\frac{B_z}{B_s}\right\rvert$	$\left\lvert\frac{H_z}{H_s}\right\rvert$	$\frac{\cosh(y/b)(t/\delta) - \cos(y/b)(t/\delta)}{\cosh(t/\delta) - \cos(t/\delta)}$
p		q_r		$(\delta/2b)\left[\frac{\sinh 2b/\delta - \sin 2b/\delta}{\cosh 2b/\delta + \cos 2b/\delta}\right]$
q		p_r		$(\delta/2b)\left[\frac{\sinh 2b/\delta + \sin 2b/\delta}{\cosh 2b/\delta + \cos 2b/\delta}\right]$

We shall see a similar symmetry for cylinders.

Chapter 9

Induction heating of cylinders

This chapter uses theory similar to that of Chapter 4. We shall start from first principles to establish the mechanism of induction heating, but we shall not repeat the Bessel discussions of Sections 4.1.1–4.1.3. Instead, we shall refer back. By now the reader will be expecting a duality relationship with Chapter 4, and this will emerge.

9.1 Basic principles

Figure 9.1 shows a long cylindrical workpiece excited by a close-coupled axial coil (solenoid), producing H_s at the surface. Since H acts in the axial (z) direction, E and J are in the peripheral (θ) direction only. As in Section 4.1, we take $\oint H\, dl = I$ and $\oint E\, dl = -d\Phi/dt$. Taking $\oint H\, dl$ for the shaded area of Fig. 9. 1b:

$$-H_{zr}\, l + (H_{zr} + \delta H_{zr})\, l = J_{\theta r}\, \delta r\, l$$

(ignoring second-order differentials), Thus

$$dH_{zr}/dr = J_{\theta r} \tag{9.1}$$

Taking $\oint E\, dl$ for the shaded area of Fig. 9. 1a:

$$-\rho\, J_{\theta r}\, 2\pi r + \rho\, (J_{\theta r} + \delta J_{\theta r})\, (2\pi r) = -(d/dt)\mu\, 2\pi r \delta r\, H_{zr}$$

$$\rho\, \delta J_{\theta r} = -\mu j\omega\, H_{zr}\, \delta r$$

$$\rho\, (dJ_{\theta r}/dr) = -\mu j\omega\, H_{zr} \tag{9.2}$$

Combining eqns 9.1 and 9.2 gives

$$\frac{d^2H}{dr^2} + \frac{dH}{dr}\frac{1}{r} = 2j\alpha^2\, H \tag{9.3}$$

where $\alpha^2 = \omega\mu/2\rho$, as before. The direction of H has been omitted from this equation as it is unambiguous. Equation 9.3 is identical to eqn 4.3, except that H has taken the place of J; it follows that the solution is identical.

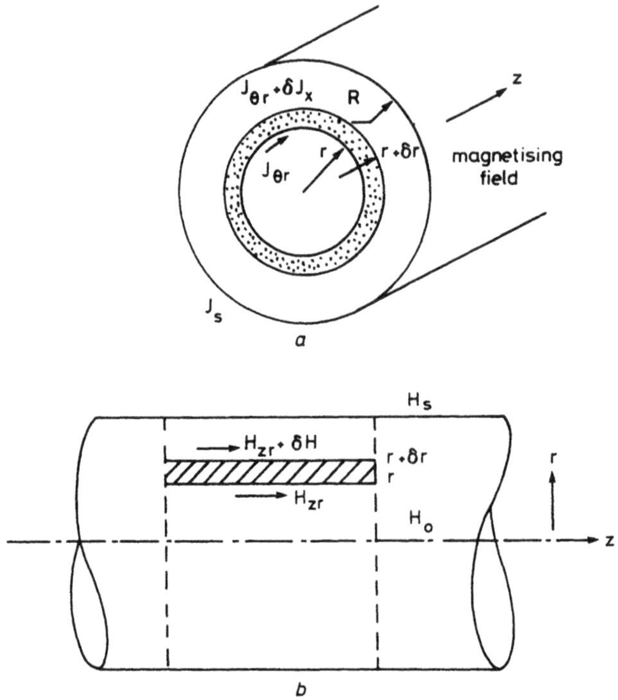

Fig. 9.1 *Cylindrical workpiece, defining symbols*

9.2 Magnetic field distribution

Following Section 4.2, the H distribution is given by

$$H = A\, I_0\, [\sqrt{(2j)}\ \alpha r) + B\, K_0\, [\sqrt{(2j)}\ \alpha r) \tag{9.4}$$

By a similar argument that H must not be infinite at $r = 0$ and defining the surface H as $H_s = H_0\, M_0\,(mR) \angle \theta_0\,(mR)$, we get

$$\frac{H}{H_s} = \frac{M_0\,(mr)}{M_0\,(mR)}\,\frac{\angle \theta_0(mr)}{\angle \theta_0(mR)} \tag{9.5}$$

and

$$\left|\frac{H}{H_s}\right| = \frac{M_0\,(mr)}{M_0\,(mR)} \tag{9.6}$$

This is the same curve as Fig. 4.4 and is reproduced here, relabelled, as Fig. 9.2. The phase change is again uniform, except at the centre of the billet (see Section 4.2).

Whilst eqns 9.5 and 9.6 are convenient for calculation, it may be helpful to record the forms

$$\frac{H}{H_s} = \frac{\text{ber}\sqrt{(2)}r/\delta + \text{j bei}\sqrt{(2)}r/\delta}{\text{ber}\sqrt{(2)}R/\delta + \text{j bei}\sqrt{(2)}R/\delta} \tag{9.5a}$$

and

$$\left|\frac{H}{H_s}\right| = \sqrt{\left[\frac{\text{ber}^2\sqrt{(2)}r/\delta + \text{bei}^2\sqrt{(2)}r/\delta}{\text{ber}^2\sqrt{(2)}R/\delta + \text{bei}^2\sqrt{(2)}R/\delta}\right]} \tag{9.6a}$$

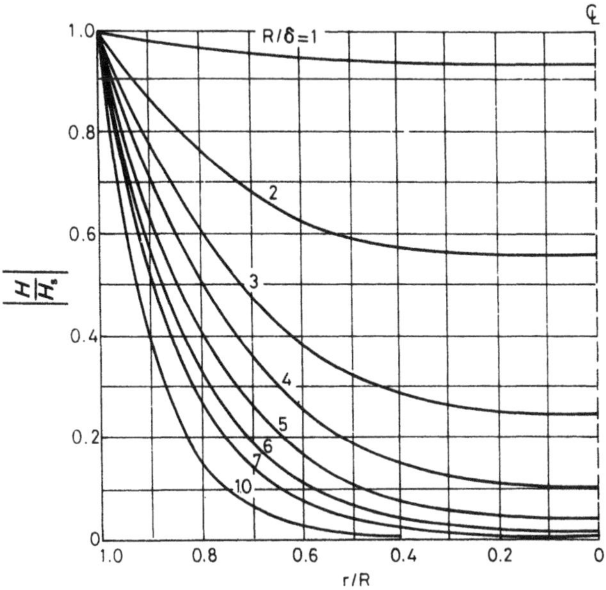

Fig. 9.2 *Fall-off of magnetic field with radius in cylinder, for various ratios of radius to depth of penetration, expressed as the modulus of H/H_s (given by $M_0(mr)/M_0(mR)$)*

9.3 Flux distribution

The total flux Φ_r lying inside radius r is given by

$$\Phi_r = \int_0^r 2\pi r\, \mu H\, \mathrm{d}r$$

$$= 2\pi\mu H_s \int_0^r r\, \frac{\text{ber}\sqrt{(2)}r/\delta + \text{j bei}\sqrt{(2)}r/\delta}{\text{ber}\sqrt{(2)}R/\delta + \text{j bei}\sqrt{(2)}R/\delta}\, \mathrm{d}r$$

$$= 2\pi\mu H_s \frac{\delta R}{\sqrt{2}} \frac{\text{bei}'\sqrt{(2)}r/\delta - \text{j ber}'\sqrt{(2)}r/\delta}{\text{ber}\sqrt{(2)}R/\delta + \text{j bei}\sqrt{(2)}R/\delta} \tag{9.7}$$

Hence, the total flux Φ_R lying inside radius R is

$$\Phi_R = 2\pi\mu H_s \frac{\delta R}{\sqrt{2}} \frac{\text{bei}'\sqrt{(2)}R/\delta - \text{j ber}'\sqrt{(2)}R/\delta}{\text{ber}\sqrt{(2)}R/\delta + \text{j bei}\sqrt{(2)}R/\delta}$$

This can be rationalised to give

$$\Phi_R = \mu H_s A_w (q - \text{j}p) \tag{9.8}$$

where

$$p = \frac{\sqrt{(2)}\delta}{R} \frac{\text{ber}\sqrt{(2)}R/\delta \ \text{ber}'\sqrt{(2)}R/\delta + \text{bei}\sqrt{(2)}R/\delta \ \text{bei}'\sqrt{(2)}R/\delta}{\text{ber}^2\sqrt{(2)}R/\delta + \text{bei}^2\sqrt{(2)}R/\delta}$$

$$q = \frac{\sqrt{(2)}\delta}{R} \frac{\text{ber}\sqrt{(2)}R/\delta \ \text{bei}'\sqrt{(2)}R/\delta - \text{bei}\sqrt{(2)}R/\delta \ \text{ber}'\sqrt{(2)}R/\delta}{\text{ber}^2\sqrt{(2)}R/\delta + \text{bei}^2\sqrt{(2)}R/\delta}$$

and A_w is the cross-section of the workpiece (πR^2). These p and q expressions have the same forms as the q_r and p_r expressions of eqns 4.15 and 4.17. They also appear in the equations for power (see Section 9.5).

9.4 Current-density distribution

Using $\boldsymbol{J}$ = curl $\boldsymbol{H}$ in cylindrical coordinates:

$$\boldsymbol{J} = \text{curl } \boldsymbol{H} = \begin{vmatrix} \boldsymbol{a}_r & \boldsymbol{a}_\psi & \boldsymbol{a}_z \\ \partial/\partial r & \partial/\partial\psi & \partial/\partial z \\ H_r & rH_\psi & H_z \end{vmatrix}$$

However, $\boldsymbol{H}$ only exists in the z direction and there is no variation of $\boldsymbol{H}$ in the z and ψ directions, so

$$\boldsymbol{J} = \text{curl } \boldsymbol{H} = \begin{vmatrix} \boldsymbol{a}_r & \boldsymbol{a}_\psi & \boldsymbol{a}_z \\ \partial/\partial r & 0 & 0 \\ 0 & 0 & H_z \end{vmatrix}$$

$$= -\boldsymbol{a}_\psi (\partial/\partial r) H_z$$

$$J = -\frac{\partial}{\partial r} \frac{\text{ber}\sqrt{(2)}r/\delta + \text{j bei}\sqrt{(2)}r/\delta}{\text{ber}\sqrt{(2)}R/\delta + \text{j bei}\sqrt{(2)}R/\delta} H_s \tag{9.9}$$

$$= -\frac{\text{ber}'\sqrt{(2)}r/\delta + \text{j bei}'\sqrt{(2)}r/\delta}{\text{ber}\sqrt{(2)}R/\delta + \text{j bei}\sqrt{(2)}R/\delta} \frac{\sqrt{(2)}H_s}{\delta} \tag{9.10}$$

But, at $r = R$, $J = J_s$:

$$J_s = -\frac{\text{ber}'\sqrt{(2)}R/\delta + \text{j bei}'\sqrt{(2)}R/\delta}{\text{ber}\sqrt{(2)}R/\delta + \text{j bei}\sqrt{(2)}R/\delta} \tag{9.11}$$

So

$$\frac{J}{J_s} = \frac{\text{ber}'\sqrt{(2)}r/\delta + \text{j bei}'\sqrt{(2)}r/\delta}{\text{ber}'\sqrt{(2)}R/\delta + \text{j bei}'\sqrt{(2)}R/\delta} \tag{9.12}$$

and

$$\left|\frac{J}{J_s}\right| = \sqrt{\left[\frac{\text{ber}'^2\sqrt{(2)}r/\delta + \text{bei}'^2\sqrt{(2)}r/\delta}{\text{ber}'^2\sqrt{(2)}R/\delta + \text{bei}'^2\sqrt{(2)}R/\delta}\right]} \tag{9.13}$$

This modulus is also given by $M_1\,(mr)/M_1\,(mR)$, which is easier to calculate, using tables (see Appendix A.3). It is plotted on Fig. 9.3, which is similar to Fig. 8.3 in form. It could be relabelled as $|H/H_s|$ in Chapter 4. Note that eqn 9.12 can also be written as

$$\frac{J}{J_s} = \frac{M_1\,(mr)\exp \text{j}\theta_1\,(mr)}{M_1\,(mR)\exp \text{j}\theta_1\,(mR)} \tag{9.14}$$

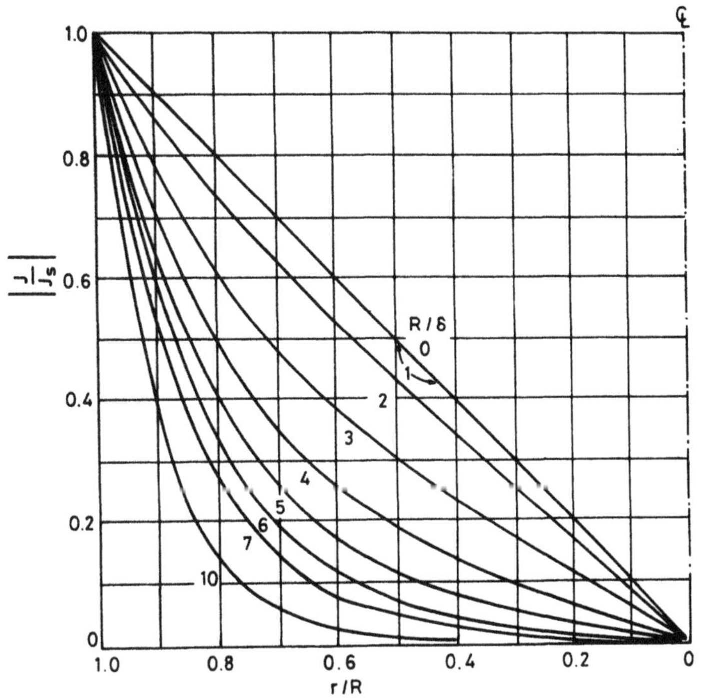

Fig. 9.3 *Fall-off of current density with depth in cylinder, for various ratios to depth of penetration, expressed as the modulus of the J/J_s ratio (given by $M_1(mr)/M_1(mR)$)*

Example 9.1

Examine the current-density distribution for the following data:

Metal	steel
Frequency f	50 Hz
Resistivity ρ	1 $\mu\Omega$ m
Radius R	7.5 mm
Magnetising field H	10 000 A/m (μ_r = 20)

Solution
The depth of penetration is

$$\delta = \sqrt{(2\rho/\mu\omega)} = \sqrt{\{(2)(10^{-6})/(20)(4\pi \times 10^{-7})(2\pi 50)\}} = 15.9\,\text{mm}$$

Thus $R/\delta = 7.5/15.9 = 0.47$. From Fig. 9.3, for $R/\delta = 0.47$, the current density drops to half its surface value at $r/R = 0.5$, so that, at this low frequency, the skin effect is not very pronounced.

Example 9.2
Examine the current-density distribution for the following data:

Metal	stainless steel
Frequency f	450 kHz
Resistivity ρ	1 $\mu\Omega$ m
Relative permeability μ_r	1
Radius R	7.5 mm

Solution
The depth of penetration is now

$$\delta = \sqrt{(2\rho/\mu\omega)} = \sqrt{\{(2)(10^{-6})/(4\pi \times 10^{-7})(2\pi\, 450\,000)\}} = 0.75\,\text{mm}$$

So $R/\delta = 7.5/0.75 = 10$. From Fig. 9.3, for $R/\delta = 10$, the current density drops to half its surface value at $r/R \simeq 0.92$, so that the skin effect is very pronounced, due to the high frequency.

$$\text{ber}\, x \simeq G \cos [(x/\sqrt{2}) - (\pi/8)]$$

$$\text{bei}\, x \simeq G \sin [(x/\sqrt{2}) - (\pi/8)]$$

$$\text{ber}'x \simeq G \cos [(x/\sqrt{2}) + (\pi/8)]$$

$$\text{bei}'x \simeq G \sin [(x/\sqrt{2}) + (\pi/8)]$$

where $G = \exp (x/\sqrt{2})/\sqrt{(2\pi x)}$, Thus

$$\frac{J}{J_s} = \sqrt{(R/r)} \exp \{(r - R)/\delta\} \frac{\cos (r/\delta + \pi/8) + \text{j} \sin (r/\delta + \pi/8)}{\cos (R/\delta + \pi/8) + \text{j} \cos (R/\delta + \pi/8)} \tag{9.15}$$

As r and R themselves become very large, the substitution $r = R - y$ reduces this to

$$\frac{J}{J_s} = \exp(-y/\delta)\exp(-jy/\delta) \tag{9.16}$$

which is the same as for a thick slab. This shows that the Bessel functions are really modifications of the basic slab equations to allow for the cylindrical shape.

9.5 Power

Using Poynting again, to get the real and reactive components of power:

$$\mathscr{P} + j\,\mathscr{Q} = \boldsymbol{E} \times \boldsymbol{H}^* \qquad \mathrm{VA/m^2}$$

Substitute eqn 9.10 for J:

$$\mathscr{P} + j\,\mathscr{Q} = \frac{\rho\sqrt{(2)}H_s}{\delta}\,\frac{\mathrm{ber}'\sqrt{(2)}/\delta + j\,\mathrm{bei}'\sqrt{(2)}R/\delta}{\mathrm{ber}\sqrt{(2)}R/\delta + j\,\mathrm{bei}\sqrt{(2)}R/\delta}\,H_s$$

$$= \frac{\rho\sqrt{(2)}H_s^2}{\delta}\,\frac{(\mathrm{ber}'\sqrt{(2)}R/\delta + j\,\mathrm{bei}'\sqrt{(2)}R/\delta)(\mathrm{ber}\sqrt{(2)}R/\delta - j\,\mathrm{bei}\sqrt{(2)}R/\delta)}{\mathrm{ber}^2\sqrt{(2)}R/\delta + \mathrm{bei}^2\sqrt{(2)}R/\delta}$$

$$= \frac{\rho R H_s^2}{\delta^2}[p + jq] \tag{9.17}$$

where p and q are defined in Section 9.3. If we use the RMS value of H_s, this gives the mean value.

The power in the workpiece P_w is given by $\mathscr{P}$ × surface area:

$$P_w = (\rho R H_s^2/\delta^2)\,p\,(2\pi R l)$$

$$= (\pi R^2 l)H_s^2\,\mu 2\pi f\,p$$

$$= (\text{billet volume})\;2\pi\mu f\,pH_s^2 \tag{9.18}$$

where $\mu = \mu_r\mu_0$.

p and q are plotted for solid cylinders in Fig. 9.4. Equations 9.17 and 9.18 and Fig. 9.4 are very important in billet heating. The remarks about frequency in Section 8.5, modified for the new values of p and q, also apply to billets.

This section has linked power with surface magnetic intensity H_s, which is what is needed in the design of induction heaters.

Example 9.3
The following data are given for the induction heating of a cylinder;

Metal	aluminium
Required work power P_w	100 kW
Frequency f	50 Hz

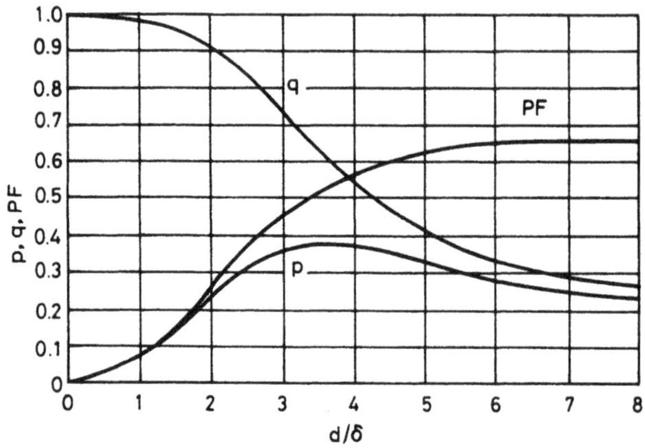

Fig. 9.4 *p and q functions for a solid cylinder, together with power factor. When $d/\delta > 8$, $q = 2/(d/\delta)$; $p = 2/(1.23 + d/\delta)$*

Resistivity ρ	$0.057\ \mu\Omega\,\mathrm{m}$
Relative permeability μ_r	1
Radius R	40 mm
Length	500 mm

What value of H_s is needed?

Solution

We have

$$\delta = \sqrt{(2\rho/\mu\omega)} = \sqrt{[(2)(5.7 \times 10^{-8})/(4\pi \times 10^{-7})(2\pi \times 50)]} = 0.017\,\mathrm{m}$$

Hence $d_w/\delta = 80/17 = 4.71$, so $p = 0.33$ from Fig, 9.4.

From eqn 9.18,

$$H_s^2 = P_w/(\text{volume})\ 2\pi\mu f\, p$$

$$= (100\,000)/(\pi \times 0.04^2 \times 0.5)\ (2\pi \times 4\pi \times 10^{-7} \times 50)(0.33)$$

$$H_s = 552\,000 \quad \mathrm{A/m}$$

At a fixed frequency, there is an optimum diameter for maximum power per unit volume, which occurs where p is maximum, i.e. at $d/\delta = 3.5$. In the above example, the optimum diameter is $3.5 \times 0.017\,\mathrm{m} \simeq 60\,\mathrm{mm}$ at 50 Hz. It will be seen later that other factors affect optimum diameter.

It is never disadvantageous to increase frequency when considering loss. Equation 9.18 gives P_w proportional to fp, where p is also a function of f. If conditions are arranged to give p maximum, then doubling f reduces p from 0.37 to 0.32, but the increased f offsets this and increases the loss by $(2 \times 0.32/0.37) = 1.73$. In the limit, as f increases, $p \rightarrow 2/(d/\delta)$ and P_w varies as

$f/\sqrt{f} = \sqrt{f}$. Below the maximum p, it can be shown that p is approximately proportional to $(d/\delta)^{1.6}$, that p varies as $f^{0.8}$ and that P_w varies as $f^{1.8}$; increasing the frequency under these conditions rapidly improves heating.

Brown, Hoyler and Bierwirth[8] discuss the question of frequency from the viewpoint of the power absorbed in the cylinder compared with that absorbed in the exciting coil. They show that little benefit is obtained by making $d/\delta > 4.5$. From this, a 'critical' frequency can be defined:

$$f_c = \frac{\rho}{\mu_0 \mu_r d^2} \frac{4.5^2}{\pi} = \frac{6.45\rho}{\mu d^2} \tag{9.19}$$

Where a free choice of frequency exists, it should be set equal to or just greater than f_c. (The term 'critical frequency' is misleading; it only indicates a frequency *below* which there is a loss of heating.)

9.6 Power factor

If the surface magnetic intensity H_s is taken as the reference phasor, the relative angles of the flux and current are shown in Fig. 9.5.

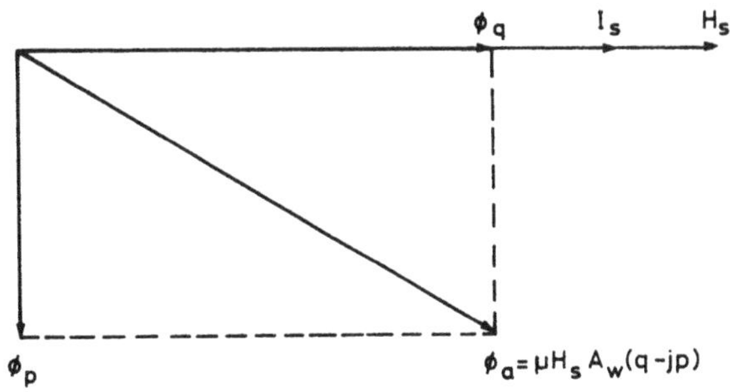

Fig. 9.5 *Phasor diagram of total work*

The total current is obtained by integrating J through the billet, but its value is not very useful. As in Section 8.6, we can get the power-factor angle from cos $[\tan^{-1}(q/p)]$. This is plotted on Fig. 9.4; again, the numbers are low. At the peak value of p, the power factor is about 0.5, rising to 0.65 at $d/\delta = 6$ and staying fairly constant at this. Even at $d/\delta = 20$, it has only risen to 0.69, i.e. it still has not reached 0.707. Compared with the values for the thin slab, where 0.7 is reached by $t/\delta = 3$ (Fig. 8.4), the cylinder is much worse and more PF correction is needed.

Since

$$p = \frac{\sqrt{(2)}\delta}{R} \frac{\text{ber}\sqrt{(2)}R/\delta \ \text{ber}'\sqrt{(2)}R/\delta + \text{bei}\sqrt{(2)}R/\delta \ \text{bei}'\sqrt{(2)}R/\delta}{\text{ber}^2\sqrt{(2)}R/\delta + \text{bei}^2\sqrt{(2)}R/\delta}$$

$$q = \frac{\sqrt{(2)}\delta}{R}\,\frac{\mathrm{ber}\sqrt{(2)}R/\delta\ \mathrm{bei}'\sqrt{(2)}R/\delta - \mathrm{bei}\sqrt{(2)}R/\delta\ \mathrm{ber}'\sqrt{(2)}R/\delta}{\mathrm{ber}^2\sqrt{(2)}R/\delta + \mathrm{bei}^2\sqrt{(2)}R/\delta}$$

and

$$\mathrm{bei}'\,x\ \mathrm{ber}\,x - \mathrm{ber}'\,x\ \mathrm{bei}\,x = M_0(x)M_1(x)\sin(\theta_1 - \theta_0 - \pi/4)$$

$$\mathrm{ber}'\,x\ \mathrm{ber}\,x + \mathrm{bei}'\,x\ \mathrm{bei}\,x = M_0(x)M_1(x)\cos(\theta_1 - \theta_0 - \pi/4)$$

So

$$q/p = \tan(\theta_1 - \theta_0 - \pi/4)$$

$$\cos[\tan^{-1}(q/p)] = \cos(\theta_1 - \theta_0 - \pi/4)$$

This gives us a straightforward method to calculate PF. The values of θ are tabulated in Reference 4. A similar method can be used to calculate the PF curve of Fig. 4.5.

9.7 Duality

The duality of the conduction and induction solutions is also true for cylinders:

Induction		*Conduction*		*Value*
$\frac{B}{B_s}$	$\frac{H}{H_s}$	$\frac{J}{J_s}$	$\frac{E}{E_s}$	$\frac{M_0(mr)\exp \mathrm{j}\,\theta_0(mr)}{M_0(mR)\exp \mathrm{j}\,\theta_0(mR)}$
$\frac{J}{J_s}$	$\frac{E}{E_s}$	$\frac{B}{B_s}$	$\frac{H}{H_s}$	$\frac{M_1(mr)\exp \mathrm{j}\,\theta_1(mr)}{M_1(mR)\exp \mathrm{j}\,\theta_1(mR)}$
$\left\lvert\frac{B}{B_s}\right\rvert$	$\left\lvert\frac{H}{H_s}\right\rvert$	$\left\lvert\frac{J}{J_s}\right\rvert$	$\left\lvert\frac{E}{E_s}\right\rvert$	$\frac{M_0(mr)}{M_0(mR)}$
$\left\lvert\frac{J}{J_s}\right\rvert$	$\left\lvert\frac{E}{E_s}\right\rvert$	$\left\lvert\frac{B}{B_s}\right\rvert$	$\left\lvert\frac{H}{H_s}\right\rvert$	$\frac{M_1(mr)}{M_1(mR)}$
p		q_r		$\frac{\sqrt{(2)}\delta}{R}\frac{\mathrm{ber}\sqrt{(2)}R/\delta\ \mathrm{ber}'\sqrt{(2)}R/\delta + \mathrm{bei}\sqrt{(2)}R/\delta\,\mathrm{bei}'\sqrt{(2)}R/\delta}{\mathrm{ber}^2\sqrt{(2)}R\delta + \mathrm{bei}^2\sqrt{(2)}R/\delta}$
q		p_r		$\frac{\sqrt{(2)}\delta}{R}\frac{\mathrm{ber}\sqrt{(2)}R/\delta\ \mathrm{bei}'\sqrt{(2)}R/\delta - \mathrm{bei}\sqrt{(2)}R/\delta\ \mathrm{ber}'\sqrt{(2)}R/\delta}{\mathrm{ber}^2\sqrt{(2)}R/\delta + \mathrm{bei}^2\sqrt{(2)}R/\delta}$

Chapter 10

Induction heating of tubes

The basic theory of induction heating of tubes[4] follows that of Chapter 9, but with different boundary conditions at the inner diameter. The solution is akin to that of conduction heating of tubes, as will be expected by now, but there is an added complication due to the flux inside the tube. We shall derive it to bring out the principles. Let the outer radius be R and the inner radius be b.

10.1 Relationship between the constants A and B in the field distribution

The general solution of eqn 9.3 is:

$$H = A(\text{ber } mr + \text{j bei } mr) + B(\text{ker } mr + \text{j kei } mr) \tag{10.1}$$

where $m = (\sqrt{2})/\delta$. At $r = R$,

$$H_s = A(\text{ber } mR + \text{j bei } mR) + B(\text{ker } mR + \text{j kei } mR) \tag{10.2}$$

At $r = b$,

$$H_b = A(\text{ber } mb + \text{j bei } mb) + B(\text{ker } mb + \text{j kei } mb) \tag{10.3}$$

To get the relationship between A and B, we use the fact that, at $r = b$, the induced voltage equals the time derivative of the flux inside the tube. Because the bore is air, the flux density is uniformly distributed and proportional to the MMF at the inner surface, i.e. $\mu_0 H_b$:

$$E\,2\pi b = \rho J_b\,2\pi b = -\text{j}\omega(B\pi b^2) = -\text{j}\omega(\mu_0 H_b)\,\pi b^2$$

$$J_b = -\frac{\text{j}\omega\mu_0 H_b \pi b^2}{2\pi\rho\, b} = -\frac{\text{j}\, bH_b}{\mu_r \delta^2} \tag{10.4}$$

μ_r is 'the relative permeability of the tube material'. It is often difficult, in practice, to attach physical meaning to those words but, for the moment, let us assume that it has a constant value throughout the tube material. Note, too, that we have introduced δ, which is a somewhat artificial concept in this problem: it is retained for comparison with previous work and can be regarded as a numeric.

We also know that $J = -\partial H/\partial r$, so

$$J_b = -(\partial H/\partial r)_{r=b}$$

$$= -m[A(\text{ber}' \, mb + \text{j bei}' \, mb) + B(\text{ker}' \, mb + \text{j kei}' \, mb)] \qquad (10.5)$$

and

$$J = -m[A(\text{ber}' \, mr + \text{j bei}' \, mr) + B(\text{ker}' \, mr + \text{j kei}' \, mr)] \qquad (10.6)$$

Equating J_b from eqns 10.4 and 10.5:

$$\frac{\text{j } bH_b}{\mu_r \delta^2} = m[A(\text{ber}' \, mb + \text{j bei}' \, mb) + B(\text{ker}' \, mb + \text{j kei}' \, mb)]$$

$$\text{j } mb[A(\text{ber } mb + \text{j bei } mb) + B(\text{ker } mb + \text{j kei } mb)]$$

$$= 2\mu_r[A(\text{ber}' \, mb + \text{j bei}' \, mb) + B(\text{ker}' \, mb + \text{j kei}' \, mb)$$

Grouping terms in A and B,

$$A[\text{j } mb(\text{ber } mb + \text{j bei } mb) - 2\mu_r(\text{ber}' \, mb + \text{j bei}' \, mb)]$$

$$= B[2\mu_r(\text{ker}' \, mb + \text{j kei}' \, mb) - \text{j } mb(\text{ker } mb + \text{j kei } mb)]$$

$$\frac{B}{A} = -\frac{2\mu_r(\text{ber}' \, mb + \text{j bei}' \, mb) - \text{j } mb(\text{ber } mb + \text{j bei } mb)}{2\mu_r(\text{ker}' \, mb + \text{j kei}' \, mb) - \text{j } mb(\text{ker } mb + \text{j kei } mb)} \qquad (10.7)$$

This may be compared with the corresponding eqn 5.4 for the conduction heating of tubes:

$$\frac{B}{A} = -\frac{\text{ber}' \, mb + \text{j bei}' \, mb}{\text{ker}' \, mb + \text{j kei}' \, mb}$$

The extra terms have crept in because of the flux in the central hole in the tube, for which there is no equivalent in conduction. There is no precise duality between the two tube solutions, although the induction solution can always be reduced to the conduction model by putting $\mu_r = 0$.

10.2 Magnetic field distribution

From eqns 10.1 and 10.2,

$$\frac{H}{H_s} = \frac{A(\text{ber } mr + \text{j bei } mr) + B(\text{ker } mr + \text{j kei } mr)}{A(\text{ber } mR + \text{j bei } mR) + B(\text{ker } mR + \text{j kei } mR)}$$

$$= \frac{(\text{ber } mr + \text{j bei } mr) + (B/A)(\text{ker } mr + \text{j kei } mr)}{(\text{ber } mR + \text{j bei } mR) + (B/A)(\text{ker } mR + \text{j kei } mR)}$$

$$= \frac{(\text{ber}\, mr + \text{j bei}\, mr)}{(\text{ber}\, mR + \text{j bei}\, mR)}$$

$$\times \left[\frac{1 + (B/A)(\text{ker}\, mr + \text{j kei}\, mr)/(\text{ber}\, mr + \text{j bei}\, mr}{1 + (B/A)(\text{ker}\, mR + \text{j kei}\, mR)/(\text{ber}\, mR + \text{j bei}\, mR)} \right] \quad (10.8)$$

The first term is the same as eqn 9.5a for a solid billet; the complex [] term modifies it for tubes. This reduces to the solid billet when the ker and kei terms go to zero or when B/A is small. The induction heating expressions are more complicated than those for conduction heating because of the flux inside the tube; if that flux is negligible, eqn 10.4 goes to zero and H/H_s becomes the analogue of eqn 4.8.

10.3 Current-density distribution

From eqn 10.6, we get:

$$\frac{J}{J_s} = \frac{A(\text{ber}'\, mr + \text{j bei}'\, mr) + B(\text{ker}'\, mr + \text{j kei}'\, mr)}{A(\text{ber}'\, mR + \text{j bei}'\, mR) + B(\text{ker}'\, mR + \text{j kei}'\, mR)}$$

This can be rearranged to conform with eqn 9.12:

$$\frac{J}{J_s} = \frac{(\text{ber}'\, mr + \text{j bei}'\, mr)}{(\text{ber}'\, mR + \text{j bei}'\, mR)}$$

$$\times \left[\frac{1 + (B/A)(\text{ker}'\, mr + \text{j kei}'\, mr)/(\text{ber}'\, mr + \text{j bei}'\, mr)}{1 + (B/A)(\text{ker}'\, mR + \text{j kei}'\, mR)/(\text{ber}'\, mR + \text{j bei}'\, mR)} \right] \quad (10.9)$$

Again, the first term is the same as eqn 9.12 for a solid billet; the complex [] term modifies it for tubes. Equation 10.9 reduces to the solid billet when the ker′ and kei′ terms go to zero or when B/A is small. Again, the induction heating expressions are complicated by the flux inside the tube; if the flux is negligible, eqn 10.4 goes to zero and J/J_s becomes the same as eqn 9.12.

After manipulation, eqn 10.9 has the form $(A + \text{j}\, B)/(C + \text{j}\, D)$, the modulus of which is $\sqrt{\{(A^2 + B^2)/(C^2 + D^2)\}}$. This modulus is the ratio of the absolute value of current density at any radius to its value at the surface. It is plotted on Figs 10.1, 10.2 and 10.3 for μ_r values of 1, 5 and 10. Only the active section is shown (there is no current in the hole) and the other side is obviously the mirror image. All are plotted for $R/\delta = 2, 4, 5, 6, 7, 8, 9$ and 10. The b/R ratios are 0.9, 0.85, 0.8, 0.6, 0.4 and 0.2, equivalent to t/d ratios of 5, 7.5, 10, 20, 30 and 40%, respectively (parts (a) to (f) of each figure).

It is interesting to compare these results with the equivalent curves for *conduction* heating in tubes (Fig. 5.1). As in the solid cylinders, the current is

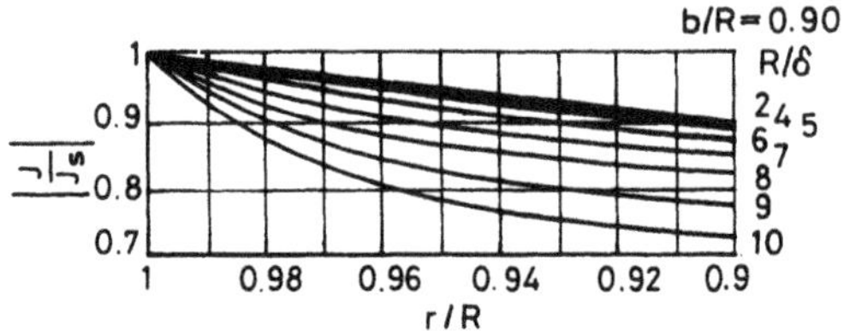

a

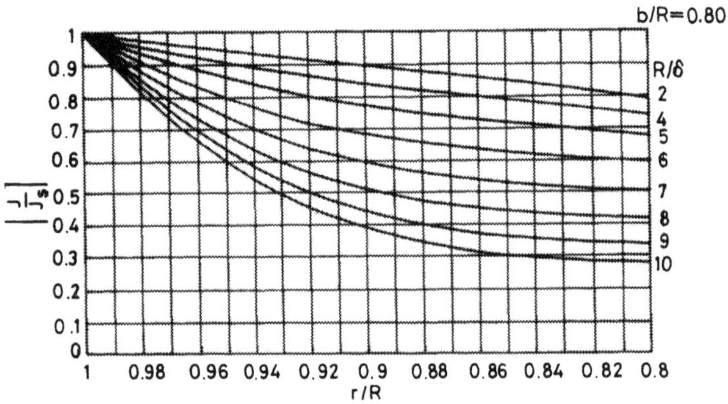

c

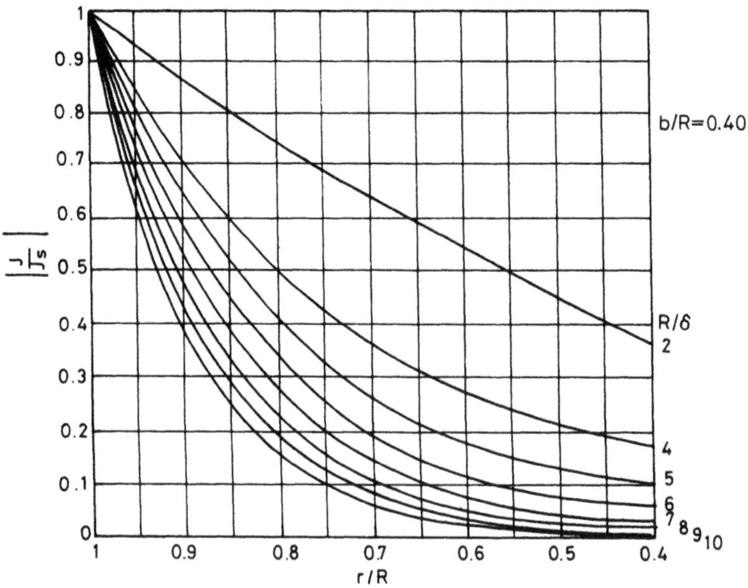

e

Fig. 10.1 *Fall-off of current density for tubes: relative $\mu = 1$. Expressed as the modulus of J/J_s, for various ratios of radius to depth of penetration, and for selected values of internal radius b to external radius R*

b

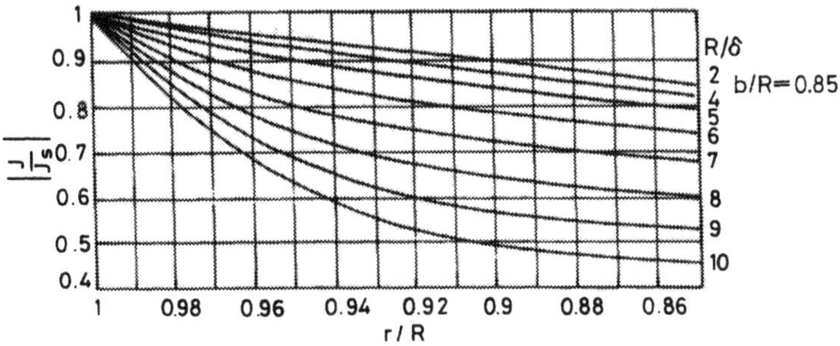

d

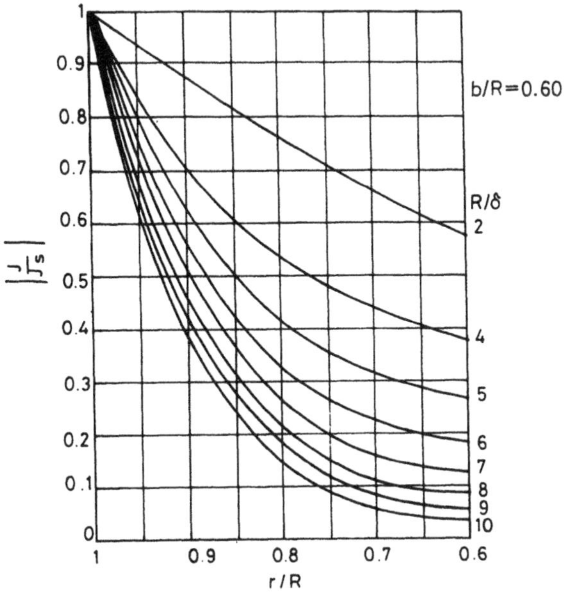

f

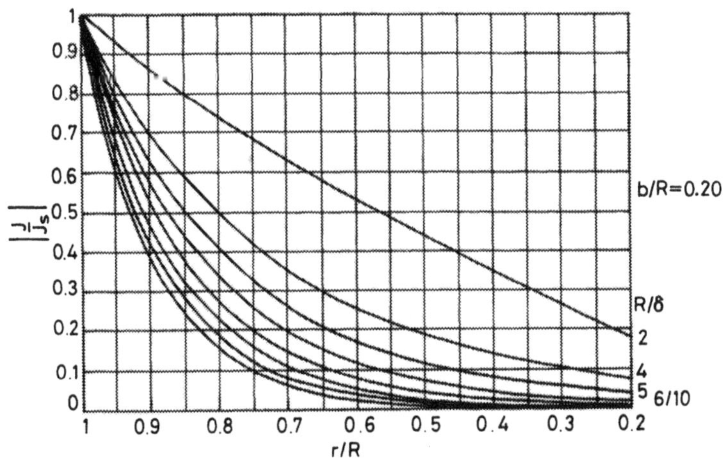

Fig. 10.1 *(Cont.)*

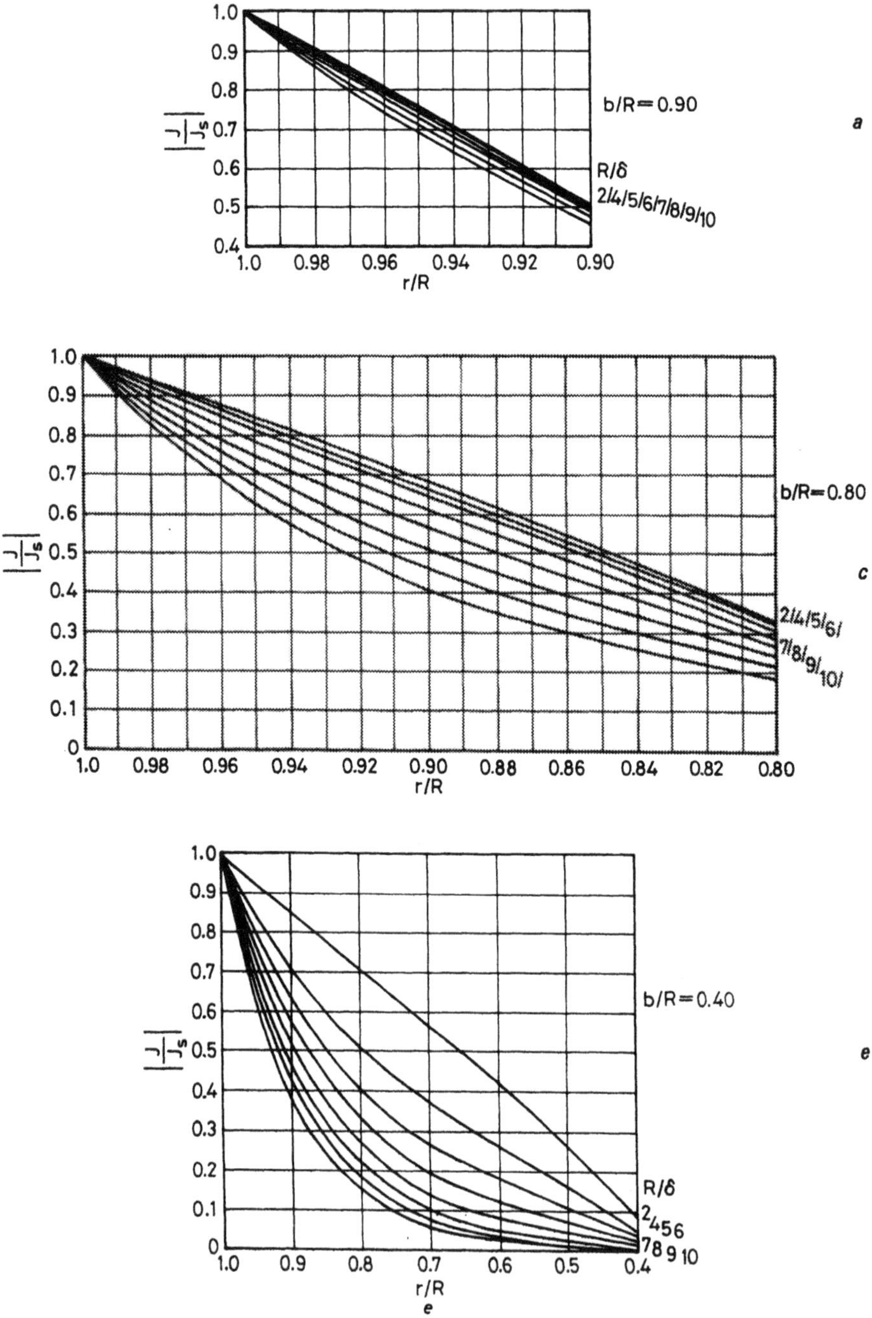

Fig. 10.2 *Fall-off of current density for tubes: relative μ = 5. Expressed as the modulus of J/J_s, for various ratios of radius to depth of penetration, and for selected values of internal radius b to external radius R*

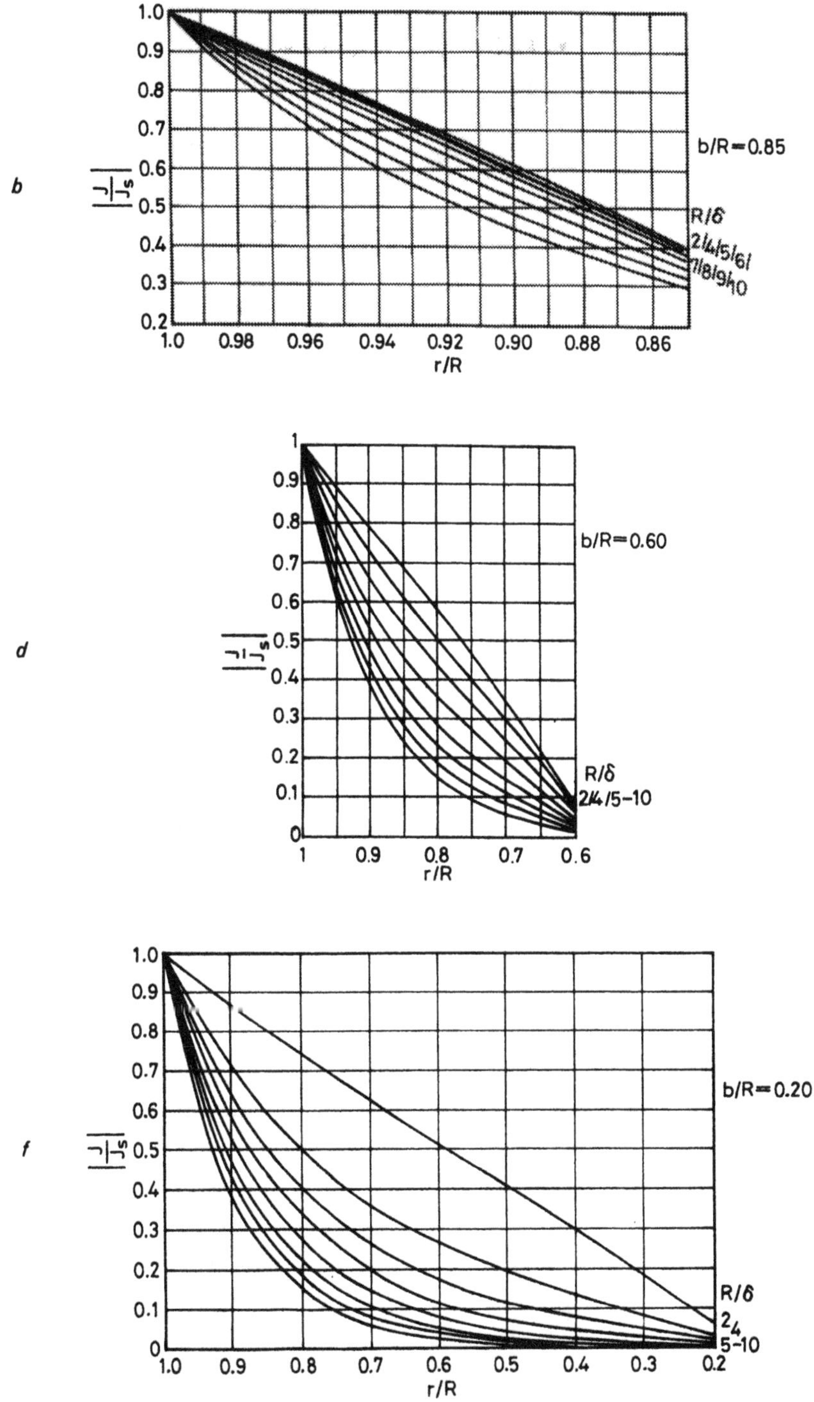
b
|J/Js|
1.0
0.9
0.8
0.7
0.6
0.5
0.4
0.3
0.2
1.0 0.98 0.96 0.94 0.92 0.90 0.88 0.86
r/R
b/R=0.85
R/δ
2/4/5/6/
7/8/9/10
d
|J/Js|
1
0.9
0.8
0.7
0.6
0.5
0.4
0.3
0.2
0.1
0
1 0.9 0.8 0.7 0.6
r/R
b/R=0.60
R/δ
2/4/5-10
f
|J/Js|
1.0
0.9
0.8
0.7
0.6
0.5
0.4
0.3
0.2
0.1
0
1.0 0.9 0.8 0.7 0.6 0.5 0.4 0.3 0.2
r/R
b/R=0.20
R/δ
2
4
5-10

Fig. 10.2 *(Cont.)*

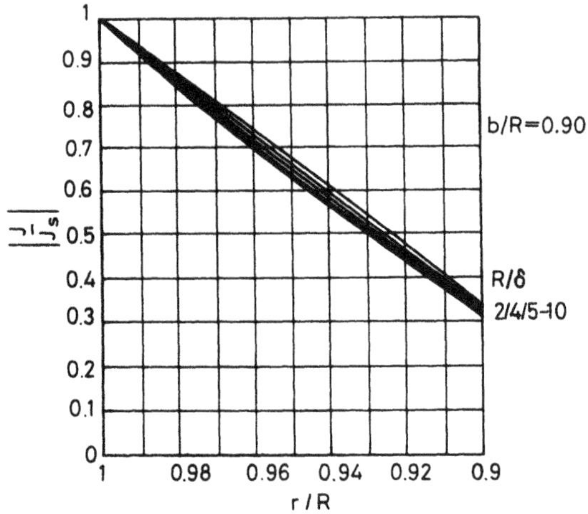
b/R=0.90
R/δ
2/4/5-10
|J/Js|
r / R
1
0.9
0.8
0.7
0.6
0.5
0.4
0.3
0.2
0.1
0
1
0.98
0.96
0.94
0.92
0.9

a

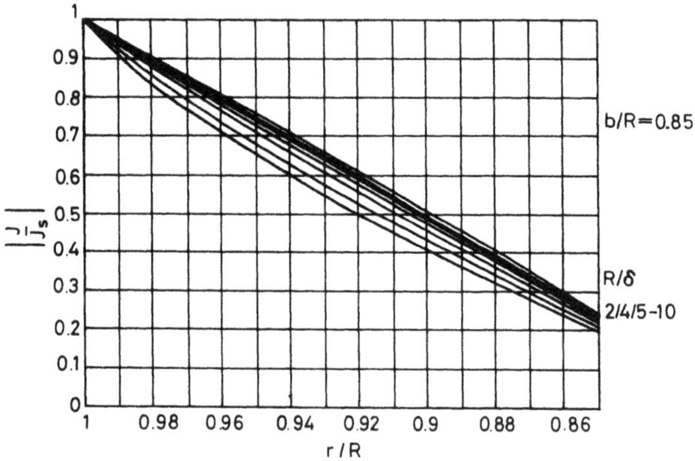
b/R=0.85
R/δ
2/4/5-10
|J/Js|
r / R
1
0.9
0.8
0.7
0.6
0.5
0.4
0.3
0.2
0.1
0
1
0.98
0.96
0.94
0.92
0.9
0.88
0.86

b

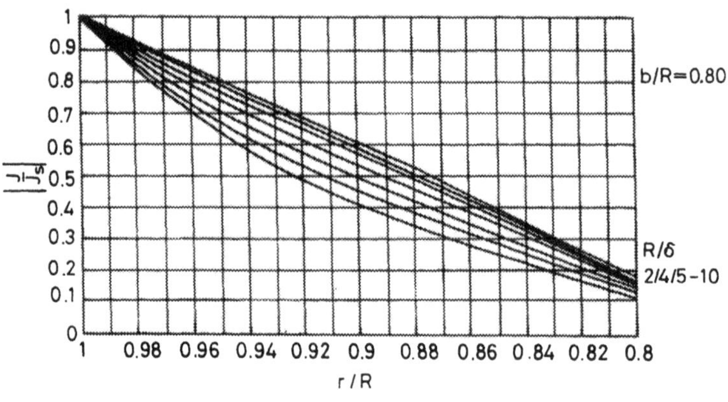
b/R=0.80
R/δ
2/4/5-10
|J/Js|
r / R
1
0.9
0.8
0.7
0.6
0.5
0.4
0.3
0.2
0.1
0
1
0.98
0.96
0.94
0.92
0.9
0.88
0.86
0.84
0.82
0.8

c

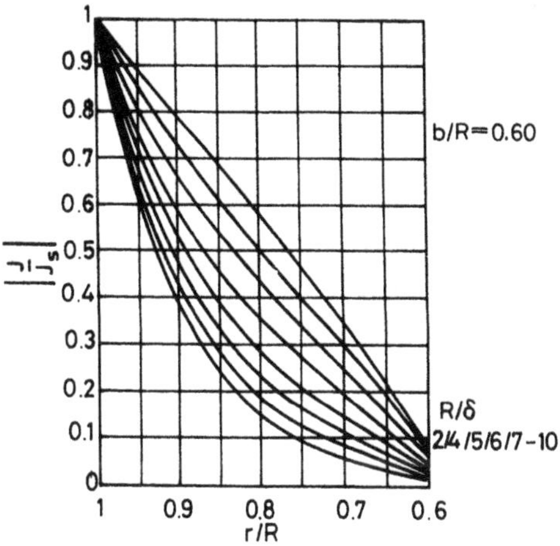

d

Fig. 10.3 *Fall-off of current density for tubes: relative μ = 10. Expressed as the modulus of J/J_s, for various ratios of radius to depth of penetration, and for selected values of internal radius b to external radius R*

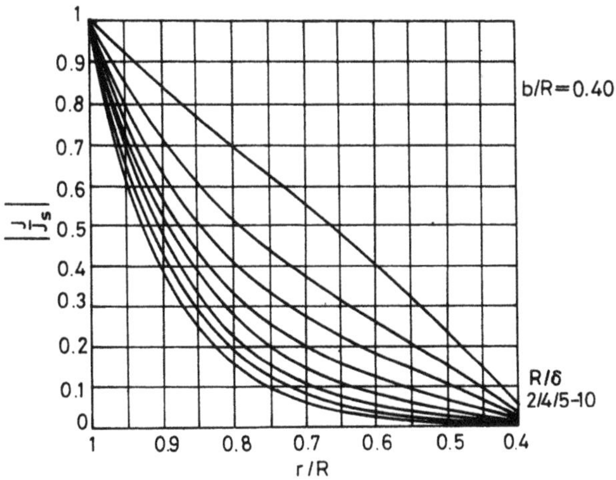

e

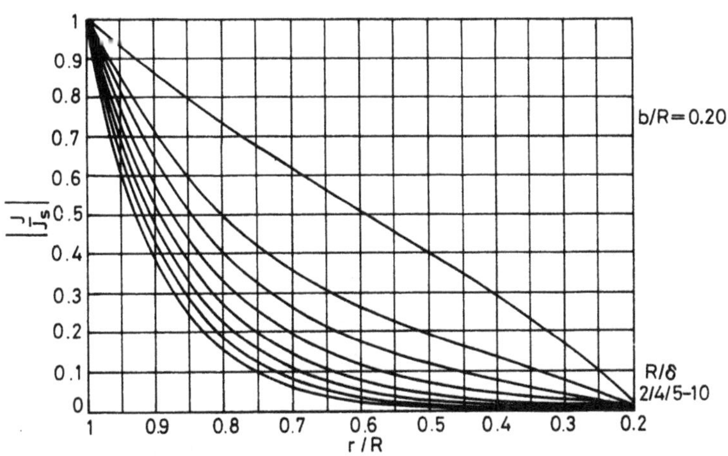

f

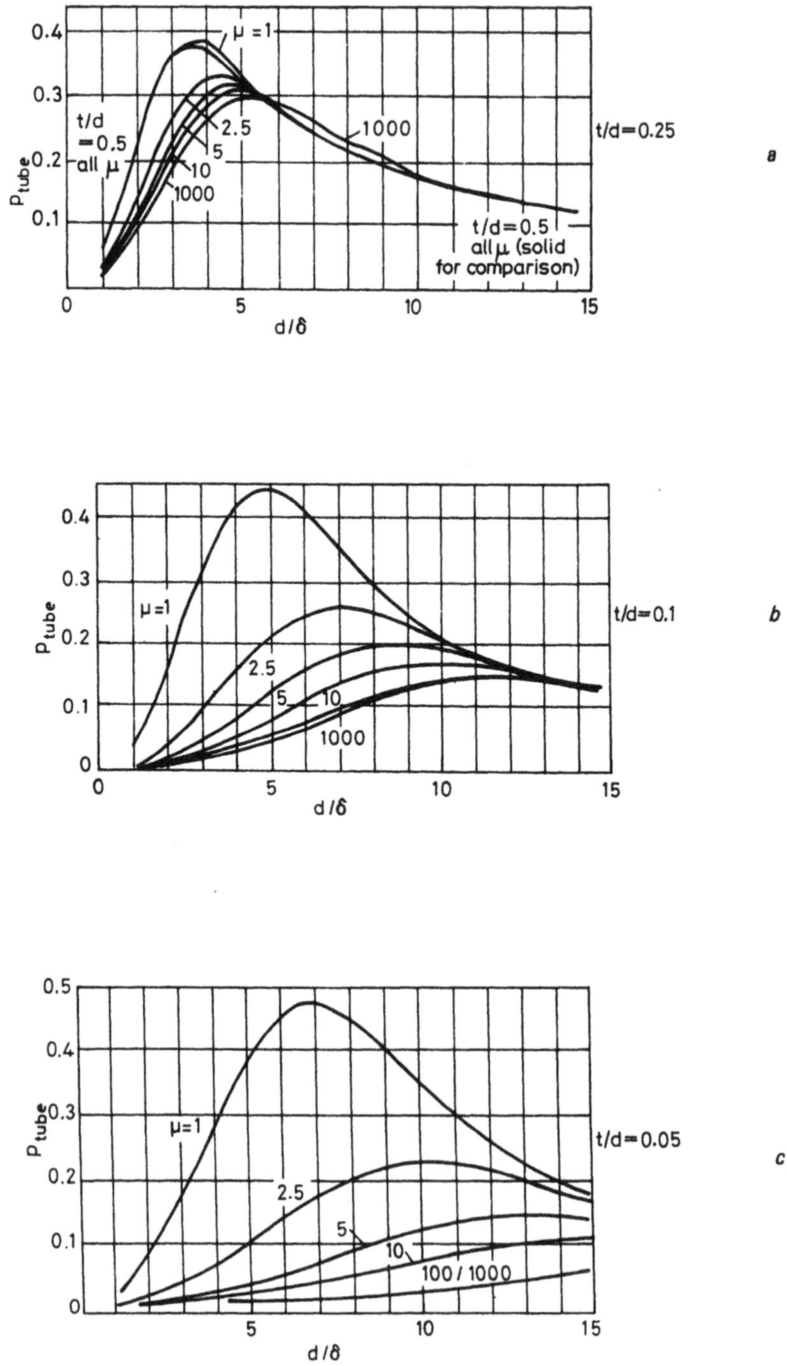

Fig. 10.4 *Curves of p_{tube} for various values of t/d and μ_r, plotted against d/δ*

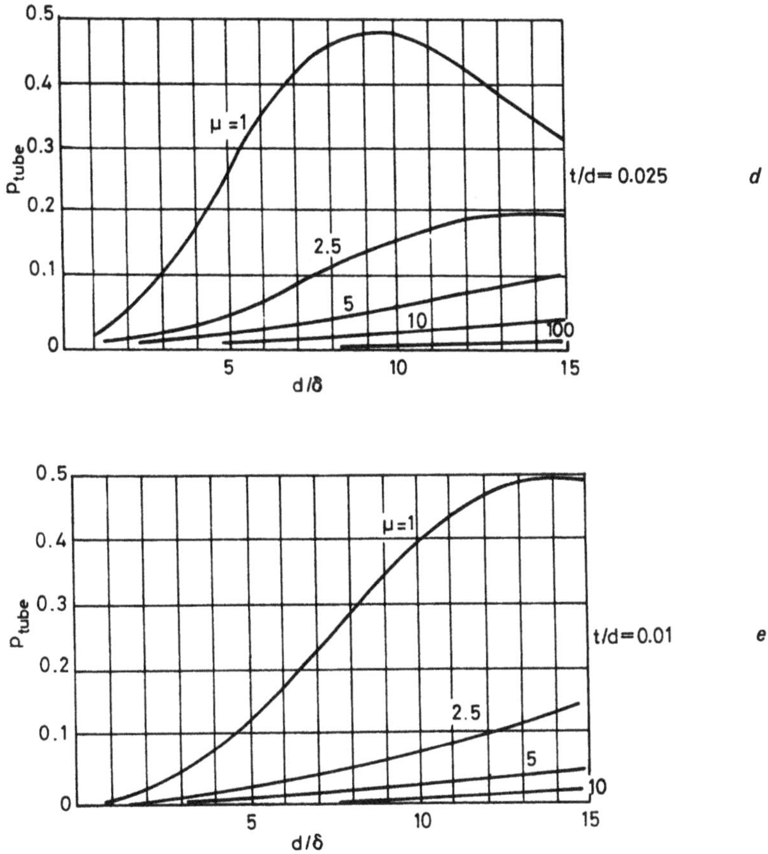

Fig. 10.4 *(Cont.)*

always worse distributed for induction: whereas, for R/δ small, the current tends to uniform (conduction) in thin tubes (say, down to $b/R = 0.8$), it tends to a linear fall-off at best, when $\mu_r = 1$, for all values of b/R. When $\mu_r > 1$, the fall-off is much more rapid: for example, for μ_r as low as 5, the fall-off rate for $b/R = 0.8$ is about 3.5 times as fast and, for $\mu_r = 10$, it is about 4 times. This is very important for our understanding, as it is impossible to visualise a practical heater of ferromagnetic tubing below Curie temperature with $\mu_r = 1$. The fall-off results in a dramatic drop in heating, since loss $\propto J^2$,

These curves must be interpreted with care, to ensure that like things are being compared. For example, for the same values of R and b/R, μ_r is implicit in R/δ ($\propto \sqrt{\mu}$), so that one has to compare, say, the $R/\delta = 6$ curve for $\mu_r = 1$ with the $R/\delta = 6\sqrt{10} = 19$ curve for $\mu_r = 10$, which makes matters much worse for heating. As the tube gets thicker ($\mu_r = 1$), the current distribution approaches that of the solid, truncated at the appropriate b/R.

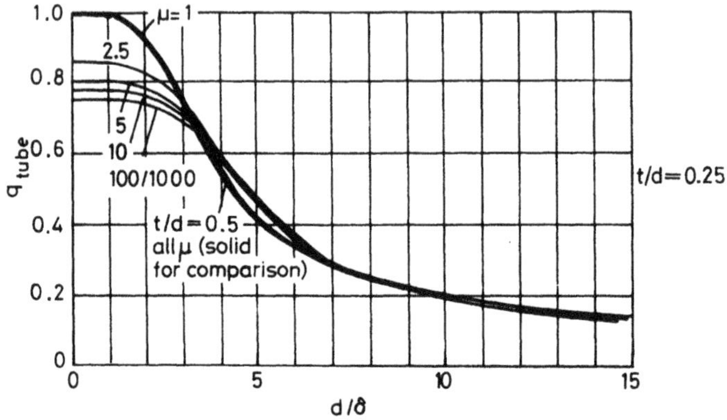

a

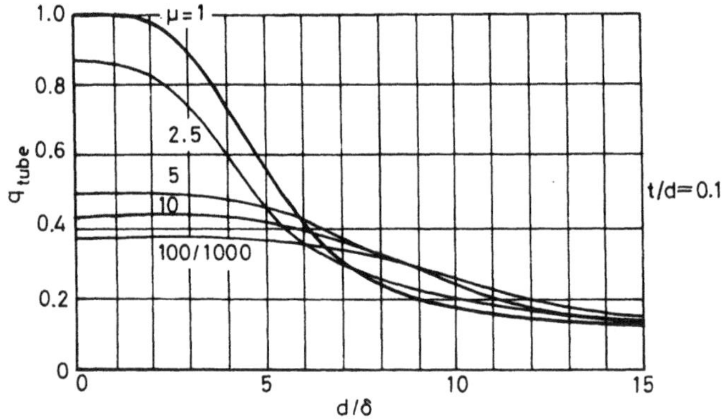

b

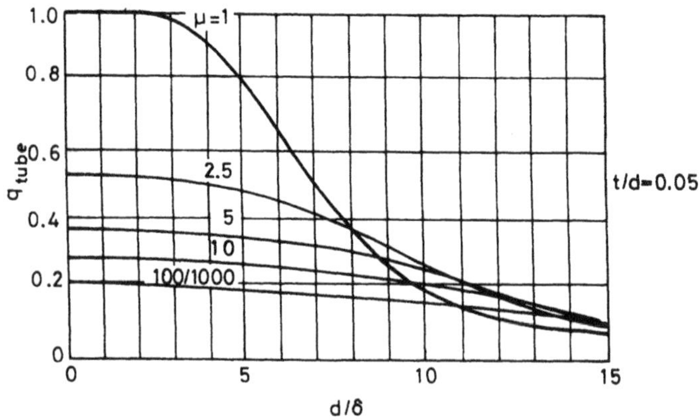

c

Fig. 10.5 *Curves of q_{tube} for various values of t/d and μ_r, plotted against d/δ*

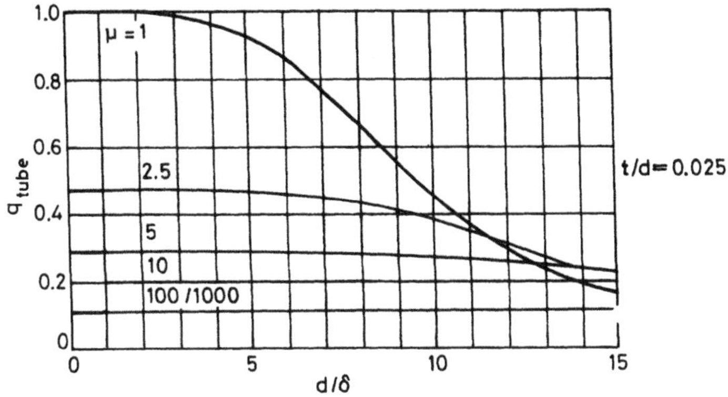

d

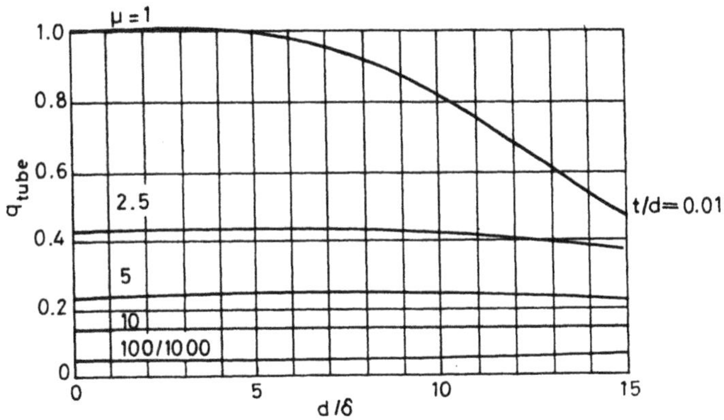

e

Fig. 10.5 *(Cont.)*

10.4 Loss-density distribution

The loss-density distribution relative to the surface is given by the square of the modulus used in Section 10.3.

10.5 Power

As before, from Poynting,

$$\mathscr{P} + j\mathscr{L} = -\mathbf{E}_s \times \mathbf{H}_s^* \qquad \text{VA/m}^2$$

$$= -\rho \mathbf{J}_s \times \mathbf{H}_s^*$$

$$= \rho m[A(\text{ber}' \, mR + j \, \text{bei}' \, mR) + B(\text{ker}' \, mR + j \, \text{kei}' \, mR)]H_s$$

But, from eqn 10.2,

$$H_s = A\,(\text{ber}\, mR + \text{j bei}\, mR) + B\,(\text{ker}\, mR + \text{j kei}\, mR)$$

So

$$\mathscr{P} + \text{j}\mathscr{Q} = \rho m\, H_s^2 \frac{A(\text{ber}'\, mR + \text{j bei}'\, mR) + B(\text{ker}'\, mR + \text{j kei}'\, mR)}{A(\text{ber}\, mR + \text{j bei}\, mR) + B(\text{ker}\, mR + \text{j kei}\, mR)}$$

$$= \rho m H_s^2 \frac{(\text{ber}'\, mR + \text{j bei}'\, mR)}{(\text{ber}\, mR + \text{j bei}\, mR)}$$

$$\times \left[\frac{1 + (B/A)(\text{ker}'\, mR + \text{j kei}'\, mR)/(\text{ber}'\, mR + \text{j bei}'\, mR)}{1 + (B/A)(\text{ker}\, mR + \text{j kei}\, mR)/(\text{ber}\, mR + \text{j bei}\, mR)}\right]$$

$$= \frac{\rho R H_s^2}{\delta^2}\,[p_{\text{tube}} + \text{j}q_{\text{tube}}] \tag{10.10}$$

where $[p_{\text{tube}} + \text{j}q_{\text{tube}}]$ is the quantity defined in Section 9.5, multiplied by the value of [ ]. Note that eqn 10.10 is in the same form as eqn 9.17, so that they become identical when $b = 0$.

To get the power loss in the tube, we multiply the Poynting expression by the surface area $2\pi Rl$:

$$P_w = (\rho R H_s^2/\delta^2)\, p_{\text{tube}}\, (2\pi R l)$$

$$= (\pi R^2 l)\, H_s^2\, \mu_r \mu_0\, 2\pi f\, p_{\text{tube}} \tag{10.11}$$

This has the same shape as eqn 9.18 though, in this context, the 'billet volume' $(\pi R^2 l)$ is not really germane. It shows, however, that the effects of the 'tubeness' have all been absorbed into $[p_{\text{tube}} + \text{j}q_{\text{tube}}]$.

10.5.1 Tube power loss constant

The value of $p_{\text{tube}} + \text{j}q_{\text{tube}}$ is given by

$$p_{\text{tube}} + \text{j}\, q_{\text{tube}} = \frac{2}{mR} \frac{(\text{ber}'\, mR + \text{j bei}'\, mR)}{(\text{ber}\, mR + \text{j bei}\, mR)}$$

$$\times \left[\frac{1 + (B/A)(\text{ker}'\, mR + \text{j kei}'\, mR)/(\text{ber}'\, mR + \text{j bei}'\, mR)}{1 + (B/A)(\text{ker}\, mR + \text{j kei}\, mR)/(\text{ber}\, mR + \text{j bei}\, mR)}\right] \tag{10.12}$$

where, as in eqn 10.7,

$$\frac{B}{A} = -\frac{2\mu_r(\text{ber}'\, mb + \text{j bei}'\, mb) - \text{j}\, mb(\text{ber}\, mb + \text{j bei}\, mb)}{2\mu_r(\text{ker}'\, mb + \text{j kei}'\, mb) - \text{j}\, mb(\text{ker}\, mb + \text{j kei}\, mb)}$$

and $m = (\sqrt{2})/\delta$. p_{tube} and q_{tube} are the real and imaginary parts of eqn 10.12, with eqn 10.7 substituted for B/A. It will be appreciated that this is a complex calculation, in more ways than one. It is best solved by computer.

Note that μ appears explicitly in eqn 10.7 and implicitly in δ and hence m. We are interested in the effect of changes in μ_r for a fixed mR. This is shown in Fig. 10.4. The calculations from which these curves were plotted are given in Appendix A.4, which shows the component parts of the equation. These are very illuminating to the reader who is interested in the detail of tube heating. Figures 10.4a–e have been plotted for a limited number of values of μ_r to show the effect of the change for various tube thickness/diameter (t/d) ratios. Obviously, as t/d approaches 0.5, the result tends to the solid, so $t/d = 0.5$ is included for comparison. Although there is a fall-off in performance with increasing μ_r for $t/d = 0.25$, this is not too serious, but thereafter the effect of μ_r rapidly becomes important, causing the heating effectiveness to diminish – sometimes to the point of ineffectiveness. For example, at $t/d = 0.01$, i.e. very thin tube, there is negligible heating even with μ_r as low as 10. If the frequency is increased to compensate, then d/δ also increases, so very little is gained.

At low values of t/d, consideration should be given to a preheat up to Curie temperature by, say, conduction heating. When μ_r has become unity, the p_{tube} curves are potentially better than for solid cylinders, assuming that t/δ can be made right.

These curves are also valuable for the calculation of when a cylinder does *not* heat. In induction-heating (and other) apparatus, it is often desirable to insert a thin cylinder for structural purposes, such as sealing, without it getting hot. Good practice will split the tube so that there is no path for peripheral currents but, if this is not possible because, for example, the cylinder must be leakproof, the above theory allows good design.

10.5.2 Tube reactive power loss constant

Curves for q_{tube} corresponding to Fig. 10.4 are shown in Fig. 10.5. As μ increases, q_{tube} falls, so improving the power factor, as expected.

For a more extensive study of losses in shells, the reader is referred to Sundberg.[14]

Chapter 11

Induction heating of hollow cylinder from inside

This topic has been included because of its possible practical importance in the preheating of the 'containers' of extrusion presses. These very thick hollow cylinders have to be brought to starting temperature before work commences or kept at temperature during breaks; at present, this is done very ineffectively using surface heaters on the external surface of the cylinder, so that the heat has to find its way through the thick metal. This could be done much more efficiently, using an induction coil temporarily suspended from the ram to heat the active surface, where it is needed.

The solution is obviously the 'inside-out' version of Chapter 9, so we shall not repeat unnecessarily.

11.1 Magnetic field distribution

As before, eqn 9.3 applies:

$$\frac{d^2H}{dr_2} + \frac{dH}{dr}\frac{1}{r} = 2j\alpha^2 H$$

where $\alpha^2 = \omega\mu/2\rho$. The direction of H has been omitted from this equation as it is unambiguous.

Let the bore radius be R. Assume that the cylinder is effectively infinitely thick.

The H distribution is given by eqn 9.4:

$$H = AI_0[\sqrt{(2j)}\alpha r] + BK_0[\sqrt{(2j)}\alpha r]$$

H must not be infinite at the outer radius, $r = \infty$, so $A = 0$ and

$$H = B(\ker mr + j\,\mathrm{kei}\,mr) \tag{11.1}$$

where $m^2 = 2\alpha^2$. Let $H = H_s$ at the inner radius, $r = R$, so that

$$B = H_s/(\ker mR + j\,\mathrm{kei}\,mR)$$

$$H = H_s \frac{\ker mr + \mathrm{j}\,\mathrm{kei}\, mr}{\ker mR + \mathrm{j}\,\mathrm{kei}\, mR} \tag{11.2}$$

and

$$\left|\frac{H}{H_s}\right| = \frac{N_0(mr)}{N_0(mR)} \tag{11.3}$$

11.2 Current-density distribution

From eqn 11.1 and $J = -\partial H/\partial r$,

$$J = -mH_s \frac{\ker' mr + \mathrm{j}\,\mathrm{kei}' mr}{\ker mR + \mathrm{j}\,\mathrm{kei}\, mR} \tag{11.4}$$

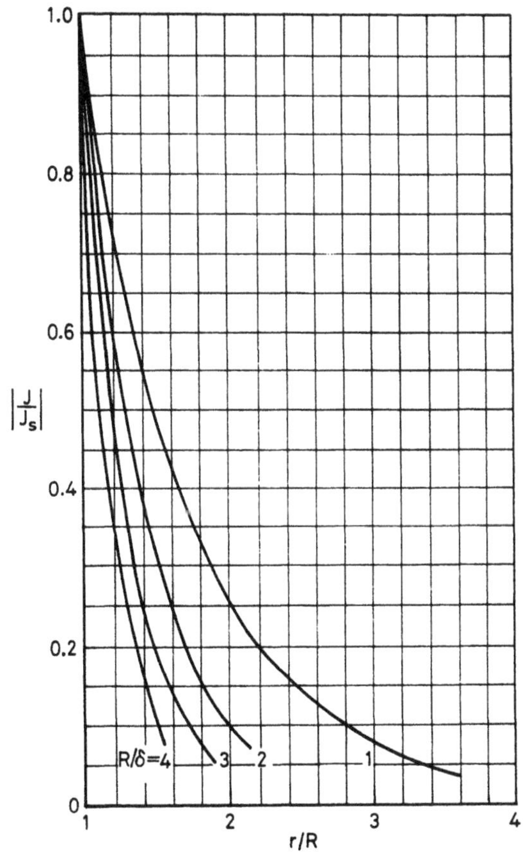

Fig. 11.1 *Fall-off of current density with radius in a hollow cylinder heated from inside, for various ratios of radius to depth of penetration, expressed as the modulus of J/J_s*

J is always in the peripheral direction. Eliminating H gives

$$\frac{J}{J_s} = \frac{\mathrm{ker}'mr + \mathrm{j}\,\mathrm{kei}'mr}{\mathrm{ker}'mR + \mathrm{j}\,\mathrm{kei}'mR} \tag{11.5}$$

$$\left|\frac{J}{J_s}\right| = \left[\frac{\mathrm{ker}'^2 mr + \mathrm{kei}'^2 mr}{\mathrm{ker}'^2 mR + \mathrm{kei}'^2 mR}\right]^{1/2} = \frac{N_1(mr)}{N_1(mR)} \tag{11.6}$$

This is plotted on Fig. 11.1.

11.3 Power loss

From Poynting, at $r = R$ (see Chapter 9):

$$\mathscr{P} + \mathrm{j}\mathscr{Q} = \boldsymbol{E} \times \boldsymbol{H}^* \qquad \mathrm{VA/m^2}$$

$$= \sqrt{(2)}\alpha\rho H_s^2 \frac{\mathrm{ker}'mR + \mathrm{j}\,\mathrm{kei}'mR}{\mathrm{ker}\,mR + \mathrm{j}\,\mathrm{kei}\,mR} \tag{11.7}$$

which can be expanded to

$$\mathscr{P} + \mathrm{j}\mathscr{Q} = \frac{\rho\sqrt{(2)}H_s^2}{\delta} \times \frac{(\mathrm{ker}'\sqrt{(2)}R/\delta + \mathrm{j}\,\mathrm{kei}'\sqrt{(2)}R/\delta)(\mathrm{ker}\sqrt{(2)}R/\delta - \mathrm{j}\,\mathrm{kei}\sqrt{(2)}R/\delta)}{\mathrm{ker}^2\sqrt{(2)}R/\delta + \mathrm{kei}^2\sqrt{(2)}R/\delta} \tag{11.8}$$

As before, we use RMS H_s to get the mean value. Multiplying by surface area and rearranging gives power:

$$P_w = \frac{\rho(2\pi R^2 l)H_s^2}{\delta^2}[p_{ext} + \mathrm{j}q_{ext}] \tag{11.9}$$

where

$$p_{ext} = \frac{\sqrt{(2)}\delta}{R}\,\frac{\mathrm{ker}\sqrt{(2)}R/\delta\,\mathrm{ker}'\sqrt{(2)}R/\delta + \mathrm{kei}\sqrt{(2)}R/\delta\,\mathrm{kei}'\sqrt{(2)}R/\delta}{\mathrm{ker}^2\sqrt{(2)}R/\delta + \mathrm{kei}^2\sqrt{(2)}R/\delta}$$

$$q_{ext} = \frac{\sqrt{(2)}\delta}{R}\,\frac{\mathrm{ker}\sqrt{(2)}R/\delta\,\mathrm{kei}'\sqrt{(2)}R/\delta - \mathrm{kei}\sqrt{(2)}R/\delta\,\mathrm{ker}'\sqrt{(2)}R/\delta}{\mathrm{ker}^2\sqrt{(2)}R/\delta + \mathrm{kei}^2\sqrt{(2)}R/\delta} \tag{11.10}$$

p_{ext} and q_{ext} are plotted for solid external cylinders in Figs 11.2. It is interesting to compare these curves with those for internal heating. These have no maxima because there is no influence from the 'other side', so we have straightforward curves, with both p and q falling off in an orderly way. Because of this, the q curve is represented by $2/(d/\delta)$ over a much greater range than in Fig. 9.5; in fact, it is within 2% for $d/\delta \geqslant 2$ and $\frac{1}{2}$% for $d/\delta \geqslant 4$.

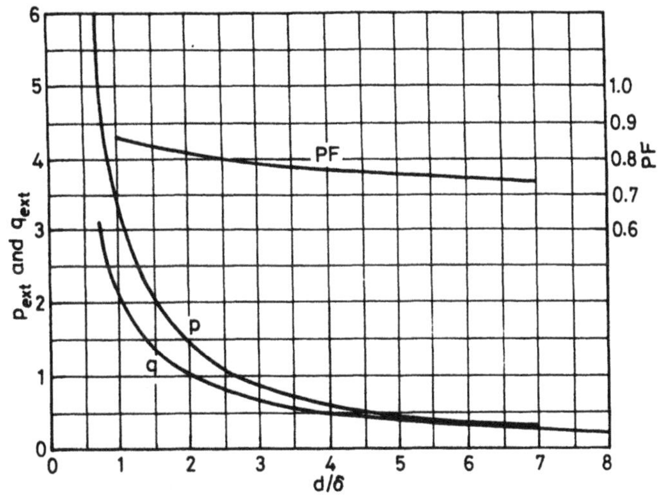

Fig. 11.2 *p_{ext} for various values of d/δ*

11.4 Power factor

Since

$$\mathrm{ker}\,x\,\mathrm{kei}'x - \mathrm{kei}\,x\,\mathrm{ker}'x = N_0(x)N_1(x)\sin(\phi_1 - \phi_0 - \pi/4)$$

$$\mathrm{ker}'x\,\mathrm{ker}\,x + \mathrm{kei}\,x\,\mathrm{kei}'x = N_0(x)N_1(x)\cos(\phi_1 - \phi_0 - \pi/4)$$

so

$$q/p = \tan(\phi_1 - \phi_0 - \pi/4)$$

$$\cos[\tan^{-1}(q/p)] = \cos(\phi_1 - \phi_0 - \pi/4)$$

This gives us a straightforward method to calculate PF. The values of ϕ are tabulated in Reference 4. A curve of power factor is included on Fig. 11.2.

11.5 Practical example

Calculate the power density at the bore of a container having the following specification, if a coil producing 5×10^5 A/m is suspended close to the surface. The inside of the coil is filled with suitably-oriented flux-guides:

Material	tool steel Ni-Cr
Frequency	50 Hz
Resistivity	1.2 $\mu\Omega$m approx.
Relative permeability	unity
Bore	125 mm

$$\delta = \sqrt{[(2)(1.2 \times 10^{-6})/(4\pi \times 10^{-7})(2\pi\, 50)]} = 78\,\text{mm}$$

$D/\delta = 125/78 = 1.60$. From Fig. 11.2, this gives $p_{ext} = 2$.
Using eqn 11.8:

$$\begin{aligned}\mathscr{P} &= R\rho H_s^2 p_{ext}/\delta^2 \\ &= (62.5/1000)\,(1.2 \times 10^{-6})\,(5 \times 10^5)^2(2)/(0.078)^2 \\ &= 481\,\text{kW/m}^2\end{aligned}$$

In Chapter 15, we shall see that the rate-of-rise of temperature is approximately $\mathscr{P}/(c\gamma)b$, where b is the thickness. If we assume that the heat is produced in one skin depth, this would give $481\,000/(5\,000\,000 \times 0.078) \simeq 1.2$ K/s, for $(c\gamma) = 5$ MJ/m^{3}K — roughly 1°C/s. The container would therefore take 5 minutes or so to get to working temperature.

Chapter 12

Effects of changing permeability and resistivity

12.1 Introduction

A method of dealing with permeability in design is discussed in Chapter 13. In this chapter, permeability μ and resistivity ρ are treated as numbers in the depth-of-penetration formula $\delta = \sqrt{(2\rho/\mu\omega)}$ and the consequences of changing them are noted, using the theory that has been developed in earlier chapters. μ and ρ both change during heating, so this is important. However, before we consider them combined, let us examine simple changes of each for the semi-infinite slab. Throughout this chapter, we shall only treat the constant E condition for conduction; variable E was discussed in Chapter 6.

12.2 Semi-infinite slab

12.2.1 Change μ: conduction, semi-infinite slab

If we take two materials (1, 2) having the same resistivity but different μ, then $\delta_1/\delta_2 = \sqrt{(\mu_2/\mu_1)}$. For example, if we compare two materials with relative μ of 1 and 100, $\delta_1/\delta_2 = 10$ and eqn 2.13 ($\mathscr{P} = \delta E_s^2/2\rho$) gives, for a semi-infinite slab, a *decrease* of loss to 10% for the higher μ, since $\mathscr{P} \propto \delta$. Thus, for two stainless steels – non-magnetic and magnetic – the magnetic will have much lower loss and heating.

For a simple increase of μ, the total current also *decreases* as δ or $1/\sqrt{\mu}$ (eqn 2.8), whilst the surface current-density remains constant, by definition, since we are applying a constant E to the purely-resistive surface.

The magnetic field intensity at the surface, for a given J_s, is proportional to $1/\alpha$, from eqn 2.14, i.e. it too is proportional to $\delta \propto 1/\sqrt{\mu}$. If δ increases, the surface magnetic flux density μH_s increases as $\sqrt{\mu}$. Both H and B fall off as $\exp(-\alpha y) \propto \exp[-(\sqrt{\mu})y]$, so that, if μ increases, both will fall faster.

In Section 2.2.7, we showed that the total flux per metre width is constant and independent of both μ and ρ.

Summarising, for μ increased by a factor x:

- E_s stays constant, by definition.
- J_s stays constant, by definition.
- Loss decreases by $\sqrt{x}$.
- Total current decreases by $\sqrt{x}$.
- H_s decreases by $\sqrt{x}$.
- B_s increases by $\sqrt{x}$.
- B, H, E and J all fall off faster, by $\exp\sqrt{x}$.
- Total flux stays constant.

Broadly speaking, the effect of saturation is equivalent to a reduction of μ. In the limit, the totally-saturated material has a relative μ of unity, as if it were non-magnetic. The conclusions of this section are valid, in qualitative terms, for x *decreasing*.

12.2.2 Change μ: induction, semi-infinite slab

The changes in the value of δ are independent of the process – conduction or induction – so we do not have to repeat that part of the discussion. It is how those changes affect the loss that differs. Another difference is the method of producing current in the metal. In conduction, a fixed voltage is applied to the surface, which we have chosen to identify by the vector quantity $\boldsymbol{E}$; the current density at the surface must also be constant, from $\boldsymbol{E} = \rho \boldsymbol{J}$, until ρ changes with temperature. On the other hand, in induction, a fixed $\boldsymbol{H}$ is imposed at the surface. In Section 7.6 it was shown that this produces a constant total current, from $\oint H \mathrm{d}l$, for a semi-infinite slab. The value of this current is independent of resistivity, although its distribution is not. It is interesting to note that here the total current is constant whereas, for conduction, it was the total flux that stayed constant and independent of μ and ρ.

Since the total current is constant, when δ falls due to an increase of μ, the current density has to rise in order to get the same current in a smaller space.

The loss density $\mathscr{P}$, from eqn 7.9, is given by $\rho H_s^2/\delta \propto 1/\delta \propto \sqrt{\mu}$, for constant H and ρ. If we assume that μ increases 16 times, a more practical value for induction, then $\mathscr{P}$ increases 4 times. This accords with experience, in that it is easier to heat magnetic steels by induction.

The total flux per metre width Φ is given by eqn 7.10 as $\mu H_s/\alpha\sqrt{2}$. Since $\alpha \propto \sqrt{\mu}$, Φ is proportional to $\sqrt{\mu}$, i.e. for μ increased by a factor of 16, the total flux increases by a factor of 4.

Summarising, for μ increased by a factor x:

- H_s is constant, by definition.
- B_s increases by x.
- Loss increases by $\sqrt{x}$.
- Total flux increases by $\sqrt{x}$.
- B, H, E and J all fall off faster, by $\exp\sqrt{x}$.
- Total current stays constant.

12.2.3 Change ρ: conduction, semi-infinite slab
Increasing ρ increases δ as $\sqrt{\rho}$. So, keeping E constant as before, we can summarise the effects of increasing ρ by a factor x:

- E_s stays constant, by definition.
- J_s decreases by x.
- Loss decreases by $\sqrt{x}$, from eqn 2.13.
- Total current increases by $x^{3/2}$, from eqn 2.8.
- H_s increases by $x^{3/2}$, from eqn 2.14.
- B_s increases by $x^{3/2}$, from eqn 2.14.
- B, H, E and J all fall off slower, by $\exp(1/\sqrt{x})$.
- Total flux stays constant.

12.2.4 Change ρ: induction, semi-infinite slab
Similarly, for induction, for ρ increased by a factor x:

- H_s is constant, by definition.
- B_s is also constant.
- Loss increases by $\sqrt{x}$.
- Total flux decreases as $1/\sqrt{x}$.
- B, H, E and J all fall off slower, by $\exp(1/\sqrt{x})$.
- Total current stays constant.

12.3 Thin slabs and cylinders

In Chapters 3, 4, 8 and 9, the losses were expressed in terms of p quantities. For conduction, from eqns 3.6 and 4.15,

$$P_w = (1/\rho)(\text{volume})E_s^2 p_r$$

For induction, in eqns 8.10 and 9.18,

$$P_w = (2\mu\pi f)(\text{volume})H_s^2 p$$

The p expressions absorbed the terms containing δ and were themselves plotted as functions of t/δ or R/δ; the two conduction plots (Figs 3.3 and 4.5) are similar in form, as are the two induction plots (Figs 8.4 and 9.4). They are grouped on Fig. 12.1 to simplify discussion.

Similarly, the moduli $|J/J_s|$ are given as functions of αy etc. in eqns 3.2, 4.10, 8.4 and 9.13 and plotted as functions of t/δ or r/δ on Figs 3.2, 4.4, 8.3 and 9.3; again, the two conduction plots (Figs 3.2 and 4.4) are similar in form, as are the two induction plots (Figs 8.3 and 9.3). Because of this similarity, it is convenient to discuss conduction first for both thin slabs and cylinders, and then to repeat for induction.

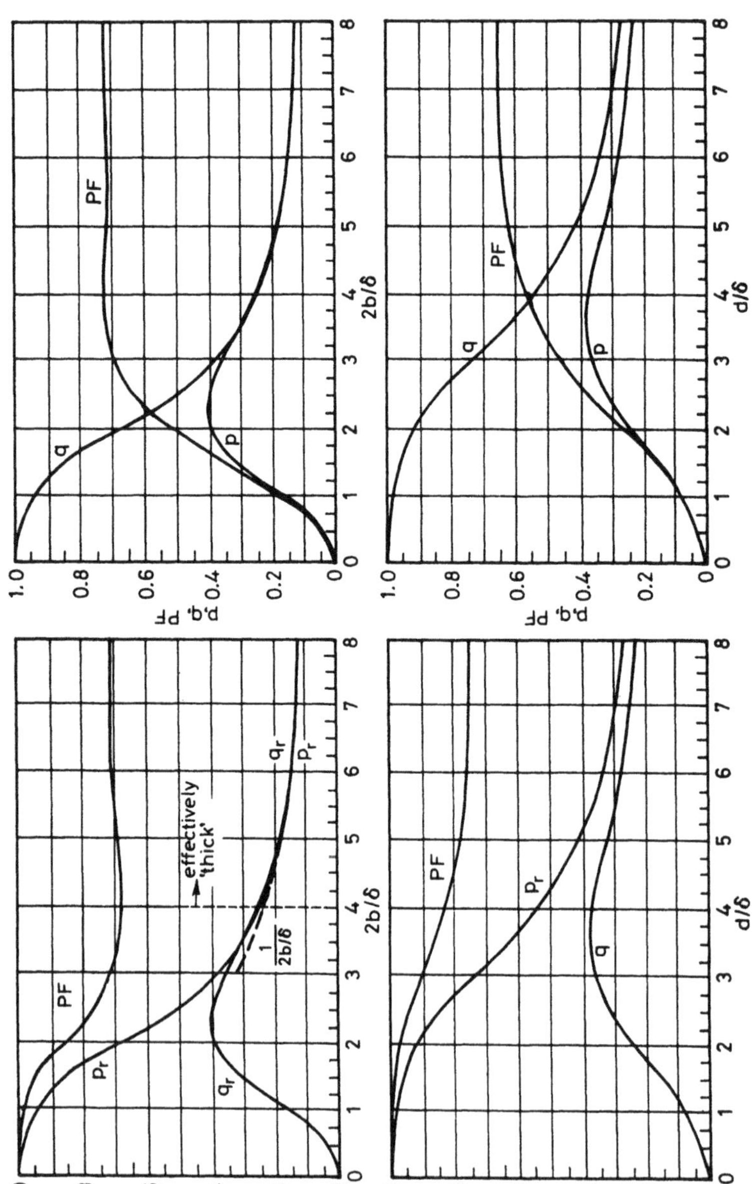

Fig. 12.1 *Functions p and q for slab and cylinder, conduction and induction*

12.3.1 Change μ: conduction, thin slabs and cylinders
For thin slabs, if μ increases by x, δ decreases by $\sqrt{x}$ and t/δ increases by $\sqrt{x}$. As expected, the current-density distribution is cramped towards the surface. The total current decreases by $\sqrt{x}$, as above.

The loss P_w, given by eqn 3.6, falls. If t/δ is chosen to give a high value of p_r (Fig. 3.3), say $t/\delta \leqslant 1.5$, x does not have to be very high for its square root to bring p_r down to the 'effectively-thick' condition for which p_r has the value $1/(t/\delta)$. For example, for $t/\delta = 1$ and $x = 25$, the new value of t/δ is 5, bringing p_r down to 0.2 from its original value of near unity. At the same time, the power factor falls from unity to 0.7. It was shown that the relation between AC and DC power is p_r, so again the drop is disadvantageous. The same arguments apply for cylinders as for thin slabs, except that eqn 4.15 and Fig. 4.5 are used.

12.3.2 Change ρ: conduction, thin slabs and cylinders
If ρ increases by x, δ increases by $\sqrt{x}$ and t/δ decreases by $\sqrt{x}$. The current penetrates deeper into the metal. The total current decreases by $\sqrt{x}$.

The loss P_w, given by eqn 3.6, increases from the improved p_r (Fig. 3.3) but is reduced linearly with ρ. The power factor improves.

The same arguments apply for cylinders, except that eqn 4.15 and Fig. 4.5 are used.

12.3.3 Change μ: induction, thin slabs and cylinders
As in Section 12.3.1, for thin slabs, if μ increases by x, δ decreases by $\sqrt{x}$ and t/δ increases by $\sqrt{x}$. As expected, the current-density distribution is cramped towards the surface. The total current decreases by $\sqrt{x}$, as above.

The loss P_w, given by eqn 8.10 (see Fig. 8.4), increases linearly due to the μ, but can rise or fall due to an increase of t/δ, depending on where it started on the curve. If p was on the left of the maximum, it increases unless the new point goes over the top and down the other side; in practice, it will probably start at or near the peak and so falls with increasing t/δ.

The same arguments apply for cylinders, except that eqn. 9.18 and Fig. 9.4 are used.

12.3.4 Change ρ: induction, thin slabs and cylinders
If ρ increases by x, δ increases by $\sqrt{x}$ and t/δ decreases by $\sqrt{x}$. The current penetrates deeper into the metal. The loss P_w, given by eqn 8.10, is now only dependent on p, which in turn depends on where we start, as discussed above. The same arguments apply for cylinders, except that eqn 9.18 and Fig. 9.4 are used.

12.4 Combined resistivity and permeability changes

12.4.1 Change ρ and μ: conduction, semi-infinite slab
For a semi-infinite slab, we can combine the effects of resistivity and permeabil-

ity, from eqn 2.13, as proportional to $\delta/\rho = 1/\sqrt{(\rho\mu)}$. As a magnetic material heats, its resistivity rises but its relative permeability remains substantially constant up to the Curie point, at which it drops fairly suddenly to unity. For constant E, the loss falls as $1/\sqrt{\rho}$ up to Curie, and falls rapidly at that point. For steel, roughly speaking, the resistivity rises by about 5 times up to Curie. If we *assume* a relative μ of 10, then the loss density falls by $1/\sqrt{5} = 0.447$, then falls by a factor of $\sqrt{10}$ as it passes through the Curie temperature.

It is important, at this stage, to remember that we are dealing with *constant* E conditions. The figures are different for strip heating, where the *current* is constant. See Chapter 6.

12.4.2 Change ρ and μ: conduction, thin slab

Here the behaviour is controlled by p_r (Fig. 3.3) and ρ. If the slab is working at $p_r \simeq 1$ at 20°C and the magnetic steel defined above then rises to the Curie temperature, δ increases by $\sqrt{(5 \times 10)}$ and $2b/\delta$ falls by $1/\sqrt{50}$ to a point where $p_r = 1$. The loss density is given by eqn 3.5:

$$\mathscr{P} \propto \rho J_s^2 p_r$$

But J varies with ρ and is given by E/ρ, E being constant, so

$$\mathscr{P} \propto p_r/\rho \qquad (12.1)$$

So $\mathscr{P}$ drops to $1/5 = 0.2$ of the 20°C value. Note that, for the starting p_r chosen, there is virtually no change of $\mathscr{P}$ on passing through the Curie point.

A thicker slab, starting at, say, $2b/\delta = 5$ and $p_r = 0.2$ and hence $\mathscr{P} \propto 0.2$, will go to $2b/\delta = 0.22$ and $p_r = 1$ above Curie, i.e. $5/\sqrt{50}$. Thus $\mathscr{P}$, which had decreased to 60% of the 20°C value just below Curie, jumps back to 100% of that figure.

As the slab heats, the increase of resistivity with temperature always acts to increase p_r. This effect combines with the decrease of $\mathscr{P}$ due to $1/\rho$. At Curie temperature, the change of relative μ to unity always lowers the value of $2b/\delta$; whether or not this affects the associated p_r depends on where one starts on Fig. 3.3.

12.4.3 Change ρ and μ: conduction, cylinder

Making the same assumptions as in Section 12.4.2 and referring to Fig. 4.5, the same arguments apply. Here, again, the loss density is given by $\mathscr{P} \propto E_s^2 p_r/\rho$ (from eqn 4.15), so the rest of the section applies. There is sufficient similarity between the two p_r curves to allow the same general conclusions, although the same numbers obviously do not apply.

12.4.4 Change ρ and μ: induction

The logic for induction is similar to conduction (Sections 12.4.1 to 12.4.3), except that Chapters 7 to 9 apply, so we leave this as an exercise for the reader. The value of μ to be used is likely to be smaller than expected. This is expanded in Chapter 13, but for now see Fig. 7.3. The normal value of H used in induction heating is $> 100\,000$ A/m, giving μ of 20 or less (right-hand scale).

Chapter 13

Non-linear theory

Ferromagnetic material has a non-linear relationship between flux density B and the magnetic field intensity H, which is described by the magnetisation curve for the material. Because of this, it is not permissible to substitute $B = \mu H$ in eqn 7.2 when dealing with steel, and the rest of the theory is then inaccurate. The non-linearity also distorts the waveforms of the flux density; if the field H is held sinusoidal, the corresponding B wave will be non-sinusoidal, and vice versa.

Much effort has been spent on theories to take this non-linearity into account.[10,11] One common simplification is to take the value of μ corresponding to the RMS value of H at the surface[12] and use this in a constant μ theory, as given in earlier chapters. Similarly, Baker[13] uses $\mu = (1.8B_s/H_{0m}) + 1$ in CGS units, where B_s is the saturated flux density and H_{0m} is the peak magnetising force. With a digital computer, is also possible to include a variable μ fairly easily, taking different values of μ with depth to correspond to the changing value of H with depth.

The approach to be used in this book has proved successful in a similar non-linear context, i.e. eddy-current couplings,[14–18] and has the advantage of giving an analytic solution, so that the effects of non-linearity are clear. The theory has been thoroughly verified as a good approximation.[19,101]

From eqn 7.9,

$$\mathscr{P} = \rho H_s^2/\delta$$

But $\delta = \sqrt{(2\rho/\mu\omega)}$, so that

$$\mu^{1/4} H_s = \sqrt[4]{(2\mathscr{P}^2/\rho\omega)} \tag{13.1}$$

Figure 13.1 shows $\mu^{1/4} H$ plotted against H for low-carbon steel on log/log scales. The plot is linear over a wide range of H, wider than the linearity of the B–H curve on a similar scale. This means that the relation

$$\mu^{1/4} H_s = kH_s^m \tag{13.2}$$

is applicable over a very wide range of H. Substituting eqn 13.2 in 13.1, and rearranging,

$$\mathscr{P} = k^2 H_s^{2m} \sqrt{(\omega\rho/2)} \tag{13.3}$$

The slope of Fig. 13.1 is 0.77. Using this, and values of $\mu^{1/4}H$ over the plotted ranges, gives

$$\mu^{1/4}H = 0.98H^{0.77} \qquad (13.4)$$

(taking a value of the constant that fits the data well from $H = 500$ to $H = 500\,000$). The index of variation of $\mathscr{P}$ with H (eqn 13.3) then becomes

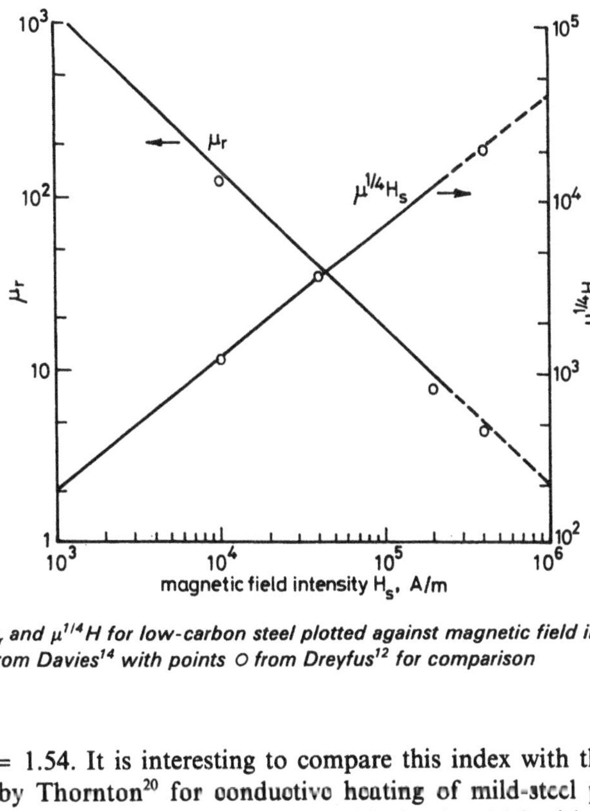

Fig. 13.1 *μ_r and $\mu^{1/4}H$ for low-carbon steel plotted against magnetic field intensity H (data from Davies[14] with points O from Dreyfus[12] for comparison*

$2 \times 0.77 = 1.54$. It is interesting to compare this index with the test results published by Thornton[20] for conductive heating of mild-steel pipes and for heating of vessels similar to those described in Section 24.3 with close-coupled coils. For both, Thornton gives an index of $(\text{current})^{1.57}$, which agrees well with 1.54. Ellet[21] also measures $(\text{current})^{1.6}$, using a closely-coupled coil, and Bowden[19] measures $H^{1.57}$ for EN1A steel. (It must always be remembered that the index will only agree for closely-coupled coils, as it is modified by the presence of magnetising current.)

Equation 13.3 is derived directly from eqn 7.9, so it implies a linear theory up to that point. To improve on this, Bowden[19,101] took the diffusion equation and substituted $B = aH^b$ from the start. This gives

$$\mathscr{P} = H_s^{(b+3)/2}\,\frac{\beta_n K_b}{2}\sqrt{\left(\frac{\omega\rho}{2}\right)}\sqrt{a} \qquad (13.5)$$

It can easily be shown that $2m = (b + 3)/2$ and that $k^2 = \sqrt{a}$, so that Bowden's equation modifies to

$$\mathscr{P} = \frac{k^2}{2}\sqrt{\left(\frac{\rho\omega}{2}\right)}\, H_s^{2m}(\beta_n K_b) \qquad (13.6)$$

This is the same as eqn 13.3 except for $\beta_n K_b$, where $\beta_n = 2/(1 - b)$ and

$$K_b = \sqrt{\left[\frac{(1 - b)^2\sqrt{2}}{(3 + b)(1 + b)^{1/2}}\right]}$$

so that

$$\beta_n K_b = \sqrt{\left[\frac{1}{m\sqrt{(2m - 1)}}\right]}$$

giving $\beta_n K_b = 1.33$ for $m = 0.77$. This means that the non-linear theory gives 33% more loss than eqn 13.3, as well as predicting the index of H correctly. Equation 13.6 accounts for the change in permeability with depth in the workpiece, as well as for the change in the level of saturation.

Equation 13.6 has the right ingredients, but Bowden[19,101] also considered the flux-density harmonics due to the magnetic non-linearity. By plotting the actual B waves caused by sinusoidal H and by analysing them, he obtained a revised expression, $B = AH^b$, where the index b was found to have the same value as that taken from B–H curves, but where the constant A is 25% greater than a. When this is used in the theory, the calculated loss gives good agreement with practice.

This theory has been verified over a wide range of H and the plot of $\mu^{1/4}H$ versus H is linear up to $H = 200\,000$ A/m, so there is no reason to doubt the extrapolation to high values of H. Nevertheless, Bowden's verification is for lower values of H and some caution is needed.

Also, the theory has been developed from the flat slab and it has not been verified for a billet, where d/δ is approximately 3.5.

It is recommended that the values obtained from eqn 7.9 be multiplied by 1.47, which is Bowden's experimental value, taking μ from Fig. 13.1, using the RMS value of H. This has been done in Section 22.8.2.

Chapter 14

Proximity heating

Heating in scanning and single-shot hardening applications is usually caused by heavy currents flowing in single conductors near the surface of the part. Because the conductor is near the surface (compared with the radius of the work), it is usually valid to treat this as a line current at a distance from a flat plane (Fig. 14.1a).

Using the method of images,[8] it can be shown that the current density J at a point distant x from the centre-line due to a current I at height h above the plane is given by

$$J = \frac{I}{\pi h} \frac{1}{1 + (x/h)^2} \tag{14.1}$$

from which the ratio of the power density at x to the maximum power density (at $x = 0$) can easily be obtained:

$$\frac{\mathcal{P}_x}{\mathcal{P}_{max}} = \left[\frac{1}{1 + (x/h)^2}\right]^2 \tag{14.2}$$

where

$$\mathcal{P}_{max} = \frac{\rho}{\delta} \frac{I^2}{\pi^2 h^2} \tag{14.3}$$

This oversimplified approach does not take into account the 'reaction' effect of the load current on H, but it clearly illustrates the effect of various parameters.

First, the maximum loss, directly under the conductor, falls off as the square of the height h. This shows the importance of the closest possible coupling between coil and work.

Secondly, the power density falls away rapidly with distance from the maximum loss point.

Thirdly, by integrating eqn 14.2 with respect to x, the total power (in watts per metre of axial length) between x and $-x$ can be shown to be[8]

$$P_{\pm x} = \frac{\rho I^2}{\pi^2 \delta h} \left[\tan^{-1} \frac{x}{h}\right]_0^x \tag{14.4}$$

giving the total power input to the workpiece (also in watts per metre of axial length), when $x = \pm\infty$, as

$$P_{\text{total}} = \frac{\rho I^2}{\pi^2 \delta h}\frac{\pi}{2} = \frac{I^2}{h}\sqrt{\left(\frac{\rho f \mu}{4\pi}\right)} \qquad (14.5)$$

Hence

$$\frac{P_{\pm x}}{P_{\text{total}}} = \frac{2}{\pi}\left[\tan^{-1}\frac{x}{h}\right] \qquad (14.6)$$

The parameters in eqns 14.2 and 14.6 are quantified in Table 14.1.

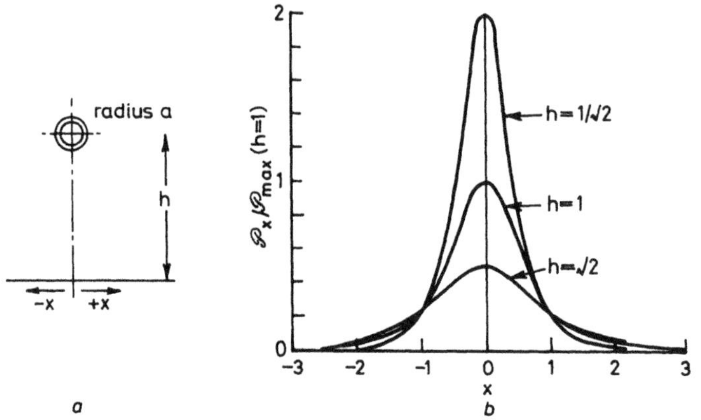

Fig. 14.1 *Proximity effect: (a) position of conductor above workpiece (flat plate) (b) power distribution in workpiece*

Figure 14.1b shows $\mathscr{P}_{\pm x}/\mathscr{P}_{\text{max}}$ in terms of the value of $\mathscr{P}_{\text{max}}$ when $h = 1$, i.e. $\mathscr{P}_{\text{max}}(h = 1) = \rho I^2/\pi^2\delta$, for $h = \sqrt{2}$, 1 and $1/\sqrt{2}$. It illustrates the power distribution in the same workpiece, as the current is moved away from the work.

A practical heater will usually have a copper coil, which itself has skin effect. Brown, Hoyler and Bierwith[8] show that the power in the sheet and the power in the coil are linked by

$$\frac{P_{\text{work}}}{P_{\text{coil}}} = \frac{a}{h}\sqrt{\left\{\frac{\mu_w \rho_w}{\mu_c \rho_c}\right\}} \qquad (14.7)$$

Table 14.1

x/h	0	0.25	0.5	0.75	1	2	3	4	5
$\mathscr{P}_x/\mathscr{P}_{\text{max}}$	1	0.89	0.64	0.41	0.25	0.04	0.01	–	–
$P_{\pm x}/P_{\text{total}}$	0	0.16	0.31	0.41	0.50	0.70	0.80	0.84	0.87

giving the coil efficiency η as

$$\eta = P_{work}/[P_{work} + P_{coil}] = 1/[1 + (h/a)\sqrt{(\rho_c\mu_c/\rho_w\mu_w)}] \quad (14.8)$$

where a is the radius of the tubular conductor. This increases with increasing a/h, i.e. closer coupling.

If copper is being induction-heated from cold, the square root is unity at the start, so that even with $a = h$ the best efficiency is only 50%. As the copper workpiece heats, the efficiency improves; for Cu at 1000°C, $\rho_{1000}/\rho_{20} = 5.7$, so $\eta \to 1/[1 + 1/\sqrt{(5.7)}]$, i.e. 70%.

For ferromagnetic workpieces, the combination of higher μ and high ρ_w/ρ_c gives good efficiencies for closely-coupled work. For example, if $\mu_r = 20$ and $\rho_w/\rho_c \simeq 10$, $\eta \to 1/[1 + 1/\sqrt{(200)}]$, i.e. 93%.

From the method of images, the effective height of a tube whose centre is h above the sheet is

$$h' = h/[1 - (a/h)^2] \quad (14.9)$$

The reader is recommended to study Chapter 8 of Reference 8 for an excellent discussion of the proximity problem. The case of a circular coil facing a conducting slab is discussed by Hammond.[22]

14.1 Scanning heating

Brown, Hoyler and Bierwith[8] also discuss the scanning coil, where the heating pattern derived above is complicated by the movement of the workpiece in the x direction. They use Duhamel's theorem to superpose the solutions; the derivations are tedious and will not be repeated here. The temperature at the surface, as a function of time t', resulting from all the power inputs that have occurred up to that time, is given by

$$\theta(t')_{surface} = \frac{(0.44 \times 10^4)P_{total}}{\sqrt{(vh')}} f\left(\frac{vt'}{h'}\right) \quad (14.10)$$

where P_{total} is the total power input per axial length (W/m) (from eqn 14.5), v is velocity (m/s), and h' is apparent height (m) (from eqn 14.9). The constant contains the thermal constants for iron ('average' values).

The function $f(vt'/h')$ is given by Table 14.2. Note that the peak value is reached at $vt'/h' = 0.5$, i.e. the peak temperature will occur when the workpiece

Table 14.2

vt'/h'	−2	−1	0	0.5	1	2	3	4	5
$f(vt'/h')$	0.03	0.08	0.87	1.0	0.84	0.6	0.48	0.4	0.36

After Brown, Hoyler and Bierwirth.[8]

has moved a distance of $h'/2$ from the centre-line of the coil. This delay should be taken into account in the placing of the quench coil.

Again, the reader with a special interest in the case-hardening of steel is recommended to read Chapter 16 of Reference 8.

Part III

Heat transfer for electroheat

Chapter 15

Basic heat transfer

15.1 Introduction

The designer and user of heating equipment is inevitably involved in heat-flow calculations. These may be explicit, starting from first principles, or implicit, using formulae based on theory but codified into design 'rules' or apparently empirical expressions. Heat transfer is a highly-developed art, based on much experimental verification and sound mathematics, so a clear understanding of the mechanisms of heat flow is possible. In this and subsequent chapters, no attempt has been made to cover all possible heat-transfer theory (many specialised textbooks do so[23,24]) only certain problems are considered that are closely associated wtih conduction and induction heating[25,27].

We have to introduce the concept of thermal conductivity k, with the units of W/m K. It is helpful to rearrange these units as follows:

$$\frac{\text{watts}}{\text{metres kelvins}} = \frac{\text{watts}}{\text{metres}^2} \bigg/ \frac{\text{kelvins}}{\text{metres}} \tag{15.1}$$

Thus

thermal conductivity = power density/thermal gradient

This rearrangement shows clearly that k, multiplied by thermal gradient, gives the power density flowing through the medium. This follows directly from Fourier's law:

$$\mathscr{P} = -k\,\partial\theta/\partial n$$

where $\partial\theta/\partial n$ is the thermal gradient in the direction normal to the area over which $\mathscr{P}$ is being measured.

By analogy, electrical conductivity σ has units $1/\Omega\text{m}$, and these can be rearranged as A/V m = $(\text{A/m}^2)/(\text{V/m})$. Thus electrical conductivity is current density/voltage gradient.

In Section 1.3 we introduced the symbol $(c\gamma)$ for volume specific heat ($\text{J/m}^3\,\text{K}$), and this can be expected to occur frequently wherever heat is stored.

In this chapter, we shall examine further the simple storage of heat, and then present introductory ideas on heat flow. The idealised solutions with all the heat entering at the surface, which we have seen to be a reasonable first assumption, will be derived for slabs and cylinders. In Chapter 18, these are modified for more practical conditions.

15.2 Heat storage: relationship between power, mean temperature and time

Assuming no losses, energy entering the material by any means whatsoever will be stored as heat. During heating, the temperature distribution in the volume is non-uniform as thermal gradients have to exist to cause heat to flow from the source regions to the rest of the metal. This space temperature-distribution will be dealt with later.

At any instant, there is a certain heat energy (joules) in the metal. If the heat sources are switched off and time is allowed for soaking (still with no losses), this heat distributes itself until the temperature is constant throughout; this notional temperature-rise is the mean temperature-rise θ_m (K). It appears in all equations as a function of time and power input, since the energy input is the product of energy-input rate (J/s) and time t (s).

There are obviously transient changes at the start and during soaking.

In induction and conduction heating, it is assumed, initially, that heat is produced uniformly in a shallow surface layer. This is idealised by defining a constant surface power density $\mathscr{P}$ (W/m^2). Consider the three configurations in Fig. 15.1. In general, energy in is equal to energy stored. In the three cases, the

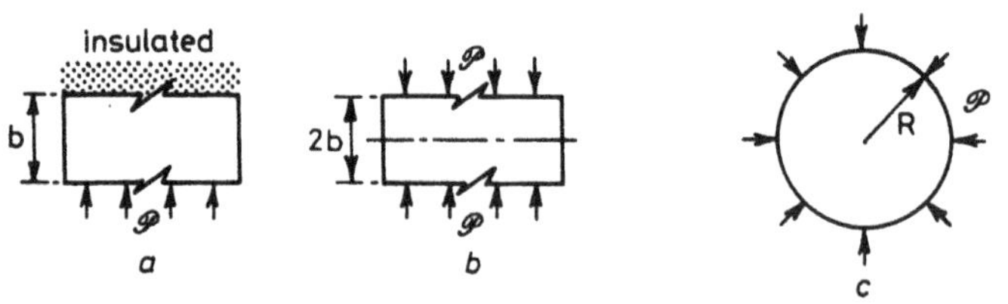

Fig. 15.1 *Configurations for heat storage problem: (a) slab heated one side only (b) slab heated both sides (c) cylinder with uniformly heated surface*

energy in is given by

(a) $\mathscr{P}t$ (surface area) (b) $2\mathscr{P}t$ (surface area) (c) $\mathscr{P}t(2\pi Rl)$

The energy stored is given by θ_m (volume) $(c\gamma)$, which in the three cases is

(a) θ_m (surface area)$b(c\gamma)$ (b) θ_m (surface area)$2b(c\gamma)$ (c) $\theta_m(\pi R^2 l)(c\gamma)$

Equating for θ_m, we have

$$\text{(a)}\ \theta_m = \frac{\mathscr{P}t}{b(c\gamma)} \qquad \text{(b)}\ \theta_m = \frac{\mathscr{P}t}{b(c\gamma)} \qquad \text{(c)}\ \theta_m = \frac{2\mathscr{P}t}{R(c\gamma)} \qquad (15.2)$$

The relationships for (a) and (b) are independent of surface area, and that for (c) is independent of length. Note that the only property of the material that appears in eqns 15.2 is its ability to store heat, given by $(c\gamma)$.

Note that it is obviously impossible to heat one side of a semi-infinite slab by conduction heating, since the appropriate current-flow conditions cannot be created. It is possible to approximate to one-sided heating using normal induction, though a certain amount of imagination is needed to picture the physical configuration. It is certainly practical if travelling-wave induction heating is used.

Example 15.1

A mild-steel slab 25 mm thick is heated on both faces at a power density of 1 MW/m². Calculate the time taken to raise the mean temperature from 20°C to 1000°C. Assume $(c\gamma) = 5 \times 10\,\mathrm{J/m^3\,K}$.

Solution

Rearranging eqn 15.2b,

$$t = \frac{\theta_m b(c\gamma)}{\mathscr{P}}$$

$$= \frac{(1000 - 20)(12.5/1000)(5 \times 10^6)}{(1 \times 10^6)} = 61.25 \quad \mathrm{s}$$

This gives the rough rule that 25 mm slabs, heated from both sides at 1 MW/m², reach 1000°C in about 1 minute. However, the calculation is a considerable oversimplification, as $(c\gamma)$ varies with temperature and it is unlikely that a uniform power density can be applied over the whole temperature range. It gives an indication of the kind of heating time to be expected.

Example 15.2

The conditions are the same as in Example 15.1, but this time for a steel billet of radius 50 mm.

Solution

Rearranging eqn 15.2c,

$$t = \frac{\theta_m R(c\gamma)}{2\mathscr{P}}$$

$$= 122.5 \quad \mathrm{s}$$

Note that this is a linear function of R, so a similar rough rule applies that a 25 mm radius billet heated at 1 MW/m² will reach 1000°C in about 1 minute. The same caveats also apply.

Example 15.3
For non-ferrous metals, the possible power densities are lower (see Chapter 9). In this example, 150 mm diameter aluminium billet is heated through 500°C with a power density of 0.3 MW/m^2. Calculate the time to reach the mean temperature, using an average value of $(c\gamma) = 2.7 \times 10^6\,J/m^3K$.

Solution
Again using eqn 15.2c,

$$t = \frac{\theta_m R(c\gamma)}{2\mathcal{P}}$$

$$= \frac{(500)(75/1000)(2.7 \times 10^6)}{2(0.3 \times 10^6)} = 169 \quad s$$

Thus, very roughly, the heating times for aluminium are the same as those for steel at the power densities possible.

15.3 Temperature distribution during heating

15.3.1 Qualitative treatment

In the discussion of heat storage (Section 15.2), θ_m was shown to be the temperature rise derived from energy considerations. What happens to a slab/billet when the power is turned on can be considered without recourse to mathematics.[25]

Transient stage: When power is applied to the surface the temperature starts to rise immediately, but none of the rest of the metal heats until a thermal gradient is established to force the energy through the material. Mathematically the calculus looks after this, but it is perhaps easier to understand if the material is mentally divided into layers. Before each layer can produce the temperature rise needed to transfer energy to the next layer, it must store energy in its own layer proportional to the temperature rise. This produces a transient temperature rise, which is very important.

The transient condition persists until every part of the material is receiving energy at the same rate. This takes a finite time.

Steady temperature rise: From the earlier discussion of heat storage, it will be obvious that the mean temperature rises linearly with time, whatever the actual temperature distribution in the material. When the transient is finished, and energy is flowing at the same rate into the whole volume, i.e. the watts per cubic metre are constant, then the whole volume rises at a constant rate (K/s) (see Section 15.3.3). The temperature across the material will adjust itself to meet this condition, and this will be shown to be parabolic for both slabs and cylinders.

The temperature pattern is then fixed *in shape* at all instants of time, i.e. it is a function of space, but not of time.

After switch-off: If the power at the surface is switched off, then another transient condition appears whilst the internal conditions adjust. The ultimate temperature rise must be the θ_m that existed at the instant of switching, and the initial internal temperature gradients that force the changes are associated with the parabolic shape at switch-off. This is known as the 'soaking period', and is needed to get approximately uniform temperatures throughout the volume, before the metal is worked. The changes are exponential and uniform heating is obtained after infinite time, so decisions have to be made about the allowable temperature difference between the surface and centre (see Chapter 16).

15.3.2 Mathematical treatment

In Section 15.2, we considered the long-term effect of a power density $\mathscr{P}$ on solids, where the heat was allowed to soak through the material. We now extend this to look at the temperature distribution in those same solids. For this, we add coordinates to the sketches in Fig. 15.1 to form Fig. 15.2.

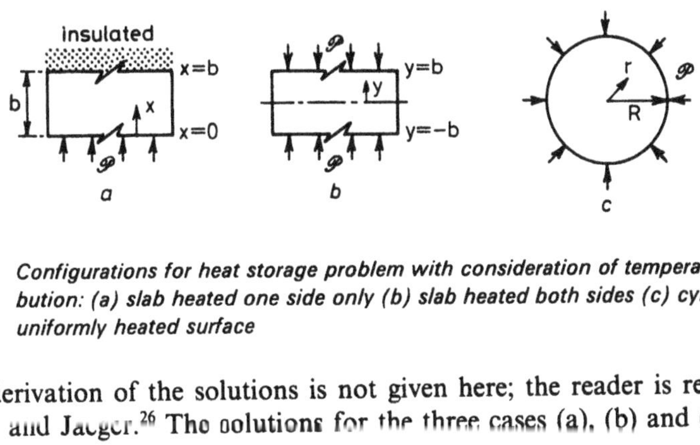

Fig. 15.2 *Configurations for heat storage problem with consideration of temperature distribution: (a) slab heated one side only (b) slab heated both sides (c) cylinder with uniformly heated surface*

The derivation of the solutions is not given here; the reader is referred to Carslaw and Jaeger.[26] The solutions for the three cases (a), (b) and (c) are as follows:

$$\theta = \frac{\mathscr{P}t}{b(c\gamma)} + \frac{\mathscr{P}b}{2k}\left[\left(1-\frac{x}{b}\right)^2 - \frac{1}{3} - \frac{4}{\pi^2}\sum_{n=1}^{\infty}\frac{(-1)^n}{n^2}\exp(-n^2\pi^2\tau)\cos n\pi\left(1-\frac{x}{b}\right)\right] \tag{15.3a}$$

$$\theta = \frac{\mathscr{P}t}{b(c\gamma)} + \frac{\mathscr{P}b}{2k}\left[\left(\frac{y}{b}\right)^2 - \frac{1}{3} - \frac{4}{\pi^2}\sum_{n=1}^{\infty}\frac{(-1)^n}{n^2}\exp(-n^2\pi^2\tau)\cos n\pi\left(\frac{y}{b}\right)\right] \tag{15.3b}$$

$$\theta = \underbrace{\frac{2\mathscr{P}t}{R(c\gamma)}}_{\theta_m} + \frac{\mathscr{P}R}{2k}\left[\underbrace{\left(\frac{r}{R}\right)^2 - \frac{1}{2}}_{\text{Steady distribution}} \underbrace{-4\sum_{n=1}^{\infty}\exp(-\beta_n^2\tau)\frac{J_0[\beta_n(r/R)]}{\beta_n^2 J_0(\beta_n)}}_{\text{Transient temperature change}}\right] \tag{15.3c}$$

The shape of all three of eqns 15.3 is the same; each has a transient, a θ_m term, and a steady distribution.

All three equations have introduced a new symbol τ in the transient solution. This is defined as *normalised time* and is dimensionless:

$$\tau = \underset{\text{Slabs}}{\frac{kt}{(c\gamma)b^2}} \quad \text{or} \quad \underset{\text{Cylinders}}{\frac{kt}{(c\gamma)R^2}} \quad \left[\frac{(\text{W/m K})\,\text{s}}{(\text{Ws/m}^3\,\text{K})\,\text{m}^2}\right] \tag{15.4}$$

The mathematical shape of the transient will not be dealt with here; suffice it to say that it becomes negligible for $\tau \geqslant 0.25$ in all three forms. For surface heating, hardening etc. this term becomes dominant; this will be discussed in Chapter 19. Very few values of n are needed to calculate the transient term, because n^2 appears in both the denominator and the exponential. With a calculator it is easy to work out the transient term for eqns 15.3a or 15.3b using a tabular method, and to plot it; this gives a feel for the relative importance of the various terms, as well as their contribution to the shape.

The first term, for θ_m, has been discussed in Section 15.2.

The steady distribution is given by the constant outside the square bracket (a function of k) multiplied by the non-transient terms inside. All three non-transient terms are parabolic, having their maximum values at the heated surface and their minimum values at $x = b$, $y = 0$ or $r = 0$. The mean temperature rise θ_m occurs where these terms are zero, namely:

(a) $1 - \dfrac{x}{b} = \dfrac{1}{\sqrt{3}} \qquad \dfrac{x}{b} = 1 - \dfrac{1}{\sqrt{3}} = 0.423$

(b) $\dfrac{y}{b} = \pm\dfrac{1}{\sqrt{3}} \qquad \dfrac{y}{b} = \pm 0.578$

(c) $\dfrac{r}{R} = \dfrac{1}{\sqrt{2}} \qquad \dfrac{r}{R} = +0.707$

Finally, $J_0(x)$ is a Bessel function of the first kind and zero order; β_n are the positive roots of $J_1(\beta) = 0$.

Note that $k/c\gamma$ varies over the temperature range, with typical values as follows:

- For steel, 14.25×10^{-6} at 20°C to 4×10^{-6} at 760°C (linear variation), rising thereafter
- For copper, 116.5×10^{-6} at 20°C to 82.7×10^{-6} at 1000°C (linear variation), rising thereafter
- For aluminium, 83.7×10^{-6} at 20°C to 69.2×10^{-6} at 600°C (linear variation), rising thereafter.

The variation for steel is very linear wtih temperature. For copper, a straight-line variation would be a reasonable approximation. For aluminium, the value is fairly constant up to 300°C and thereafter falls linearly. For the theoretical discussion, $k/c\gamma$ has been treated as constant, but the variation must be taken into account in practical calculations.

15.3.3 Generalised values in terms of surface–centre temperature difference

It is always helpful to have expressions and curves which are general and expressed in non-dimensional terms. This can be done for eqns 15.3 by considering the temperature difference between the surface and the back for (a), and the surface and the centre for (b) and (c), after the transient:

$$\left.\begin{array}{llll} x = 0 & \theta_s = \theta_m + \dfrac{\mathscr{P}b}{2k}\dfrac{2}{3} \\ x = b & \theta_b = \theta_m - \dfrac{\mathscr{P}b}{2k}\dfrac{1}{3} \end{array}\right\} \quad \theta_s - \theta_b = \frac{\mathscr{P}b}{2k} \qquad (15.5a)$$

$$\left.\begin{array}{llll} y = \pm b & \theta_s = \theta_m + \dfrac{\mathscr{P}b}{2k}\dfrac{2}{3} \\ y = 0 & \theta_c = \theta_m - \dfrac{\mathscr{P}b}{2k}\dfrac{1}{3} \end{array}\right\} \quad \theta_s - \theta_c = \frac{\mathscr{P}b}{2k} \qquad (15.5b)$$

$$\left.\begin{array}{llll} r = R & \theta_s = \theta_m + \dfrac{\mathscr{P}R}{2k}\dfrac{1}{2} \\ r = 0 & \theta_c = \theta_m - \dfrac{\mathscr{P}R}{2k}\dfrac{1}{2} \end{array}\right\} \quad \theta_s - \theta_c = \frac{\mathscr{P}R}{2k} \qquad (15.5c)$$

Thus, in all three forms, the term outside the square bracket in eqns 15.3 represents $\theta_s - \theta_b$ or $\theta_s - \theta_c$ as appropriate.

The following equations can also be written:

$$\theta_m = \frac{\mathscr{P}t}{b(c\gamma)} = \frac{\mathscr{P}b}{k}\tau = 2\tau(\theta_s - \theta_b) \qquad (15.6a)$$

$$\theta_m = \frac{\mathscr{P}t}{b(c\gamma)} = 2\tau(\theta_s - \theta_c) \qquad (15.6b)$$

$$\theta_m = \frac{2\mathscr{P}t}{R(c\gamma)} = 4\tau(\theta_s - \theta_c) \qquad (15.6c)$$

Equations 15.3 can then be normalised in terms of $\theta_s - \theta_b$ or $\theta_s - \theta_c$ as appropriate, together with normalised time τ and normalised depth:

$$\frac{\theta}{\theta_s - \theta_b} = 2\tau + \left(1 - \frac{x}{b}\right)^2 - \frac{1}{3} - \frac{4}{\pi^2}\Sigma \text{ etc.} \qquad (15.7a)$$

$$\frac{\theta}{\theta_s - \theta_c} = 2\tau + \left(\frac{y}{b}\right)^2 - \frac{1}{3} - \frac{4}{\pi^2}\Sigma \text{ etc.} \qquad (15.7b)$$

$$\frac{\theta}{\theta_s - \theta_c} = 4\tau + \left(\frac{r}{R}\right)^2 - \frac{1}{2} - 4\,\Sigma \text{ etc.} \qquad (15.7c)$$

These are plotted in generalised form in Figs. 15.3 and 15.4. Figure 15.3 is drawn for a slab heated from one surface, but applies to both slabs, since eqns. 15.7a

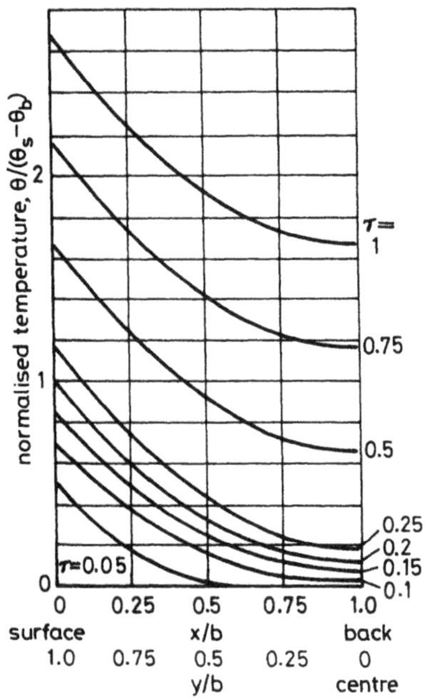

Fig. 15.3 *Temperature distribution in a slab with power input from one side: normalised temperature plotted against position for various values of normalised time τ*

and b are identical, except for the change of origin. The other half (for a slab heated from both sides) is not shown but is the mirror image. Up to $\tau = 0.25$, the heat flow is in a transient state, with the surface rising faster than the centre. The summation term then becomes zero, the temperature distribution becomes parabolic, and thereafter the temperature profile remains the same, with all parts rising at the same rate.

If the temperature of all parts of the workpiece is rising at a uniform rate, then the power input to every element of the volume must be the same. This follows from energy equations:

$$\begin{aligned} \text{energy into elemental volume} &= P_v \, dt \, dv \\ \text{energy stored in elemental volume} &= (c\gamma) \, dv \, d\theta \end{aligned} \qquad (15.8)$$

where P_v is the power per unit volume (W/m^3). Hence, if $d\theta/dt$ is constant, P_v

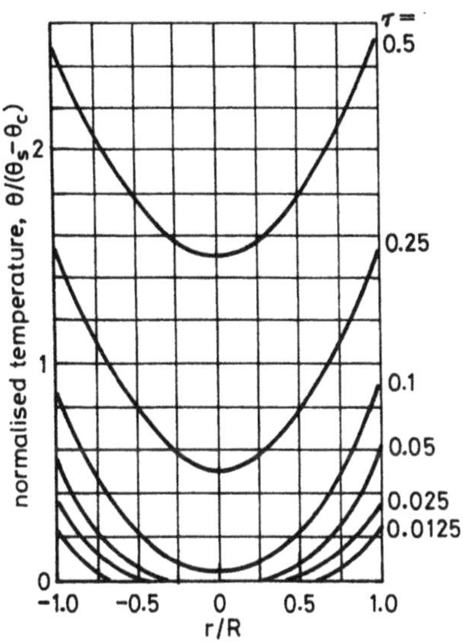

Fig. 15.4 *Temperature distribution in a cylinder with power input at the surface: normalised temperature plotted against normalised radius for various values of normalised time τ*

is the same throughout the volume and is equal to the total surface power divided by the total volume. Let $\mathscr{P}$ be the surface power-density. For the slab heated from both sides, $P_v = 2\mathscr{P}/t$; for the cylinder, $P_v = \mathscr{P}(2\pi Rl/\pi R^2 l) = 2\mathscr{P}/R$, i.e. the volume power-density is inversely proportional to the thickness t and outer radius R, respectively. For the slab, the power density at y from the centre is $\mathscr{P}_y = \mathscr{P}(2y/t)$; for the cylinder, $\mathscr{P}_r = \mathscr{P}(2r/R)$, i.e. falling linearly in both as the centre-line is approached. This power is being transferred by conduction.

15.3.4 Surface temperature

From the above equations, θ_s will always be greater than θ_m. From eqns 15.5,

$$\theta_s = \theta_m + \frac{\mathscr{P}b}{3k} \tag{15.9a}$$

$$\theta_s = \theta_m + \frac{\mathscr{P}b}{3k} \tag{15.9b}$$

$$\theta_s = \theta_m + \frac{\mathscr{P}R}{4k} \tag{15.9c}$$

The last terms are therefore the excess surface temperature when θ_m has reached the desired temperature. This is usually modified in practice by the fall-off of $\mathscr{P}$ at higher temperatures, which changes Figs 15.3–15.5.

Example 15.4
A steel billet, 200 mm in diameter, is heated with a uniform power density of 1 MW/m^2. The mean temperature rise after soaking is to be 1000°C. Assume that the thermal conductivity is 40 W/m K. Find

(a) the time to reach θ_m
(b) the surface temperature at that time
(c) the difference between surface and centre temperature at that time.

Solution
(a) Using eqn 15.3c,

$$\theta_m = \frac{2\mathscr{P}t}{R(c\gamma)}$$

So

$$t = \frac{R(c\gamma)\theta_m}{2\mathscr{P}}$$

$$= \frac{(100/1000)(5 \times 10^6)(1000)}{2(1 \times 10^6)} = 250\,\text{s}$$

(b) Using eqn 15.9c,

$$\theta_s = \theta_m + \frac{\mathscr{P}R}{4k}$$

$$= 1000 + \frac{(1 \times 10^6)(100/1000)}{(4)(40)} = 1000 + 625 = 1625°\text{C}$$

That is, the surface would have melted before θ_m was reached.
(c) Using eqn 15.5c,

$$\theta_s - \theta_c = \frac{\mathscr{P}R}{2k}$$

$$= 1250°\text{C}$$

This shows that the centre has only risen by 375°C after 250 s. These extreme figures have been quoted to bring out the thermal problem of induction heating and the importance of understanding the equations of this chapter.

Before starting (b) and (c) in this example, the validity of the assumption that the transient had passed should have been tested. From eqn 15.4,

$$\tau = \frac{kt}{(c\gamma)R^2}$$

$$= \frac{(40)(250)}{(5 \times 10^6)(0.1)^2} = 0.2$$

Since this is less than 0.25, the transient is still present and the surface–centre differential is even greater than calculated in (b) and (c).

From Appendix A.1, $(c\gamma)$ rises from 3.65 MJ/m^3 K at 20°C to a peak at 800°C and falls to 5.09 at 1000°C. The value used above is a mean value of 5.09 over the range, rounded to 5 for simplicity. In practice, there is substantial storage of energy at the Curie point, where the stored energy changes without change of temperature. The calculations of this example must be taken as indicative of the magnitudes, rather than exact figures.

Similarly, in the calculation of τ there should be calculations over the temperature ranges, using the appropriate values of k and $(c\gamma)$. However, in the light of Example 15.5 below, which uses more realistic power densities, it is not worth the effort.

This example, which oversimplifies the practical problem by its assumption of constant $\mathscr{P}$ and k, shows that the temperature difference at the end of the heating period is high and that a substantial soaking period is needed to equalise the temperatures. It also indicates that there are practical limitations on the power density that can be used without excessive temperature at the surface.

In Chapter 18 it is shown that $\theta_s - \theta_c$ is reduced when the loss is distributed, rather than being concentrated at the surface.

Example 15.5

The billet of Example 15.4 is heated to Curie temperature (760°C) from 20°C at 1 MW/m^2, but after that the power input reduces to 0.25 MW/m^2. Recalculate (a), (b) and (c).

Solution

(a) We have

$$t = \frac{R(c\gamma)}{2}\left[\frac{\theta_{m1}}{\mathscr{P}_1} + \frac{\theta_{m2}}{\mathscr{P}_2}\right]$$

$$= \frac{(0.1)(5 \times 10^6)}{2(10^6)}\left[\frac{760 - 20}{1} + \frac{1000 - 760}{0.25}\right] = 425\,\text{s}$$

(b), (c) *Conditions below Curie*

Check the normalised time:

$$\tau_1 = \frac{k_1 t_1}{(c\gamma)R^2} = \frac{(45)(185)}{(4.8 \times 10^6)(0.1)^2} = 0.17$$

so that $\tau < 0.25$ still. This means that it is inaccurate to use eqns 15.4–15.7; they are only approximately applicable. (Note that mean values have been used for k and $(c\gamma)$ over the 20–760°C range: see Appendix A.1.)

The time for heating to Curie temperature is 185 s. From eqn 15.9c,

$$\theta_{s1} = \theta_{m1} + \frac{\mathscr{P}_1 R}{4k_1}$$

$$= (760 - 20) + \frac{(10^6)(0.1)}{(4)(45)} = 740 + 556 = 1296°C$$

Since θ_s is a temperature rise, the surface temperature will be 1316°C.

From eqn 15.5c,

$$\theta_{s1} - \theta_{c1} = \frac{\mathscr{P}R}{2k}$$

$$= \frac{(10^6)(0.1)}{(2)(45)} = 1111°C$$

so that the centre temperature is only (1296 − 1111) = 185°C(!).

(b), (c) *Conditions above Curie*
Since soaking has not yet been dealt with, the detailed behaviour of the temperatures above Curie cannot be calculated. If it is assumed, without proof, that the billet would achieve a substantially uniform temperature of 760°C if allowed to soak for the remainder of the 425 s heating time (i.e. 240 s) without loss, the second stage can be superimposed on a uniform temperature of 760°C. Also, it follows logically that if, at 760°C, the thermal differential was right to cause 1 MW/m² to be carried from the surface, the surface rise will not increase above 1296°C when asked to cope with only 0.25 MW/m².

Therefore from eqn 15.9c,

$$\theta_{s2} = \theta_{m2} + \frac{\mathscr{P}_2 R}{4k_2}$$

$$= (1000 - 760) + \frac{(0.25 \times 10^6)(0.1)}{(4)(26)} = 240 + 240°C$$

Thus the surface temperature at the end of heating is 760 + 240 + 240 = 1240°C.

τ_2 should be checked:

$$\tau_2 = \frac{k_2 t_2}{(c\gamma)R^2} = \frac{(26)(240)}{(6 \times 10^6)(0.1)^2} = 0.10$$

This again makes the calculations invalid, since $\tau < 0.25$.

The diameter in these examples has deliberately been chosen to be large and therefore to give 'difficult' results. If the diameter is made smaller by only a small amount, the R^2 term will greatly increase τ.

Note that τ can be rearranged as follows:

$$\tau = \frac{kt}{(c\gamma)R^2} = \frac{k}{R}\frac{\theta_m}{2\mathscr{P}} = \frac{k\theta_m}{2R\mathscr{P}} \qquad (15.10)$$

which may be preferred for testing its validity.

15.4 Identity of slab solutions

Comparison of the a and b forms of the equations of this chapter shows that, apart from 15.3, they are identical. Equations 15.3a and b are the same equation with the origin shifted. Examination of Fig. 15.2a and b shows why this is so. By our choice of $2b$ for the thickness of the slab heated from both sides, that problem becomes two one-sided slabs back to back. In this chapter the solutions have been kept separate to emphasise this, but in Chapter 16 the identity will be taken as established.

Chapter 16

Soaking conditions

16.1 Introduction

We showed in Chapter 15 that the temperature difference between centre and surface is often too high for the workpiece to be used immediately, so that a short soaking period is needed to reduce the differential. Since we start this chapter with the end results of Chapter 15, surface heating is implicit here: this is modified in Chapter 18, where we take into account the fact that the heat is not produced at the surface.

It was shown that the temperature distribution is parabolic for all three configurations considered: two slabs and the cylinder (Fig. 15.2). As in Chapter 15, we shall take the three solutions in parallel; in practice this means two solutions, since the two slabs can be treated together by a simple change of symbol.

The parabolic temperature distributions at the end of the heating are the initial conditions for further solutions of the heat-flow equations. To distinguish temperatures during soaking, we define the temperatures when the power is switched off as $\bar{\theta}_s$, $\bar{\theta}_c$ and $\bar{\theta}_m$, still denoting temperatures at other times by θ_s, θ_c and θ_m, where $\bar{\theta}_m = \theta_m$.

16.1.1 Value of final temperature θ_m

Equations 15.5a–c can easily be rearranged. From eqn 15.5a we have

$$\bar{\theta}_b = \bar{\theta}_m - \frac{\mathscr{P}b}{2k}\frac{1}{3} \qquad \bar{\theta}_s - \bar{\theta}_b = \frac{\mathscr{P}b}{2k}$$

Substituting,

$$\bar{\theta}_b = \bar{\theta}_m - \frac{\bar{\theta}_s - \bar{\theta}_b}{3}$$

Hence

$$\bar{\theta}_m = \bar{\theta}_b + (\bar{\theta}_s - \bar{\theta}_b)/3 \qquad (16.1a)$$

Similarly,

$$\bar{\theta}_m = \bar{\theta}_c + (\bar{\theta}_s - \bar{\theta}_c)/3 \quad (16.1b)$$

$$\bar{\theta}_m = \bar{\theta}_c + (\bar{\theta}_s - \bar{\theta}_c)/2 = (\bar{\theta}_s + \bar{\theta}_c)/2 \quad (16.1c)$$

16.1.2 Solutions for temperature distribution

The general solutions for slab and cylinder for any initial temperature distribution can be found in Carslaw and Jaeger:[26]

$$\theta = \theta_m + 2\sum_{n=1}^{\infty} \exp(-n^2\pi^2\tau)\cos n\pi\frac{x}{b}\int_0^b f(x')\cos n\pi\frac{x'}{b}\,\mathrm{d}\left(\frac{x'}{b}\right) \quad (16.2a/b)$$

$$\theta = \theta_m + \frac{2}{R^2}\sum_{n=1}^{\infty} \exp(-\alpha_n^2\tau)\frac{J_0(r\alpha_n/R)}{J_0^2(\alpha_n)}\int_0^R rf(r)J_0(\alpha_n r/R)\,\mathrm{d}r \quad (16.2c)$$

If we were dealing with *any* temperature distribution, perhaps that found in Chapter 18, we should have to get an expression for $f(x)$, $f(y)$ or $f(r)$, but we have already defined it as parabolic. We shall derive the solution in detail for a slab previously heated from one side, modifying it subsequently for double-sided heating and then, by analogy, getting the cylindrical solution.

16.2 Soaking of slab previously heated from one side

Assume that the heating has been applied long enough for the temperature distribution to become parabolic. From eqns 15.5a and 15.3c:

$$\bar{\theta}_s = \bar{\theta}_m + \frac{\mathscr{P}b}{3k} \quad (16.3)$$

$$\bar{\theta}_b = \bar{\theta}_m - \frac{\mathscr{P}b}{6k} \quad (16.4)$$

$$\bar{\theta} = \bar{\theta}_m + \frac{\mathscr{P}b}{2k}\left[\left(1 - \frac{x}{b}\right)^2 - \frac{1}{3}\right] \quad (16.5)$$

From eqn 16.2a, the temperature distribution during soaking is

$$\theta = \theta_m + 2\sum_{n=1}^{\infty} \exp(-n^2\pi^2\tau)\cos n\pi\frac{x}{b}\int_0^b f(x')\cos n\pi\frac{x'}{b}\,\mathrm{d}\left(\frac{x'}{b}\right)$$

In our problem, $f(x')$ is eqn 16.5. Taking the integral first,

$$\int_0^b \left\{\bar{\theta}_m + \frac{\mathscr{P}b}{2k}\left[\left(1 - \frac{x}{b}\right)^2 - \frac{1}{3}\right]\right\}\cos n\pi\frac{x}{b}\,\mathrm{d}\left(\frac{x}{b}\right)$$

$$= \frac{\mathscr{P}b}{2k}\int_0^b \left(\frac{x^2}{b^2} - \frac{2x}{b}\right)\cos n\pi\frac{x}{b}\,\mathrm{d}\left(\frac{x}{b}\right)$$

since terms in cos $n\pi\theta \mathrm{d}\theta$ disappear between these limits. Thus the integral becomes

$$\frac{\mathscr{P}b}{2k}\left[\frac{2\dfrac{n\pi x}{b}\cos\dfrac{n\pi x}{b} + \left\{\left(\dfrac{n\pi x}{b}\right)^2 - 2\right\}\sin\dfrac{n\pi x}{b}}{n^3\pi^3} - \frac{2\left(\cos\dfrac{n\pi x}{b} + \dfrac{n\pi x}{b}\sin\dfrac{n\pi x}{b}\right)}{n^2\pi^2}\right]_0^b$$

$$= \frac{\mathscr{P}b}{2k}\left[\frac{2}{n^2\pi^2}\right] \tag{16.6}$$

Hence, eqn 16.2a reduces to

$$\theta = \bar{\theta}_m + \frac{2\mathscr{P}b}{\pi^2 k}\sum_{n=1}^{\infty}\frac{\exp(-n^2\pi^2\tau)}{n^2}\cos n\pi\frac{x}{b} \tag{16.7}$$

(Note that, for $x = 0$ and $\tau = 0$,

$$\sum_{n=1}^{\infty}\frac{1}{n^2} = \frac{\pi^2}{6}$$

so that

$$\bar{\theta}_s = \theta_m + \frac{2\mathscr{P}b}{\pi^2 k}\frac{\pi^2}{6} = \theta_m + \frac{\mathscr{P}b}{3k}$$

For $x = b$ and $\tau = 0$,

$$\sum_{n=1}^{\infty}\frac{1}{n^2}(-1)^n = -\left(\frac{\pi^2}{12}\right)$$

so that

$$\bar{\theta}_b = \theta_m + \frac{2\mathscr{P}b}{\pi^2 k}\left(-\frac{\pi^2}{12}\right) = \theta_m - \frac{\mathscr{P}b}{6k}$$

This checks that the initial conditions agree with eqns 16.3 and 16.4.)

From eqns 16.3 and 16.4 the differential temperature $\bar{\theta}_s - \bar{\theta}_b$ is $\mathscr{P}b/2k$. Eqn 16.7 can be rearranged as:

$$\frac{\theta_s - \theta_b}{\bar{\theta}_s - \bar{\theta}_b} = \frac{4}{\pi^2}\sum_1^{\infty}\frac{\exp(-n^2\pi^2\tau) - (-1)^n}{n^2}$$

$$= \frac{4}{\pi^2}\left[\sum_1^{\infty}\frac{\exp(-n^2\pi^2\tau)}{n^2}\right] + \left(\frac{1}{3}\right)$$

Similarly

$$\frac{\theta_b - \bar{\theta}_b}{\bar{\theta}_s - \bar{\theta}_b} = \frac{4}{\pi^2}\left[\sum_1^{\infty} \frac{(-1)^n \exp(-n^2\pi^2\tau)}{n^2}\right] + \left(\frac{1}{3}\right) \tag{16.8}$$

These dimensionless functions are plotted on Fig. 16.1. For large τ, the exponen-

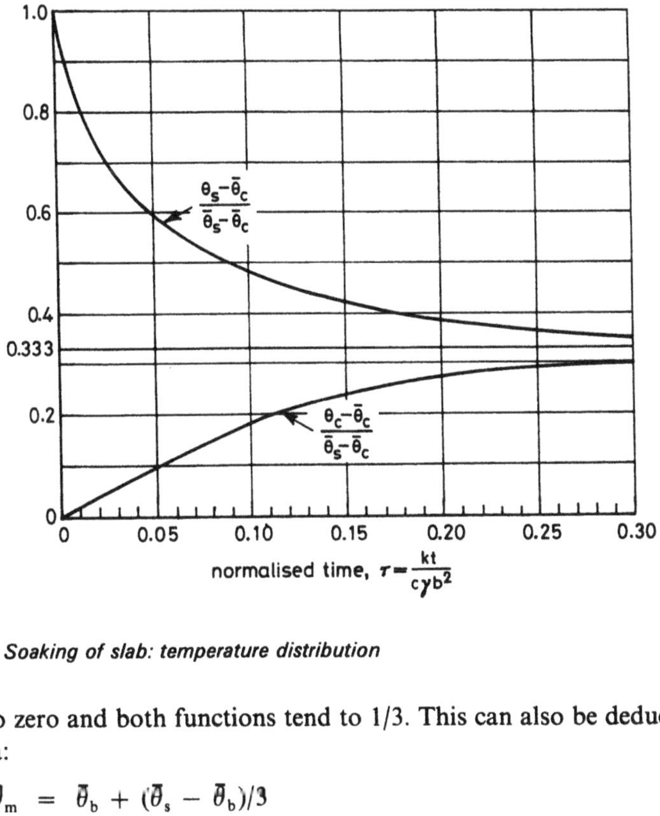

Fig. 16.1 *Soaking of slab: temperature distribution*

tials go to zero and both functions tend to 1/3. This can also be deduced from eqn 16.1a:

$$\bar{\theta}_m = \bar{\theta}_b + (\bar{\theta}_s - \bar{\theta}_b)/3$$

Since $[\exp(-n^2\pi^2\tau)]/n^2$ falls very rapidly with increasing n, it is valid to use only the $n = 1$ term for $\tau \geqslant 0.1$, giving

$$\theta \approx \bar{\theta}_m + \frac{2\mathscr{P}b}{\pi^2 k} \exp(-\pi^2\tau) \cos \pi \frac{x}{b} \tag{16.9}$$

We are interested in the instant when $\theta_s - \theta_b$ reaches a certain value, expressed either as a temperature difference θ_d, or as a proportion of the final temperature θ_m:

$$\theta_d = \theta_s - \theta_b \approx \frac{2\mathscr{P}b}{\pi^2 k}[\exp(-\pi^2\tau) + \exp(-\pi^2\tau)]$$

$$\approx \frac{4\mathscr{P}b}{\pi^2 k} \exp(-\pi^2\tau) \tag{16.10}$$

Note that this is independent of the original heating time.

If the differential temperature at the start of soaking is $\bar{\theta}_d = \bar{\theta}_s - \bar{\theta}_b = \mathscr{P}b/2k$, then

$$\tau \approx \frac{1}{\pi^2} \ln \frac{8}{\pi^2(\theta_d/\bar{\theta}_d)} \tag{16.11}$$

$$t \approx \frac{(c\gamma)b^2}{k\pi^2} \ln \frac{8}{\pi^2(\theta_d/\bar{\theta}_d)} \qquad \tau \geqslant 0.1 \tag{16.12}$$

where θ_d is the specified differential temperature. This can be expressed in terms of $\theta_d/\bar{\theta}_m$:

$$t \approx \frac{(c\gamma)b^2}{k\pi^2} \ln \frac{(4/\pi^2)(c\gamma b^2/k\bar{t})}{(\theta_d/\bar{\theta}_m)} \tag{16.13}$$

where $\bar{t}$ is the time of application of the power before soaking.

16.3 Soaking of slab previously heated from both sides

In Chapter 15, we established that we could use the solutions for one-sided heating for a slab $2b$ thick, heated from both sides. Equations 16.1 to 16.13 are still valid except that the temperature at the centre-line θ_c is substituted for θ_b and it is again convenient to measure from the axis of symmetry, as in Chapter 15, so we make the substitution $x = b - y$.

At the end of the heating period, assume that the distribution has become parabolic. From eqn 15.5b, at $y = \pm b$,

$$\bar{\theta}_s = \bar{\theta}_m + \frac{\mathscr{P}b}{3k} \tag{16.14}$$

At $y = 0$,

$$\bar{\theta}_c = \bar{\theta}_m - \frac{\mathscr{P}b}{6k} \tag{16.15}$$

Generally, from eqn 15.3b,

$$\bar{\theta} = \bar{\theta}_m + \frac{\mathscr{P}b}{2k}\left[\left(\frac{y}{b}\right)^2 - \frac{1}{3}\right] \tag{16.16}$$

The key equations 16.8 are in non-dimensional form and are unchanged, except that θ_c is substituted for θ_b. Figure 16.1 is still valid.

16.4 Soaking of cylinders

In Section 16.1.2, we quoted the general solution (eqn 16.2c) for a cylinder without heat input and with an initial temperature distribution $\theta = f(r)$. This was slightly simplified from Carslaw and Jaeger[26] to include θ_m:

$$\theta = \theta_m + \frac{2}{R^2}\sum_{n=1}^{\infty} \exp(-\alpha_n^2\tau)\,\frac{J_0(r\alpha_n/R)}{J_0^2(\alpha_n)}\int_0^R rf(r)J_0(\alpha_n r/R)\,dr$$

Baker[27] solved the transient part of eqn 16.2c, with an initial parabolic temperature distribution; the result is plotted in Fig. 16.2, which has the same form as Fig. 16.1.

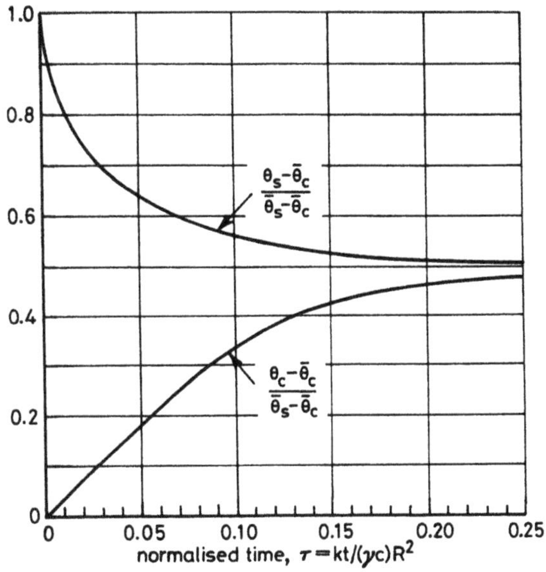

Fig. 16.2 *Soaking of cylinder: temperature distribution*

Example 16.1
In Example 15.4, it was shown that $\theta_s - \theta_c$ is 1250°C at the end of the heating period. Find the time taken for the temperature difference to fall to 10% of this value, after the heating has finished. Assume $(c\gamma) = 5 \times 10^6\,\mathrm{J/m^3K}$ and $k = 40\,\mathrm{W/m\,K}$.

Solution
From Fig. 16.2, for 10%, $\tau = 0.154$. From the definition of τ (Chapter 15):

$$\tau = kt/(c\gamma)R^2$$

$$t = \frac{0.154(c\gamma)R^2}{40}$$

$$= \frac{(0.154)(5 \times 10^6)(0.1)^2}{40}$$

$$= 192.5 \quad \mathrm{s}$$

Note that this is the time taken to reduce the differential to 10% of its initial value, whatever that initial value. Since 1250°C is too high in practice, and the effect of current depth has not yet been taken into account, the time calculated is grossly pessimistic and is probably more representative of a 50°C differential.

Chapter 17

Radiation

The radiation loss $\mathscr{P}_R$ is given by the Stefan-Boltzmann law:

$$\mathscr{P}_R = 5.67 \times 10^{-8} \varepsilon T_s^4 \qquad \text{W/m}^2 \tag{17.1}$$

where ε is the emissivity coefficient of the surface (dimensionless) (see Fig. 22.6) and T_s is the absolute surface temperature (K). The constant $5.67 \times 10^{-8}\,\text{W/m}^2\,\text{K}^4$ is known as Stefan's constant.

Note that the symbol for temperature has been changed here from θ to T, denoting that we are working in absolute temperatures in kelvin, i.e. Celsius + 273.15°. θ is used both for 'scale' temperatures, e.g. 760°C, and for temperature intervals, which can be stated in either kelvin or Celsius, since 1°C = 1 K.[28,29]

If the body is radiating to a surface at absolute temperature T_a kelvin, the radiation loss is reduced to

$$\mathscr{P}_R = 5.67 \times 10^{-8} \varepsilon (T_s^4 - T_a^4) \tag{17.2}$$

For low temperatures, $\mathscr{P}_R$ is negligible. For example, if a surface at 400 K is radiating to 300 K, the loss density is roughly 1 kW/m^2, which is small compared with the values of $\mathscr{P}_s$ input, e.g. 1 MW/m^2, that have been found in Chapter 7. Also, if there is a substantial difference between the temperature of the radiating surface and the ambient, then T_a^4 is usually negligible compared with T_s^4; for example, for a surface at 1000 K radiating to 300 K, $T_s^4 = (1000)^4 = 10^{12}$ and $T_a^4 = (300)^4 = 81 \times 10^8 < 10^{10}$. It is often sufficient to ignore the T_a term.

17.1 Time to heat a billet, including radiation

In Chapter 15, the energy into a cylinder was equated to the energy stored. If there is radiation loss, this becomes, for unit length,[30]

$$\begin{aligned} \text{power in} &= \text{power stored} + \text{power radiated} \\ \mathscr{P}t(2\pi R) &= \theta_m(\pi R^2)(c\gamma) + \mathscr{P}_R t(2\pi R) \end{aligned} \tag{17.3}$$

This can be rearranged and written in differentials:

$$\frac{2\,dt}{R(c\gamma)} = \frac{d\theta}{\mathscr{P} - \mathscr{P}_R} \tag{17.4}$$

If $\mathscr{P}$ is a constant input power, it is convenient to turn it into an artificial radiation temperature $T_{\mathscr{P}}$, i.e. the temperature that would have to exist to radiate $\mathscr{P}$. Thus

$$\mathscr{P} = CT_{\mathscr{P}}^4 \qquad \mathscr{P}_R = CT^4 \tag{17.5}$$

where C is a constant for the material ($5.67 \times 10^{-8}\varepsilon$ W/m^2 K) and T_a is ignored. Equation 17.4 becomes a function of T and can be integrated:

$$\frac{2C}{R(c\gamma)}[t]_0^t = \int_{T_1}^{T_2} \frac{dT}{T_{\mathscr{P}}^4 - T^4} \tag{17.6}$$

where T_2 is the final temperature and T_1 is the starting temperature (both in K). Hence

$$\frac{2C}{R(c\gamma)}t = \frac{1}{T_{\mathscr{P}}^3}\left[\frac{1}{2}\tan^{-1}\frac{T}{T_{\mathscr{P}}} + \frac{1}{2}\tanh^{1}\frac{T}{T_{\mathscr{P}}}\right]_{T_1}^{T_2} \tag{17.7}$$

Let the square bracket be $f(\Theta)$, where

$$\Theta_1 = T_1/T_{\mathscr{P}} \qquad \Theta_2 = T_2/T_{\mathscr{P}} \tag{17.8}$$

remembering that Θ is in radians. Then

$$t_R = \frac{(c\gamma)R}{2CT_{\mathscr{P}}^3}[f(\Theta_2) - f(\Theta_1)] \tag{17.9}$$

$f(\Theta)$ is so easily calculated, using an engineering calculator, that it is not worth plotting.

17.2 Time to heat a slab, including radiation

Following the same arguments as Section 17.1, but for a slab,[30]

$$\frac{dt}{b(c\gamma)} = \frac{d\theta}{\mathscr{P} - \mathscr{P}_R} \tag{17.10}$$

The shape of this equation is the same as that of eqn 17.4, so the rest of the solution is the same, with b substituted for $R/2$ in eqn 17.9, to give

$$t_R = \frac{(c\gamma)b}{CT_{\mathscr{P}}^3}[f(\Theta_2) - f(\Theta_1)] \tag{17.11}$$

Example 17.1
Calculate the time taken to heat a steel billet of 0.05 m diameter from 1000 to 1500 K, with a surface power density of 1 MW/m^2, (a) without radiation (b) with radiation with an emissivity ε of 0.7. Assume $(c\gamma) = 5.6 \times 10^6$ W s/m^3 K.

Solution

(a) Without radiation, from eqn 15.2c,

$$t_0 = \frac{(c\gamma)R}{2\mathscr{P}}\Delta\theta$$

$$= \frac{(5.6 \times 10^6)(0.05)}{(2)(10^6)(2)} 500 = 35.0\,\text{s}$$

(b) With radiation, from eqn 17.5,

$$T_{\mathscr{P}} = \sqrt[4]{(\mathscr{P}/C)}$$

$$= \sqrt[4]{[(10^6)/(5.67 \times 10^{-8})(0.7)]} = 2240\,\text{K}$$

From eqns 17.7 and 17.8,

$$\Theta_2 = 1500/2240 = 0.670 \qquad f(\Theta_2) = 0.7$$

$$\Theta_1 = 1000/2240 = 0.446 \qquad f(\Theta_1) = 0.45$$

The time to temperature is given by eqn 17.9:

$$t_R = \frac{(c\gamma)R}{2CT_{\mathscr{P}}^3}[f(\Theta_2) - f(\Theta_1)]$$

$$= \frac{(5.6 \times 10^6)(0.05/2)}{2(5.67 \times 10^{-8})(0.7)(2240)^3}[0.7 - 0.45] = 39.2\,\text{s}$$

$$t_R/t_0 = 39.2/35 = 1.12$$

That is, with radiation the billet heating time is increased by 12% in the temperature range above the Curie point.

17.3 Mean radiated power over a temperature interval

In Chapter 15 it was shown that, after an initial transient, the temperature rose linearly with time. By the time radiation is important, this condition will certainly be reached. This allows us to calculate the mean radiated power over the temperature range T_1 to T_2.[30] First, the total radiated power-density is given by

$$\int_{t_1}^{t_2} \mathscr{P}_R\,dt = \int_{T_1}^{T_2} k\mathscr{P}_R\,dT \qquad \text{if } dt = k\,dT$$

$$= \int_{T_1}^{T_2} kCT^4\,dT$$

$$= \frac{kC}{5}(T_2^5 - T_1^5) \tag{17.12}$$

If $\mathscr{P}_R$ were constant over the time interval at a level given by the mean radiated power-density $\mathscr{P}_{Rm}$, then the total radiated power-density would be

$$\int_{t_1}^{t_2} \mathscr{P}_{Rm}\, dt = \int_{T_1}^{T_2} k\mathscr{P}_{Rm}\, dT$$

$$= k\mathscr{P}_{Rm}(T_2 - T_1) \qquad (17.13)$$

Equating eqns 17.12 and 17.13,

$$\mathscr{P}_{Rm} = \frac{C}{5}\left(\frac{T_2^5 - T_1^5}{T_2 - T_1}\right) = \frac{5.67 \times 10^{-8}\varepsilon}{5}\left(\frac{T_2^5 - T_1^5}{T_2 - T_1}\right) \qquad (17.14)$$

Example 17.2

Calculate the mean radiated power-density for steel ($\varepsilon = 0.7$) over the ranges (a) 20°C to 750°C (i.e. up to Curie) and (b) 750°C to 1250°C.

Solution

Use eqn 17.14.

(a) From 20°C to 750°C:

$$\mathscr{P}_{Rm} = \frac{5.67 \times 10^{-8} \times 0.7}{5}\left(\frac{1023^5 - 293^5}{1023 - 293}\right) = 12.2 \text{ kW/m}^2$$

(b) From 750°C to 1250°C:

$$\mathscr{P}_{Rm} = \frac{5.67 \times 10^{-8} \times 0.7}{5}\left(\frac{1523^5 - 1023^5}{1523 - 1023}\right) = 112 \text{ kW/m}^2$$

17.4 Cooling by radiation

When $\mathscr{P} = 0$, eqn 17.6 becomes, with appropriate limits on the integral,[30]

$$\frac{2C}{R(c\gamma)}[t]_0^t = -\int_{T_2}^{T_f} \frac{dT}{T^4}$$

$$= \left[\frac{1}{3T^3}\right]_{T_2}^{T_f}$$

where T_f is the final temperature after the billet has cooled for t seconds, starting at temperature T_2. Thus

$$T_f = \sqrt[3]{\left\{1\Big/\left[\frac{1}{T_2^3} + \frac{6C}{R(c\gamma)}t\right]\right\}} = T_2\sqrt[3]{\left\{1\Big/\left[1 + \frac{6C}{R(c\gamma)}tT_2^3\right]\right\}}$$

$$= T_2\Big/\sqrt[3]{\left[1 + \frac{6C}{R(c\gamma)}tT_2^3\right]} \qquad (17.15)$$

The same equation holds for slabs, with b in place of $R/2$:

$$T_f = T_2 \bigg/ \sqrt[3]{\left[1 + \frac{3C}{b(c\gamma)} t T_2^3\right]} \tag{17.16}$$

Example 17.3
A steel billet of 0·10 m diameter, originally at 1500 K, cools by radiation for 30 s. Find the final temperature, using the constants of Example 17.1.

Solution
From eqn 17.15:

$$T_f = T_2 \bigg/ \sqrt[3]{\left[1 + \frac{6C}{R(c\gamma)} t T_2^3\right]}$$

$$= 1500 \bigg/ \sqrt[3]{\left[1 + \frac{(6)(5.67 \times 10^{-8})(0.7)}{(0.05)(5.6 \times 10^6)} (30)(1500^3)\right]}$$

$$= 1469 \text{ K}$$

Chapter 18

Effect of current depth and radiation

Chapters 15 and 16 were based on the simplifying assumption that all the heat entered at the surface. In practice, this is only true if the frequency is sufficiently high for the depth of heating to be small compared with the radius or thickness. In this chapter, we modify this to allow for the two practical conditions:

- the actual heat distribution in the metal
- the heat lost by radiation from the surface.

The loss equations can be combined with the heat-flow equations to give an accurate solution.[12] To deal with this, we need to expand the heat-flow equation to include the volumetric production of heat.

18.1 General heat-flow equation

It is convenient to derive the expressions for cartesian and cylindrical coordinates in parallel, since the same logic applies.

Consider the heat input into an element 1 m long, at depth y in cartesian coordinates and at radius r in cylindrical coordinates. Then the energy input is given by

Cartesian: $(c\gamma)x\ \mathrm{d}y\ \partial\theta$

Cylindrical: $(c\gamma)2\pi r\ \mathrm{d}r\ \partial\theta$

This is made up of two parts; the net heat conducted into the element and the heat produced in the element. The heat conducted from the element is given by

Cartesian: at y, $k(\partial\theta/\partial y)\ x\ \partial t$

Cylindrical: at r, $k(\partial\theta/\partial r)\ 2\pi r\ \partial t$

The heat conducted into the element is given by

Cartesian: at $(y + \mathrm{d}y)$, $k(\partial/\partial y)[\theta + (\partial\theta/\partial y)\mathrm{d}y]\ x\ \partial t$

Cylindrical: at $(r + \mathrm{d}r)$, $k(\partial/\partial r)[\theta + (\partial\theta/\partial r)]\,2\pi(r + \mathrm{d}r)\,\partial t$

This enables the net heat conducted to be obtained.

Now, when P_v is the power loss per unit volume (W/m^3), a variable function with depth, the heat produced in the element is given by

Cartesian: $(c\gamma)x\,\mathrm{d}y\,\partial\theta = P_v\,x\,\mathrm{d}y\,\partial t + k\,(\partial^2\theta/\partial y^2)x\,\mathrm{d}y\,\partial t$

$$(c\gamma)\,\partial\theta/\partial t - k\,(\partial^2\theta/\partial y^2) = P_v \tag{18.1}$$

Cylindrical $(c\gamma)2\pi r\,\mathrm{d}r\partial\theta = P_v 2\pi r\,\mathrm{d}r\partial t + k(\partial^2\theta/\partial r^2)2\pi r\,\mathrm{d}r\partial t$

$$+ \; k(\partial\theta/\partial r)\mathrm{d}r\,2\pi\partial t$$

$$= P_v 2\pi r\,\mathrm{d}r\partial t + 2\pi r\mathrm{d}r\,k\left(\frac{1}{r}\frac{\partial\theta}{\partial r} + \frac{\partial^2\theta}{\partial r^2}\right)\partial t$$

$$(c\gamma)\,\partial\theta/\partial t - (k/r)\,\partial/\partial r\,(r\partial\theta/\partial r) = P_v \tag{18.2}$$

ignoring second-order smallness terms.

Comparison of these equations with the diffusion equation, derived earlier in Chapters 2, 3, 4, 7, 8 and 9, shows that the equations of heat and current flow etc. are similar. In fact, the concept of 'diffusion' came in Fourier's theory of heat flow and was only later applied to electromagnetism.

18.2 Loss density

In Chapters 3, 4, 8 and 9, we obtained expressions for the current-density distribution (eqns 3.1, 8.3, 4.8 and 9.14).

Conduction, slab: $$J = \frac{\cosh\,(1 + \mathrm{j})\alpha y}{\cosh\,(1 + \mathrm{j})\alpha b}J_s$$

Induction, slab: $$J = \frac{\sinh\,(1 + \mathrm{j})\alpha y}{\sinh\,(1 + \mathrm{j})\alpha b}J_s$$

Conduction, cylinder: $$J = \frac{M_0(mr)\exp\mathrm{j}\theta_0(mr)}{M_0(mR)\exp\mathrm{j}\theta_0(mR)}J_s$$

Induction, cylinder: $$J = \frac{M_1(mr)\exp\mathrm{j}\theta_1(mr)}{M_1(mR)\exp\mathrm{j}\theta_1(mR)}J_s$$

The volumetric loss-density is given by $\rho|J|^2$ for all four conditions and the appropriate moduli can be found in eqns 3.2, 8.4, 4.10 and 9.13. These are used in the next sections.

The loss density at any depth can be plotted relative to the surface: this has been done in Figs 18.1 to 18.4, which show clearly how different the distributions are for conduction and induction.

18.3 Rectangular slab: conduction and induction heating

It is convenient to deal with the solutions for conduction and induction in parallel because of their similarity. We know that

Conduction: $$P_v = \rho \frac{\cosh 2y/\delta + \cos 2y/\delta}{\cosh 2b/\delta + \cos 2b/\delta} J_s^2 \qquad (18.3)$$

Induction: $$P_v = \rho \frac{\cosh 2y/\delta - \cos 2y/\delta}{\cosh 2b/\delta - \cos 2b/\delta} J_s^2 \qquad (18.4)$$

These can be combined into

$$P_v = \rho \frac{\cosh 2y/\delta \pm \cos 2y/\delta}{\cosh 2b/\delta \pm \cos 2b/\delta} J_s^2$$

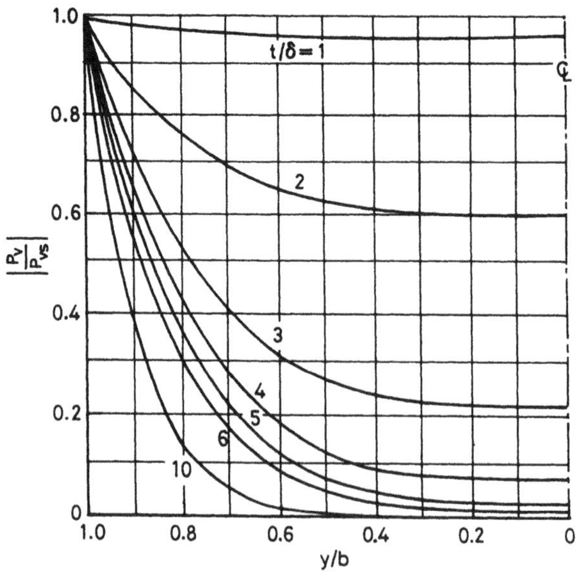

Fig. 18.1 *Loss-density distribution in a slab: conduction. Expressed as per-unit of the surface value, for various t/δ ratios*

where the top signs apply to conduction and the bottom signs to induction. Note that here P_v is a function of y.

We can combine the expressions for $\mathscr{P}$ for conduction and induction, using eqns 3.4 and 8.8:

$$\mathscr{P} = \frac{\rho\delta}{2} \frac{\sinh 2b/\delta \pm \sin 2b/\delta}{\cosh 2b/\delta \pm \cos 2b/\delta} J_s^2 \tag{18.6}$$

Combining eqns 18.5 and 18.6 gives

$$P_v = \frac{2\mathscr{P}}{\delta} \frac{\cosh 2y/\delta \pm \cos 2y/\delta}{\sinh 2b/\delta \pm \sin 2b/\delta} \tag{18.7}$$

We can now use this in eqn 18.1, in Cartesian coordinates, giving the result in terms of $\mathscr{P}$ as before. Equation 18.1 is

$$(c\gamma)\ \partial\theta/\partial t - k\ (\partial^2\theta/\partial y^2) = P_v$$

where $(c\gamma)$ is mass specific heat (J/m^{3}K), k is thermal conductivity (W/m K) and

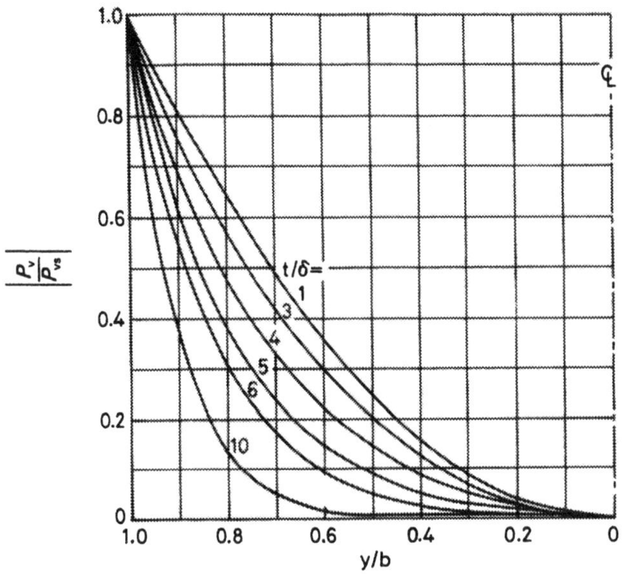

Fig. 18.2 *Loss-density distribution in a slab: induction. Expressed as per-unit of the surface value, for various t/δ ratios*

P_v is the power input per unit volume (W/m^3). We also know from eqn 15.2b that $\theta_m = \mathscr{P}t/b(c\gamma)$, which can be written

$$(c\gamma)\frac{\partial\theta_m}{\partial t} = \frac{\mathscr{P}}{b} \tag{18.8}$$

Also, since θ_m is a function of t but not of position y, its differentials with respect to y are zero, so we can write

$$\partial^2\theta_m/\partial y^2 = 0 \tag{18.9}$$

Taking eqns 18.7, 18.8 and 18.9 in conjunction with eqn 18.1,

$$(c\gamma)\frac{\partial}{\partial t}(\theta - \theta_{\rm m}) - k\frac{\partial^2}{\partial y^2}(\theta - \theta_{\rm m})$$

$$= P_{\rm v} - \frac{\mathscr{P}}{b} = \frac{2\mathscr{P}}{\delta}\frac{\cosh 2y/\delta \pm \cos 2y/\delta}{\sinh 2b/\delta \pm \sin 2b/\delta} - \frac{\mathscr{P}}{b} \quad (18.10)$$

Now, because θ and $\theta_{\rm m}$ are rising together in time, once steady-state has been reached, there is no variation of $(\theta - \theta_{\rm m})$ with time, only with y; so the first term on the left-hand side of eqn 18.10 is zero. Integrating the rest of eqn 18.10 with respect to y,

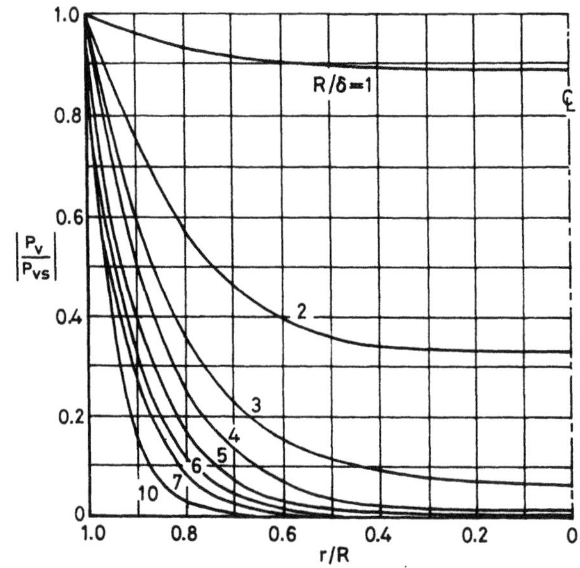

Fig. 18.3 *Loss-density distribution in a cylinder: conduction. Expressed as per-unit of the surface value, for various R/δ ratios*

$$k\frac{\partial}{\partial y}(\theta - \theta_{\rm m}) = \frac{\mathscr{P}y}{b} - \frac{2\mathscr{P}}{\delta}\frac{\delta}{2}\frac{\sinh 2y/\delta \pm \sin 2y/\delta}{Z(2b/\delta)} + \text{constant} \quad (18.11)$$

where $Z(2b/\delta) = \sinh 2b/d \pm \sin 2b/\delta$. To evaluate the constant, remember that $\partial\theta/\partial y = 0$ at $y = 0$, i.e. no temperature gradient exists across the centre-line, from symmetry. Therefore

$$\text{constant} = \text{zero} \quad (18.12)$$

Integrating again,

$$k(\theta - \theta_{\rm m}) = \frac{\mathscr{P}y^2}{2b} - \frac{\mathscr{P}\delta}{2}\frac{\cosh 2y/\delta \mp \cos 2y/\delta}{Z(2b/\delta)} + \text{constant} \quad (18.13)$$

When $y = 0$, $\theta = \theta_{\rm c}$, the centre temperature, so

$$\left.\begin{array}{lll} \text{for conduction } k(\theta_c - \theta_m) & = & \text{constant} + 0 \\ \text{for induction } \quad k(\theta_c - \theta_m) & = & \text{constant} - \mathscr{P}b/Z(2b/\delta) \end{array}\right\} \tag{18.14}$$

Subtracting eqn 18.14 from eqn 18.13,

$$\theta - \theta_c = \frac{\mathscr{P}y^2}{2bk} - \frac{\mathscr{P}\delta}{2k}\frac{\cosh 2y/\delta \mp \cos 2y/\delta \pm {}^0_2}{Z(2b/\delta)}$$

$$= \frac{\mathscr{P}b}{2k}\left[\left(\frac{y}{b}\right)^2 - \frac{\delta}{b}\frac{\cosh 2y/\delta \mp \cos 2y/\delta \pm {}^0_2}{Z(2b/\delta)}\right] \tag{18.15}$$

This gives the temperature differential between the centre and the points $\pm y$, for all y.

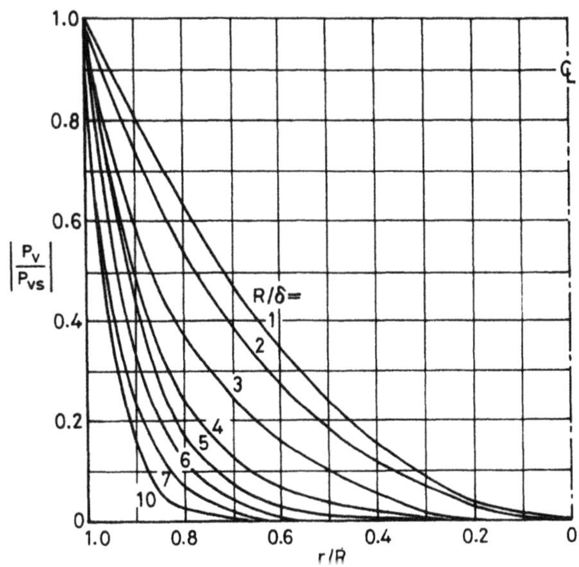

Fig. 18.4 *Loss-density distribution in a cylinder: induction. Expressed as per-unit of the surface value, for various R/δ ratios*

We are specially interested in the difference between the surface and the centre to compare with eqn 15.5b, i.e. for $y = \pm b$

$$\theta_s - \theta_c = \frac{\mathscr{P}b}{2k}\left[1 - \frac{\delta}{b}\frac{\cosh 2b/\delta \mp \cos 2b/\delta \pm {}^0_2}{Z(2b/\delta)}\right] \tag{18.16}$$

If all the heat entered at the surface, $\theta_s - \theta_c$ would be $\mathscr{P}b/2k$, as in eqn 15.5b, so the second term in the square bracket represents the effect of current depth. Therefore

$$\frac{\theta - \theta_c}{(\theta_s - \theta_c)_{\text{idealised}}} = \left(\frac{y}{b}\right)^2 - \frac{\delta}{b}\,\frac{\cosh 2y/\delta \mp \cos 2y/\delta \pm {}^{0}_{2}}{\sinh 2b/\delta \pm \sin 2b/\delta} \qquad (18.17)$$

This is plotted on Fig. 18.5 for conduction and on Fig. 18.6 for induction. The curves are shown for various ratios of thickness to depth of penetration, i.e. $2b/\delta$. When $2b/\delta \rightarrow \infty$, the curves are the same as the idealised Fig. 15.3.

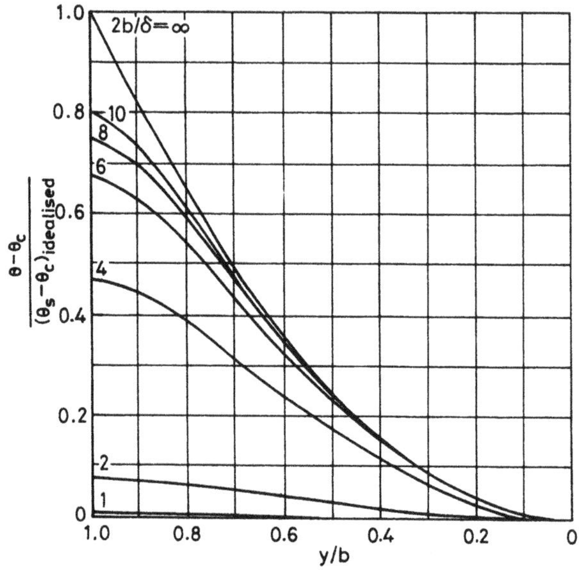

Fig. 18.5 *Variation of temperature in a slab with finite current depth: conduction. Curves based on steady-state conditions: $2b/\delta = \infty$ curve is 'idealised' condition, with all heat entering at surface*

Note that the complementary function, which would be obtained by putting $P_v = 0$, giving the transient solution, has been omitted, so that the equations above only apply when steady conditions have been reached.

The curves form an interesting contrast. For $2b/\delta = \infty$, 10, 8 and 6 they are the same, but they increasingly deviate as the ratio falls. This reflects the difference in the current-density patterns; compare Figs 18.1 and 18.2. At higher values of $2b/\delta$, the current distribution is exponential, or nearly so, approximating to the infinite slab for both DRH and induction, which explains why the curves are the same for these values.

For a slab being heated by induction at $2b/\delta = 2$, i.e. near the peak of the p curve, the temperature difference is approximately 50% of the idealised value, which helps considerably, permitting higher power-densities to be used than were found in Chapter 15. Comparison of the two sets of curves shows another advantage of using conduction heating at low frequency: in addition to the improved value of p and power factor, we have greatly improved temperature

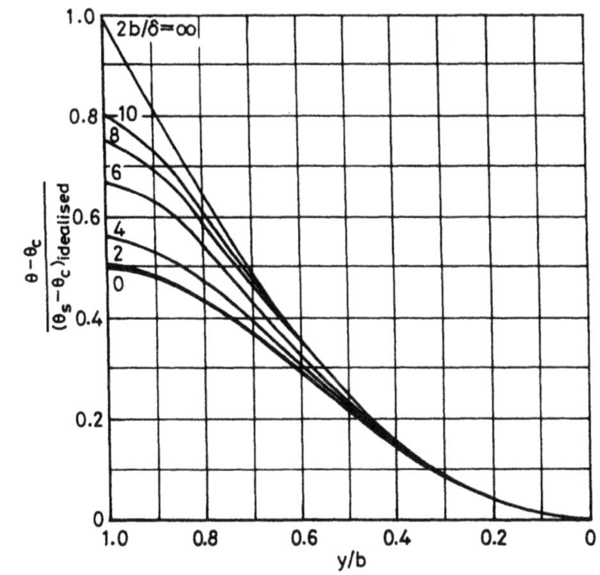

Fig. 18.6 *Variation of temperature in a slab with finite current depth: induction. Curves based on steady-state conditions: $2b/\delta = \infty$ curve is 'idealised' condition, with all heat entering at surface*

distribution. In the limit, as DC is approached, the differential disappears. The importance of the modified theory is best seen by an example.

Example 18.1
A slab, 75 mm thick, is being heated by passing current through it. Calculate the effect of finite current depth for (a) steel (b) copper at 50 Hz.

Solution
(a) For steel, from Chapter 2 the depth of penetration δ is roughly 1.5 mm ignoring saturation. Hence $2b/\delta \simeq 75/1.5 = 50$, so the $2b/\delta = \infty$ curve applies, to a good degree of approximation. Using eqn 18.17, at the surface ($y = b$),

$$\frac{\theta_s - \theta_c}{(\theta_s - \theta_c)_{\text{idealised}}} = 1 - \frac{1 \cosh 50 - \cos 50}{50 \sinh 50 + \sin 50} = 0.98$$

which is negligibly different from unity.

(b) For copper, from Chapter 2 the depth of penetration is roughly 10 mm. Hence $2b/\delta \simeq 75/10 \simeq 7.5$, and the $2b/\delta = 8$ curve is nearly correct. Thus

$$\frac{\theta_s - \theta_c}{(\theta_s - \theta_c)_{\text{idealised}}} = 0.87$$

The depth of penetration increases as $\sqrt{\rho}$, so that as the copper heats the ratio decreases; for example, at 500°C ρ has increased from 0.017 to 0.052, roughly 3 times. The value of $2b/\delta$ decreases to $75/17.3 = 4.3$ and $(\theta_s - \theta_c)/(\theta_s - \theta_c)_{\text{idealised}}$ becomes approximately 0.7, which is a substantial difference. The importance of this, in practice, is that it is the temperature differential that sets the level of power input. The user has to decide the temperature differential that he can use in his process: if the differential is reduced by the above theory, then the power input can be correspondingly increased, resulting in a shorter cycle time.

18.4 Cylindrical billets : conduction and induction heating

As in Section 18.3, it is convenient to deal with conduction and induction heating together, because they again have similar forms. $m = (\sqrt{2})/\delta$ throughout. P_v is a function of r.

For conduction, from eqn 4.10,

$$P_v = \frac{\rho}{2} \frac{\text{ber}^2\, mr + \text{bei}^2\, mr}{\text{ber}^2\, mR + \text{bei}^2\, mR} J_s^2$$

But from eqn 4.13,

$$\mathscr{P} = \frac{\rho\,\delta}{2\sqrt{2}} \frac{\text{ber}\, mR\, \text{bei}'\, mR - \text{bei}\, mR\, \text{ber}'\, mR}{\text{ber}^2\, mR + \text{bei}^2\, mR} J_s^2$$

giving

$$P_v = \mathscr{P} m \frac{\text{ber}^2\, mr + \text{bei}^2\, mr}{\text{ber}\, mR\, \text{bei}'\, mR - \text{bei}\, mR\, \text{ber}'\, mR} \tag{18.18}$$

$$= \mathscr{P} m \frac{\text{ber}^2\, mr + \text{bei}^2\, mr}{W(mR)}$$

where

$$W(mR) = \text{ber}\, mR\, \text{bei}'\, mR - \text{bei}\, mR\, \text{ber}'\, mR$$

For induction, from eqn 9.13,

$$P_v = \frac{\rho}{2} \frac{\text{ber}'^2\, mr + \text{bei}'^2\, mr}{\text{ber}'^2\, mR + \text{bei}'^2\, mR} J_s^2$$

But, from Section 9.5,

$$\mathscr{P} = \frac{\rho\,\delta}{2\sqrt{2}} \frac{\text{ber}\, mR\, \text{ber}'\, mR + \text{bei}\, mR\, \text{bei}'\, mR}{\text{ber}'^2\, mR + \text{bei}'^2\, mR} J_s^2$$

giving

$$P_v = \mathscr{P}m \frac{\mathrm{ber}'^2\, mr + \mathrm{bei}'^2\, mr}{\mathrm{ber}\, mR\, \mathrm{ber}'\, mR + \mathrm{bei}\, mR\, \mathrm{bei}'\, mR} \tag{18.19}$$

$$= \mathscr{P}m \frac{\mathrm{ber}'^2\, mr + \mathrm{bei}'^2\, mr}{Z(mR)}$$

where

$$Z(mR) = \mathrm{ber}\, mR\, \mathrm{ber}'\, mR + \mathrm{bei}\, mR\, \mathrm{bei}'\, mR$$

The equation of heat flow in cylindrical coordinates is eqn 18.2:

$$(c\gamma)\frac{\partial\theta}{\partial t} - \frac{k}{r}\frac{\partial}{\partial r}\left(r\frac{\partial\theta}{\partial r}\right) = P_v$$

where P_v is the loss per unit volume (W/m^3).

We now follow the same method that we used for the slab; unfortunately the solutions cannot be dealt with elegantly by the previous simple $\pm$ expedient, and we must keep them separate. Instead can treat them in parallel, tagging similar equation numbers with C and I, respectively, where necessary.

The mean temperature $\theta_m = 2\mathscr{P}t/(c\gamma)R$, from eqn 15.2c, giving

$$(c\gamma)\frac{\partial\theta_m}{\partial t} = \frac{2\mathscr{P}}{R} \tag{18.20}$$

This can be added to both sides of eqn 18.2 without changing it. Also, since θ_m is not a function of r, its derivatives with respect to r are zero:

$$(c\gamma)\frac{\partial}{\partial t}(\theta - \theta_m) - \frac{k}{r}\frac{\partial}{\partial r}\left[r\frac{\partial}{\partial r}(\theta - \theta_m)\right] = \mathscr{P}\left[m\frac{\mathrm{ber}^2\, mr + \mathrm{bei}^2\, mr}{W\,(mR)} - \frac{2}{R}\right] \tag{18.21C}$$

$$(c\gamma)\frac{\partial}{\partial t}(\theta - \theta_m) - \frac{k}{r}\frac{\partial}{\partial r}\left[r\frac{\partial}{\partial r}(\theta - \theta_m)\right] = \mathscr{P}\left[m\frac{\mathrm{ber}'^2\, mr + \mathrm{bei}'^2\, mr}{Z\,(mR)} - \frac{2}{R}\right] \tag{18.21 I}$$

If the temperatures everywhere have reached the condition where they are all rising at a constant rate, then $\theta - \theta_m$ is only a function of αr, not of time. So the first term is zero, and eqns 18.22 can be solved by integrating twice with respect to r:

$$\frac{\partial}{\partial r}\left[r\frac{\partial}{\partial r}(\theta - \theta_m)\right] = \frac{\mathscr{P}}{k}\left[\frac{2r}{R} - mr\frac{\mathrm{ber}^2\, mr + \mathrm{bei}^2\, mr}{W\,(mR)}\right] \tag{18.22C}$$

$$\frac{\partial}{\partial r}\left[r\frac{\partial}{\partial r}(\theta - \theta_m)\right] = \frac{\mathscr{P}}{k}\left[\frac{2r}{R} - mr\frac{\mathrm{ber}'^2\, mr + \mathrm{bei}'^2\, mr}{Z\,(mR)}\right] \tag{18.22 I}$$

The first integration for the conduction equation gives

$$r\frac{\partial}{\partial r}(\theta - \theta_m) = \frac{\mathscr{P}}{k}\left[\frac{r^2}{R} - \frac{m}{W(mR)}\int r\,(\text{ber}^2\, mr + \text{bei}^2\, mr)\,\text{d}r\right] + \text{constant} \tag{18.23C}$$

But

$$\int r\,(\text{ber}^2\, mr + \text{bei}^2\, mr)\,\text{d}r = \frac{r}{m}(\text{ber}\, mr\ \text{bei}'\, mr - \text{bei}\, mr\ \text{ber}'\, mr)$$

(see, for example, Reference 41). Therefore

$$\frac{\partial}{\partial r}(\theta - \theta_m) = \frac{\mathscr{P}}{k}\left[\frac{r}{R} - \frac{(\text{ber}\, mr\ \text{bei}'\, mr - \text{bei}\, mr\ \text{ber}'\, mr)}{W(mR)}\right] + \text{constant} \tag{18.24C}$$

The first integration for the induction equation gives

$$r\frac{\partial}{\partial r}(\theta - \theta_m) = \frac{\mathscr{P}}{k}\left[\frac{r^2}{R} - \frac{m}{Z(mR)}\int r\,(\text{ber}'^2\, mr + \text{bei}'^2\, mr)\,\text{d}r\right] + \text{constant} \tag{18.23 I}$$

From Warren[2] (p. 256), for example, we know that

$$\int_0^a r(\text{ber}'^2\, kr + \text{bei}'^2\, kr)\text{d}r = \frac{a}{k}\,Z(ka)$$

Therefore

$$\frac{\partial}{\partial r}(\theta - \theta_m) = \frac{\mathscr{P}}{k}\left[\frac{r}{R} - \frac{Z(mr)}{Z(mR)}\right] + \text{constant} \tag{18.24 I}$$

From symmetry, when r is zero there is no temperature gradient, so the integration constant is zero for both conduction and induction.

From here, the solutions differ; one is approximate, the other analytic. We shall deal with the analytic case (induction) first.

18.4.1 Induction

Integrating again (from eqn 18.24 I).

$$\theta - \theta_m = \frac{\mathscr{P}}{k}\left[\frac{r^2}{2R} - \frac{1}{2}\frac{\text{ber}^2\, mr + \text{bei}^2\, mr}{mZ(mR)}\right] + \text{constant}$$

$$= \frac{\mathscr{P}R}{2k}\left[\frac{r^2}{R^2} - \frac{\text{ber}^2\, mr + \text{bei}^2\, mr}{mR\ Z(\text{mR})}\right] + \text{constant} \tag{18.25}$$

When $r = 0$, $\theta = \theta_c$ and $(\text{ber}^2\, 0 + \text{bei}^2\, 0) = 1$:

$$\theta_c - \theta_m = \frac{\mathscr{P}R}{2k}\left[-\frac{1}{mR\ Z(mR)}\right] + \text{constant}$$

Hence

$$\theta - \theta_c = \frac{\mathscr{P}R}{2k}\left[\frac{r^2}{R^2} - \frac{X(mr) - 1}{mR\,Z(mR)}\right] \qquad (18.26)$$

where

$$X(x) = \text{ber}^2\,x + \text{bei}^2\,x \qquad (18.27)$$

If we compare eqn 18.26 with eqn 15.5c, the final term represents the reduction due to the heat being generated under the surface instead of at the surface. Equation 18.26 is plotted on Fig. 18.7; the finite depth of penetration changes

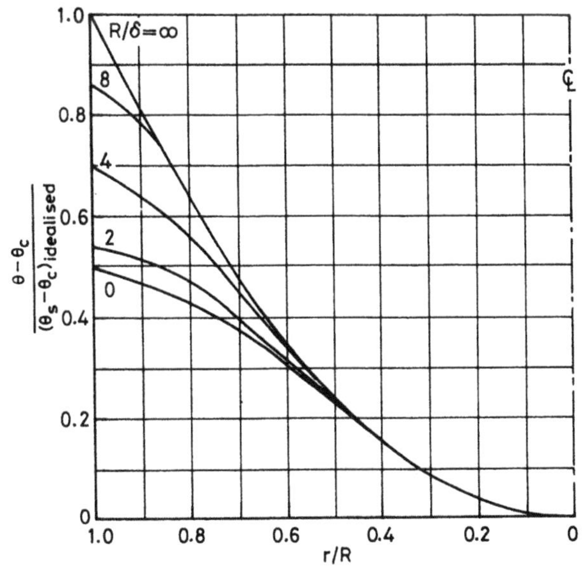

Fig. 18.7 *Variation of temperature in a cylinder with finite current depth: induction. Curves based on steady-state conditions. $R/\delta = \infty$ curve is 'idealised' condition, with all heat entering at surface*

the temperature distribution considerably. For $R/\delta = \infty$, i.e. for $\delta = 0$ or $R \rightarrow \infty$, the pattern is identical, but the temperature difference is decreased for all other values of R/δ. At first sight, it appears advantageous to reduce R/δ to get a more even temperature distribution, but this also reduces the power input.

The most important value of eqn 18.26 is for $r = R$, which gives the surface–centre temperature difference:

$$\theta_s - \theta_c = \frac{\mathscr{P}_0 R}{k}\left[1 - \frac{X(mR) - 1}{mR\,Z\,(mR)}\right] \qquad (18.28)$$

Note that the complementary function, obtained by putting $P_v = 0$ in the heat-flow equation, gives the transient term and has been omitted in this section.

If required it can be found in Dreyfus[12] (pp. 92–93) and is quoted in Chapter 15. The above solution is only valid for $\tau \geqslant 0.25$, or $t \geqslant (c\gamma)R^2/4k$, when all parts of the billet are increasing in temperature at a constant rate; this implies that the sum of heat generated and heat inflow by conduction is constant in every part of the billet.

It is interesting to compare the shape of eqn 18.26 with that of eqn 18.15. b has been replaced by R throughout; the argument of the functions has changed from $2\alpha b$ to $(\sqrt{2})\alpha R$; the dimensionless ratio y/b has become r/R; and the hyperbolic functions are now Bessel functions. With these changes, the shape of the corresponding equations matches exactly.

18.4.2 Conduction

Integrating again (from eqn 18.23C),

$$\theta - \theta_m = \frac{\mathscr{P}R}{2k}\left[\left(\frac{r}{R}\right)^2 - \frac{2}{R}\frac{\int W(mr)\,dr}{W(mR)}\right] + \text{constant} \qquad (18.29)$$

Unfortunately, as far as the writer knows, there is no standard form for $\int W(mr)\,dr$. This can be dealt with in two steps:

- From $R/\delta = \infty$ downwards. Comparison of the loss-distribution curves (Figs 18.3 and 18.4) shows that they are practically identical down to $R/\delta = 3$. If the loss distribution is the same, the temperature distribution is too, so we can copy the appropriate curves from Fig. 18.7.
- From $R/\delta = 0$ upwards. The integral can be solved using the series for ber and bei given in Chapter 4, together with the equivalent series for ber′ and bei′. This solution is given in Appendix A.5: it is not included here for space reasons. It has limited validity due to its series derivation but is certainly correct up to $R/\delta = 4$, which overlaps with the step above.

The second step is carried out as follows. When $r = 0$, $\int W(mr)dr = 0$, so

$$\theta_c - \theta_m = 0 + \text{constant} \qquad (18.30)$$

Hence

$$\theta - \theta_c = \frac{\mathscr{P}R}{2k}\left[\left(\frac{r}{R}\right)^2 - \frac{4}{mR}\,\frac{(mr)^2/8 + (mr)^6/2304 + (mr)^{10}/4.92\times10^6 + (mr)^{14}/4.17\times10^{10}}{(mR)/2 + (mR)^5/192 + (mR)^9/245\,760 + (mR)^{13}/1.49\times10^9}\right] \qquad (18.31)$$

$$= \frac{\mathscr{P}R}{2k}\left[\left(\frac{r}{R}\right)^2 - \frac{4}{mR}\,\frac{(mr)^2}{8}\,\frac{2}{mR}\,\frac{1 + (mr)^4/288 + (mr)^3/614\,400 + (mr)^{12}/5.21 \times 10^9}{1 + (mR)^4/96 + (mR)^8/122\,880 + (mR)^{12}/0.75 \times 10^9}\right]$$

But $mr = [(\sqrt{2})/\delta]r = [(\sqrt{2})R/\delta](r/R)$, so

$$\frac{\theta - \theta_c}{(\theta_s - \theta_c)_{\text{idealised}}} = \left(\frac{r}{R}\right)^2\left[1 - \frac{1 + (R/\delta)^{14}(r/R)^4/72 + (R/\delta)^8(r/R)^8/\text{38th}\,400 + (R/\delta)^{12}(r/R)^{12}/8.14 \times 10^7}{1 + (R/\delta)^4/24 + (R/\delta)^8/7680 + (R/\delta)^{12}/11.64 \times 10^6}\right] \tag{18.32}$$

Since $(r/R)^2$ is the parabolic distribution given in Chapter 15, it follows that the bracket [] is the effect of current distribution. For $R/\delta = 0$, the bracket is zero for all r/R (cf. Fig. 18.6).

Equation 18.32 is plotted on Fig. 18.8 for $R/\delta = 0$ to 4. The $R/\delta = 4$ curve

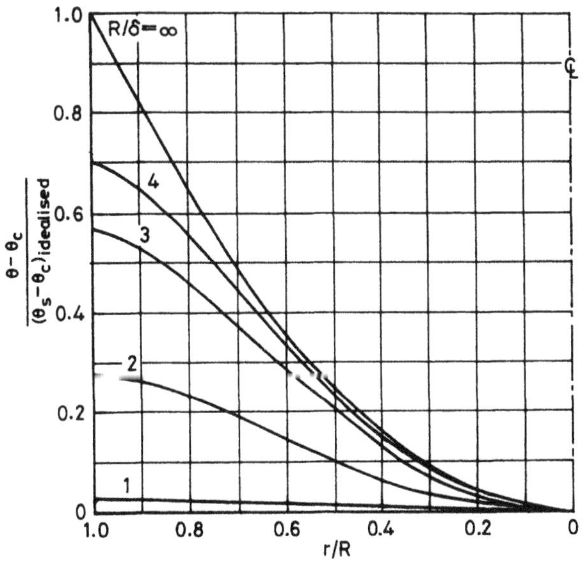

Fig. 18.8 *Variation of temperature in a cylinder with finite current depth: conduction. Based on the approximation of Appendix A.5 for R/δ = 0 to 4: above these values, the curves are the same as Fig. 18.7*

is identical to the $R/\delta = 4$ curve on Fig. 18.7 for all practical purposes, confirming the basis used to derive Fig. 18.8 above. Note that eqn 18.32 used the approximations for ber etc. in both numerator and denominator, since this allowed the limiting value when $R/\delta = 0$ to be seen easily: it would have been possible to use precise values for the denominator, but the agreement is so good up to $R/\delta = 4$ that it seemed justifiable to use it in the expanded form. The approximations are very accurate for low values of the argument, and get worse as it increases. The numerator is increasingly accurate as r/R decreases.

The same comments apply to the comparison of induction and conduction as were made for Figs 18.5 and 18.6.

To allow interpolation, without cluttering by too many closely-spaced curves, the intercepts at $y/b = 1$ and $r/R = 1$ have been plotted on Fig. 18.9.

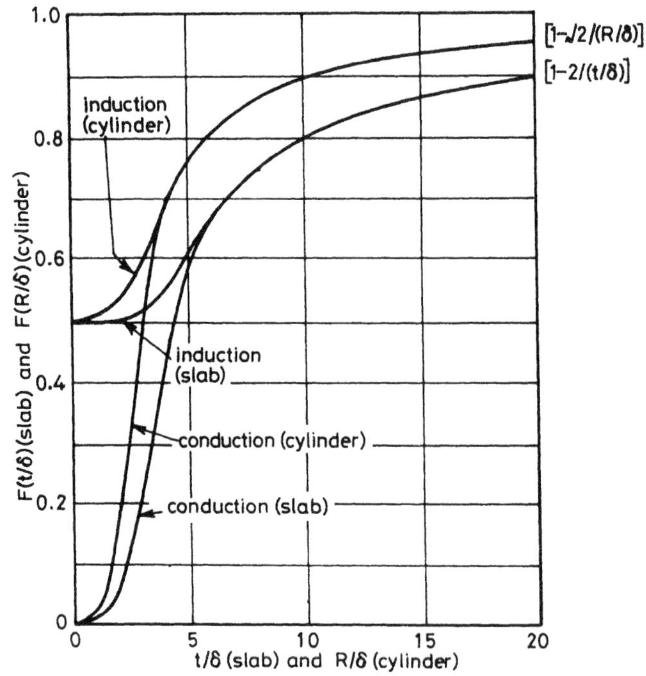

Fig. 18.9 *Correction factors for temperature distribution due to finite current depth for a cylinder and for a slab (induction and conduction)*

18.5 Effect of radiation

Radiation can be regarded as a simple surface loss subtracting from the surface power input. For the theory of Chapter 15, it is sufficient to substitute $\mathscr{P}_n$, the net power-density after subtracting radiation, instead of $\mathscr{P}_s$. The temperature distribution is still parabolic, but the value of $\theta_s - \theta_c$ is reduced by $\mathscr{P}_n/\mathscr{P}_s$.

The present chapter has been based on the assumption that the power density generated plus inflow in the billet is uniform. Radiation from the surface is obviously going to upset this assumption. As an *approximation*, the first term in eqn 18.26 can be taken as proportional to $\mathscr{P}_n$ (following the argument of the last paragraph), but the total power input $\mathscr{P}_s$ is used for the second term, giving

$$\theta - \theta_c = \frac{\mathscr{P}_n R}{2k}\left[\frac{r^2}{R^2} - \frac{X(mr) - 1}{(\mathscr{P}_n/\mathscr{P}_s)\, mR\, Z\,(mR)}\right] \tag{18.33}$$

$\mathscr{P}_s$ is the electromagnetic input (W/m^2), calculated in Chapter 9. $\mathscr{P}_n/\mathscr{P}_s$ is less than unity and the temperature differential is reduced because of the $\mathscr{P}_n$ term outside the bracket and also because the second term in the bracket is increased by $\mathscr{P}_s/\mathscr{P}_n$.

For $r = R$, the surface–centre differential is given by

$$\theta_s - \theta_c = \frac{\mathscr{P}_n R}{2k}\left[1 - \frac{\mathscr{P}_s}{\mathscr{P}_n}\frac{X(mR) - 1}{mR\, Z(mR)}\right] \tag{18.34}$$

The correction factor $F(\mathscr{P}_n/\mathscr{P}_s, mR)$, the bracketed term in eqn 18.34, is plotted in Fig. 18.10. Similar curves can be derived for conduction and for slabs. Note

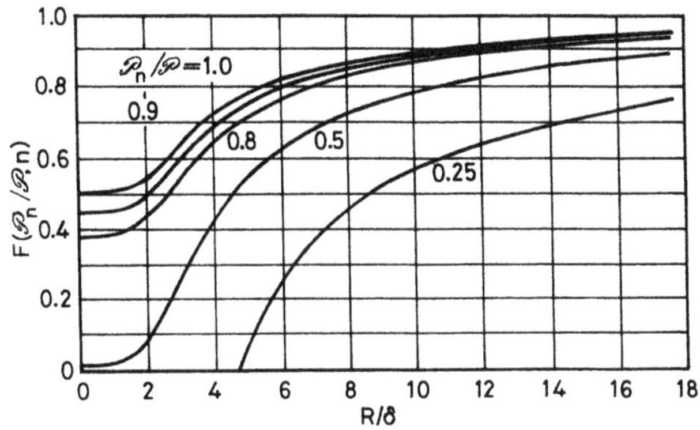

Fig. 18.10 *Radiation correction factor for cylinder under induction*

that the correction factor is used in conjunction with the net power input after radiation, $\mathscr{P}_n$. It should be used with caution because of the assumptions involved.

The calculation of radiation loss was dealt with in Chapter 17.

Chapter 19

Heat transfer during surface hardening

In surface hardening, the work surface is brought rapidly to temperature by a high power-density applied at the surface. The aim is to heat a shallow layer without affecting the rest of the workpiece. These two features – the high intensity and the thin layer – mean that different criteria from billet heating apply. Dreyfus[12] deals with this problem in great detail, and the reader is advised to study that book if he is particularly interested in the hardening process.

Since the skin depth is small, there is no need to use cylindrical coordinates or thin slab ideas; this problem can be treated as a semi-infinite slab. In practice, during hardening, the surface will be heated under one of two conditions:

- a variable temperature $\theta(t)$ applied to the surface
- a variable power-density $\mathscr{P}(t)$ applied to the surface.

These are not really separate, since one implies the other; it is a question of which quantity is known. Before discussing these cases, it is helpful to consider the problem of a semi-infinite slab suddenly placed in a new temperature environment. This is not an artificial problem, as it is what happens when a piece to be hardened is heated in an oven, but it is not the induction surface-hardening problem; it is used here to introduce the concepts.

19.1 Slab subjected to sudden increase in surface temperature

From Carslaw and Jaeger[26] (p. 60), if the face of a semi-infinite slab has its temperature suddenly raised by θ_s at $t = 0$, the temperature rise θ at any point in the slab, distant y from the surface, at any time t, is

$$\theta = \theta_s\{1 - \mathrm{erf}[y/2\surd(kt/c\gamma)]\} \qquad (19.1)$$

where $y/2\surd(kt/c\gamma)$ is a dimensionless quantity. The absolute temperatures are obtained from eqn 19.1 by adding the initial temperature to these temperature rises.

The error function $\mathrm{erf}\,x$ is important in probability and is one form of solution of the heat-flow equation. It is defined by

$$\operatorname{erf} x = \frac{2}{\sqrt{\pi}} \int_0^x \exp(-u^2)\,du \tag{19.2}$$

A curve of erf x is plotted against $y/2\sqrt{(kt/c\gamma)}$ in Fig. 19.1, together with values of θ/θ_s from eqn 19.1, which is also the curve of $1 - \operatorname{erf} x$ (curve A).

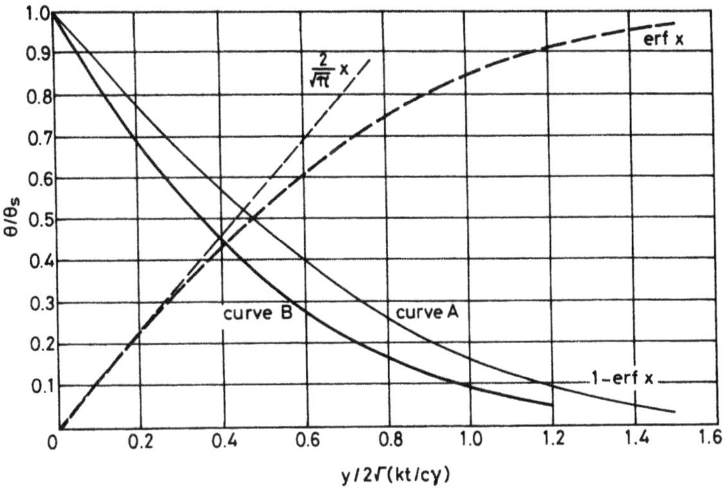

Fig. 19.1 *Temperature distribution in a workpiece. Curve A shows the surface maintained at a constant temperature θ_s above the initial temperature of the workpiece. Curve B shows the surface subjected to a constant power-density $\mathcal{P}$. The dashed curve shows erf x using the same scales. θ is the temperature rise at any point in the slab, distant y from the surface, at time t. Actual temperatures are obtained by adding the initial temperature to these rises*

Up to $x = 0.4$,

$$\operatorname{erf} x \simeq \frac{2}{\sqrt{\pi}}\left[x - \frac{x^3}{3}\right] \tag{19.3}$$

to 5% accuracy. For $x = 0$, erf $x = 0$; for $x = \infty$, erf $x = 1$. From eqn 19.1, the temperature is within 5% of its final value when erf $y/2\sqrt{(kt/c\gamma)}$ is 0.05. From the approximation of eqn 19.3, this occurs when $x = 0.05\sqrt{(\pi/2)}$. Taking values of $c = 857$ J/kg K, $k = 26$ W/m K, and $\gamma = 7850$ kg/m^3 gives

$$y/2\sqrt{\left[\frac{(26)t}{(857)(7850)}\right]} = \frac{0.05\sqrt{\pi}}{2}$$

$$\frac{y}{\sqrt{t}} = 0.05\sqrt{\left[\frac{(\pi)(26)}{(857)(7850)}\right]} = \frac{0.17}{10^3} \tag{19.4}$$

Hence, for steel, we can say that, after one second, material at 0.17 mm from the surface is near its final temperature, and that, at other depths, the time taken to arrive at 95% of the final temperature is given by $(10^3/0.17)^2 y^2$; for example, at 10 mm from the surface it takes 3460 s.

The power density is given by $-k\partial\theta/\partial y$ everywhere. We are interested in the surface power-density $\mathscr{P}_s$ needed to produce these conditions. In general,[26]

$$\frac{d}{dx}(\operatorname{erf} x) = \frac{2}{\sqrt{\pi}} \exp(-x^2)$$

Hence

$$\mathscr{P}_s = \left[\theta_s k \frac{2}{\sqrt{\pi}} \exp\left(\frac{-y^2}{t}\frac{c\gamma}{4k}\right) \frac{1}{2\sqrt{(kt/c\gamma)}}\right]_{y=0}$$

$$= \frac{\theta_s}{\sqrt{t}} \sqrt{\left(\frac{kc\gamma}{\pi}\right)} \quad \text{W/m}^2 \tag{19.5}$$

The energy input per unit area is given by

$$\int_0^t \mathscr{P}_s \, dt = \theta_s \sqrt{\left(\frac{kc\gamma}{\pi}\right)} [2\sqrt{t}]_0^t$$

$$= 2\theta_s \sqrt{\left(\frac{kc\gamma}{\pi} t\right)} \quad \text{J/m}^2 \tag{19.6}$$

The power density falls off with increasing time ($\propto \sqrt{t}$), as might be expected for these conditions, where the temperature gradient decreases as the slab heats. It is an artificial condition, as the power density would have to be infinite near $t = 0$ to produce such a sudden change of temperature.

19.2 Slab subjected to a temperature $\theta(t)$ varying with time

This solution superimposes a large number of solutions of the kind given in Section 19.1, using the final condition of each time interval as the initial condition for the next. Using Duhamel's theorem, these superpositions become an integral. The solution is again given by Carslaw and Jaeger[26] (p. 63):

$$\theta = \frac{2}{\sqrt{\pi}} \int_{\mu = y/2\sqrt{(kt/c\gamma)}}^{\mu=\infty} \phi(\lambda) \exp(-\mu^2) \, d\mu \tag{19.7}$$

where

$$\lambda = t - \frac{c\gamma}{4k}\frac{y^2}{\mu^2} \tag{19.8}$$

and $\phi(\lambda)$ is the time-varying temperature applied to the surface. The dummy variable λ is used to avoid confusion with t; the integral is treated as if t and y are constants and integrated (if possible) as an integral in μ over the given limits. θ is, again, a function of both time and depth.

Example 19.1
Obtain the slab surface-hardening solution when the applied temperature varies linearly with time as $\theta = \phi(t) = at$, where a is a constant.

Solution
Combining eqns 19.7 and 19.8,

$$\theta = \frac{2}{\sqrt{\pi}} \int_{\mu=y/2\sqrt{(kt/c\gamma)}}^{\mu=\infty} \phi\left(t - \frac{y^2 c\gamma}{4k\mu^2}\right) \exp(-\mu^2)\,\mathrm{d}\mu \tag{19.9}$$

The first bracketed term is $\phi(\lambda)$; here $\phi(\lambda) = a\lambda$. Making the substitution, we obtain

$$\theta = \frac{2a}{\sqrt{\pi}} \int_{y/2\sqrt{(kt/c\gamma)}}^{\infty} \left(t - \frac{y^2 c\gamma}{4k}\frac{1}{\mu^2}\right) \exp(-\mu^2)\,\mathrm{d}\mu \tag{19.10}$$

where the integral and limits are entirely in μ, both t and y being treated as constants. It can be shown[26] that

$$\int_x^{\infty} \exp(-\mu^2)\,\mathrm{d}\mu = \frac{\sqrt{\pi}}{2} - \int_0^x \exp(-\mu^2)\,\mathrm{d}\mu = \frac{\sqrt{\pi}}{2}(1 - \operatorname{erf} x) \tag{19.11}$$

and

$$\int_x^{\infty} \frac{\exp(-\mu^2)\,\mathrm{d}\mu}{\mu^2} = \frac{\exp(-x^2)}{x} - 2\int_x^{\infty} \exp(-\mu^2)\,\mathrm{d}\mu \tag{19.12}$$

The solution of eqn 19.10 is, therefore,

$$\begin{aligned}\theta &= \frac{2a}{\sqrt{\pi}}\left\{t\frac{\sqrt{\pi}}{2}(1 - \operatorname{erf} x) - \frac{c\gamma}{4k}y^2\left[\frac{\exp(-x^2)}{x} - 2\frac{\sqrt{\pi}}{2}(1 - \operatorname{erf} x)\right]\right\}\\ &= a\left\{\left(t + \frac{c\gamma}{2k}y^2\right)(1 - \operatorname{erf} x) - \frac{2}{\sqrt{\pi}}\frac{c\gamma}{4k}\frac{y^2\exp(-x^2)}{x}\right\}\end{aligned}$$

But $x = y/2\sqrt{(kt/c\gamma)}$. Hence

$$\theta = a\left\{\left(t + \frac{c\gamma}{2k}y^2\right)\left[1 - \operatorname{erf}\frac{y}{2\sqrt{(kt/c\gamma)}}\right] - \sqrt{(t)}\sqrt{\left(\frac{c\gamma}{k\pi}\right)}\,y\exp\left(\frac{-y^2}{t}\frac{c\gamma}{4k}\right)\right\} \tag{19.13}$$

Similarly, any other function for the applied temperature can be substituted.

19.3 Slab subjected to a constant power-density

As stated in Section 19.1, the power density associated with a given time-function of temperature at the surface is given by $(-k\partial\theta/\partial y)$, so the surface power-density associated with a given surface time-function of temperature is obtained by $-k$ times the differential of eqn 19.7 with respect to y, for $y = 0$.

If

$$\mathscr{P} = -k\frac{\partial\theta}{\partial y} \tag{19.14}$$

where $\theta = f(t, y)$, and

$$\frac{\partial^2\theta}{\partial y^2} - \frac{c\gamma}{k}\frac{\partial\theta}{\partial t} = 0 \tag{19.15}$$

then

$$\frac{\partial^2\mathscr{P}}{\partial y^2} - \frac{c\gamma}{k}\frac{\partial\mathscr{P}}{\partial t} = 0 \tag{19.16}$$

In Section 19.1 the solution for eqn 19.15 was given, with θ_s suddenly applied at the surface. Obviously the solution for eqn 19.16 is similar:

$$\mathscr{P} = \mathscr{P}_s\{1 - \operatorname{erf}[y/2\surd(kt/c\gamma)]\} \tag{19.17}$$

At $y = \infty$, the temperature rise is zero and the slab is at the initial temperature. At any other depth, the temperature rise can be found by integrating between the depth and infinity; from eqn 19.14,

$$\theta_y = \frac{1}{k}\int_y^\infty \mathscr{P}\,\mathrm{d}y$$

$$= \frac{\mathscr{P}_s}{k}\int_y^\infty \{1 - \operatorname{erf}[y/2\surd(kt/c\gamma)]\}\,\mathrm{d}y$$

Carslaw and Jaeger[26] (p. 484) show that

$$\int_x^\infty (1 - \operatorname{erf}\xi)\,\mathrm{d}\xi = \frac{\exp(-x^2)}{\surd\pi} - x(1 - \operatorname{erf} x)$$

Thus

$$\theta = 2\frac{\mathscr{P}_s}{k}\sqrt{\left(\frac{kt}{c\gamma}\right)}\int_y^\infty\left[1 - \operatorname{erf}\frac{y}{2\surd(kt/c\gamma)}\right]\frac{\mathrm{d}y}{2\surd(kt/c\gamma)} \tag{19.18}$$

$$= \frac{2\mathscr{P}_s}{k}\left\{\frac{\exp[-y^2(c\gamma)/4kt]}{\surd(\pi c\gamma/k)}\surd(t) - \frac{y}{2}\left[1 - \operatorname{erf}\frac{y}{2\surd(kt/c\gamma)}\right]\right\} \tag{19.19}$$

When $y = 0$,

$$\theta_s = \frac{2\mathscr{P}_s}{\surd(\pi c\gamma k)}\surd t \tag{19.20}$$

Note that the integral in eqn 19.18 has a mathematical symbol,

ierfc $y/2\surd(kt - c\gamma)$ and a table of values is given in Carslaw and Jaeger.[26]
The solution eqn 19.20 can be put in the form θ/θ_s:

$$\theta/\theta_s = [\exp(-u^2) - \surd(\pi)u(1 - \operatorname{erf} u)] \tag{19.21}$$

which is also plotted on Fig. 19.1 (curve B).

Remember that, whereas θ_s is a constant in eqn 19.1, here it varies with $\surd t$ according to eqn 19.20.

For small values of u, eqn 19.21 simplifies to

$$\theta/\theta_s = 1 - u\surd\pi + u^2 \ldots \tag{19.22}$$

19.4 Slab subjected to a surface power-density $\mathscr{P}_s(t)$ varying with time

By similar reasoning to Section 19.2, the varying power input can be summed. The power expression is exactly as eqn 19.17. The corresponding temperature distribution is[26]

$$\theta = \frac{1}{\surd(\pi c\gamma k)} \int_{\tau=0}^{\tau=t} f(t - \tau)\exp\left(-\frac{y^2}{\tau}\frac{c\gamma}{4k}\right)\frac{d\tau}{\tau^{1/2}} \tag{19.23}$$

19.5 Slab allowed to soak

If the slab is heated at a uniform surface power-density $\mathscr{P}_s$ for T seconds and then allowed to soak, the temperature at time t ($t > T$) is given by

$$\theta = \frac{2\mathscr{P}_s}{(c\gamma k)}\left[\surd(t)\operatorname{ierfc}\frac{y/\surd t}{2\surd(k/c\gamma)} - \surd(t - T)\operatorname{ierfc}\frac{y/\surd(t - T)}{2\surd(k/c\gamma)}\right] \tag{19.24}$$

Chapter 20

Water cooling in conductors

20.1 General

Water is widely used in induction-heating applications to cool hollow tubular conductors. Because of the inherent importance of the topic, a great deal of experimental work has been done to verify the theory, and the results can be used with confidence.

From the principles of dimensionless analysis,[24] the heat flow from the walls of a pipe carrying a fluid can be expressed in terms of three dimensionless quantities:

$$\text{Nusselt number} \qquad \text{Nu} = \mathcal{P}d/k\theta \tag{20.1}$$

$$\text{Reynolds number} \qquad \text{Re} = v\gamma d/\eta \tag{20.2}$$

$$\text{Prandtl number} \qquad \text{Pr} = c\eta/k \tag{20.3}$$

where k, γ, c, η are properties of the fluid at the mean bulk temperature, v is the velocity of the fluid, d is the pipe internal diameter, $\mathcal{P}$ is the heat transfer per unit area, and θ is the temperature difference between the wall and the fluid.

Because these quantities are dimensionless, any consistent system of units is valid. In the SI system, the following units are used:

$\mathcal{P}$	W/m^2	v	m/s	c	J/kg K
k	W/m K	γ	kg/m^3	η	N s/m^2
θ	K	d	m		

Various theoretical relationships[23] between Nu, Re, and Pr have been proposed, but these will not be discussed here. There exists a summary of a large body of experiment for liquids and gases (see Fishenden and Saunders,[24] Fig. 31) suggesting that the practical equation relating them is

$$\text{Nu} = 0.023\text{Re}^{0.8}\text{Pr}^{0.4} \tag{20.4}$$

which is valid for turbulent flow ($\text{Re} \geqslant 2100$). There should be a small correct-

Table 20.1

	Temperature (°C)					
	10	20	30	40	50	60
k	0.58	0.60	0.62	0.63	0.64	0.65
Pr	9.5	7.0	5.43	4.34	3.57	3.07
η	13.1×10^{-4}	10.1×10^{-4}	8.0×10^{-4}	6.5×10^{-4}	5.5×10^{-4}	4.7×10^{-4}
$Pr^{0.4}/\eta^{0.8}$	498	542	591	638	674	718
p.u.	0.92	1.0	1.09	1.18	1.24	1.33

Data from Ede,[23] p. 262.

ing term proportional to l/d (where l is the pipe length), but this can be ignored for l/d ratios above about 10, as the power of l/d is only about 0.05.

In passing through the pipe, the water is itself heated and the temperature difference to be used in Nu will vary. Also, the properties of the water vary with temperature. Water enters the piping at 10° to 25°C and leaves about 30°C hotter, so the range is (say) 10° to 60°C.

Over this range, γ varies from 1000 to 980 kg/m^3 and c varies from 4190 to 4180 J/kg K. These are effectively constants, and the recommended values are 995 kg/m^3 and 4180 J/kg K respectively.

By contrast, the other quantities vary over a wide range, as shown in Table 20.1. The correction factor for Nu with change of temperature (constant k times $\mathrm{Pr}^{0.4}/\eta^{0.8}$) is given, with sufficient accuracy, relative to conditions at 20°C by

$$0.84 + 0.008\theta \qquad (20.5)$$

where θ is in degrees Celsius; this is the final row in Table 20.1.

Using eqn 20.4, it is easy to plot straight lines on log-log paper relating loss per degree Celsius per metre length of pipe against water quantity or against velocity, for various pipe diameters, by multiplying $\mathscr{P}$ by the appropriate internal surface area. Rearranging eqn 20.4,

$$\begin{aligned}\frac{\text{loss}}{\theta l} &= 0.023\pi(vd)^{0.8}\gamma^{0.8}\frac{k\mathrm{Pr}^{0.4}}{\eta^{0.8}}\\ &= 0.023\pi(vd)^{0.8}(995)^{0.8}(0.6 \times 542)\\ &= 5893(vd)^{0.8} \qquad \text{W/mK}\end{aligned} \qquad (20.6)$$

This is plotted against velocity in Fig. 20.1 for various diameters. Other diameters can be interpolated.

Since water quantity $= (\pi d^2/4)v \times 60 \times 10^3 = 47\,120vd^2$ litres/min, eqn 20.6 can be rearranged as follows:

$$\frac{\text{loss}}{\theta l} = 1.076\left(\frac{L}{d}\right)^{0.8} \qquad (20.7)$$

where L is water quantity in litres/min (note that these units are not SI but are in common use). This is plotted on Fig. 20.2 for various diameters. Note that for L/d constant, the loss is constant, so the calibration line is horizontal, using the bottom scale × 10. The lines are plotted over the range of known validity. If L is too low, there is a risk of laminar flow; for L high, the pressure drop is likely to be excessive.

Figures 20.1 and 20.2 are plotted for 20°C values of the constants, and the loss values should be corrected for other temperatures using eqn 20.5.

Example 20.1

Correct the values of heat-transfer coefficients at (a) 0·8 m/s in a 6 mm pipe (Fig. 20.1) and (b) 15 litres/min in a 12 mm pipe (Fig. 20.2) to a mean bulk temperature of 35°C.

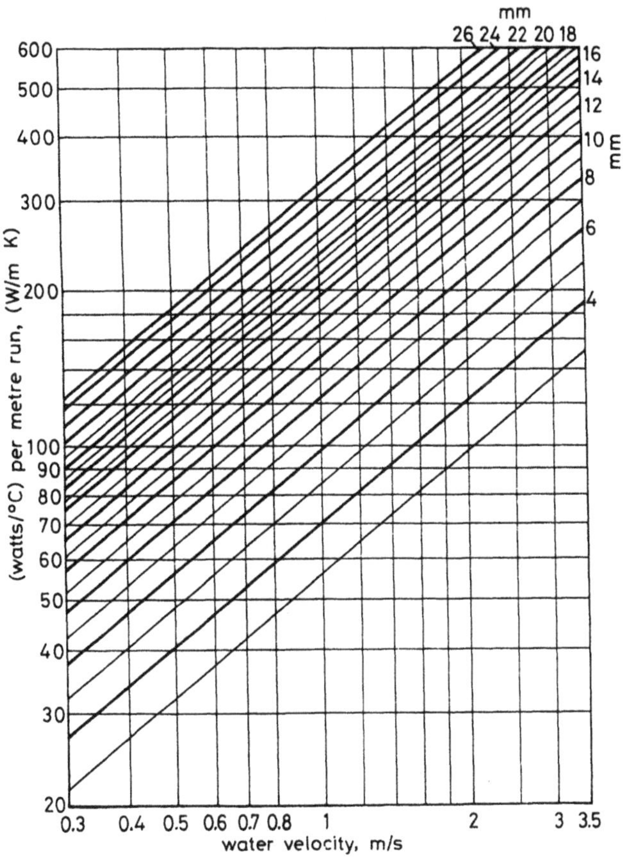

Fig. 20.1 *Heat transfer from water-cooled tubes of various internal diameters, plotted against water velocity*

Solution

(a) Using Fig. 20.1, the heat-transfer coefficient at 20°C = 82 W/m K. The correction factor, from eqn 20.5, is given by $0.84 + 0.008\theta = 0.84 + 35 \times 0.008 = 1.12$, so the corrected coefficient is $82 \times 1.12 = 92$ W/m K.
(b) Figure 20.2 gives 32 W/mK at 20°C; the corrected heat-transfer coefficient is 36 W/m K. This correction is substantial and should always be taken into account. Although it apparently gives 10% more heat transfer, it should be remembered that water at 20°C rising to 60°C has four-thirds of the heat capacity of water rising from 30 to 60°C, so it is always advantageous to start with cooler water for a given maximum outlet temperature.

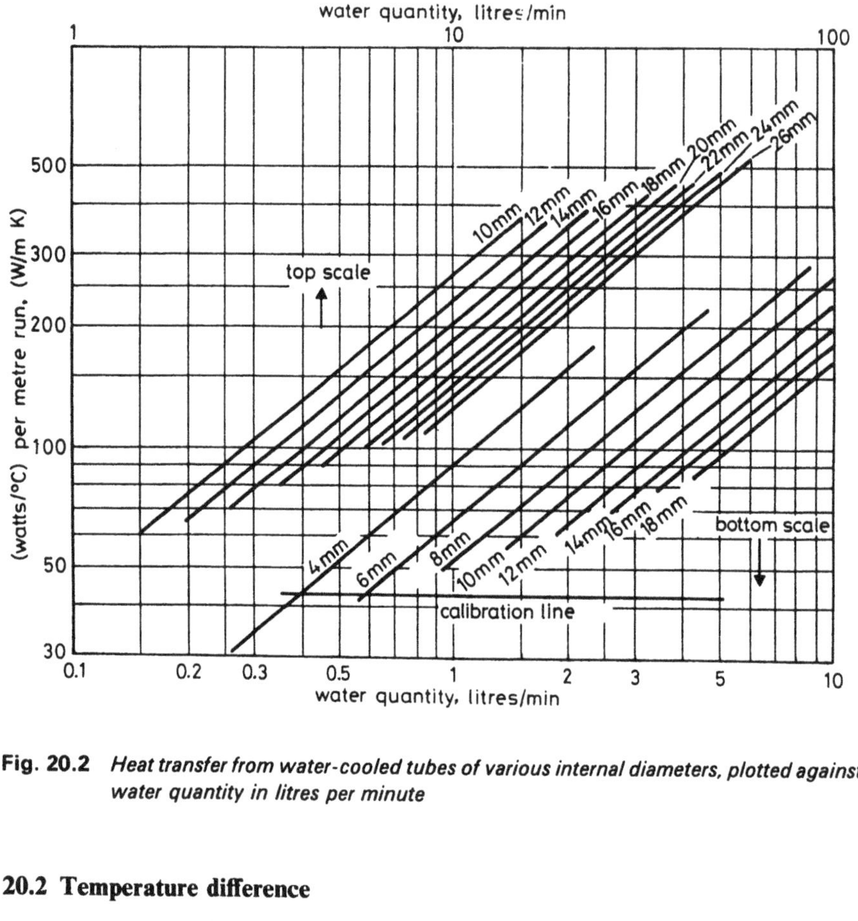

Fig. 20.2 *Heat transfer from water-cooled tubes of various internal diameters, plotted against water quantity in litres per minute*

20.2 Temperature difference

As the water goes through the pipe, it heats up and the temperature difference between pipe and water decreases. Strictly, the heat-transfer coefficient should be integrated, which is done by using the logarithmic mean temperature difference, given by $(\theta_1 - \theta_2)/\ln(\theta_1/\theta_2)$, where θ_1 and θ_2 are the temperature differences at the start and finish. This is always more accurate, but for the kind of problem being considered here, where the rise in the water temperature is small compared with the temperature of the tube, it is sufficient to take $(\theta_1 + \theta_2)/2$. This is accurate to 10% for $\theta_1/\theta_2 < 3$.

20.3 Water quantity

The loss to be removed (copper loss plus radiation gain) will be calculable, as described earlier. If the temperature rise of the water is to be θ°C and the water quantity is Q cubic metres per second, then

$$\text{heat removed} = (c\gamma)Q\theta \quad \text{W} \tag{20.8}$$

It is usually more convenient to give the water quantity as L litres per minute and to calculate the heat removed in kilowatts, using $\gamma = 995\,\text{kg/m}^3$ and $c = 4180\,\text{J/kg K}$:

$$\text{heat removed} = \frac{4180 \times 995}{1000} \frac{L}{60 \times 1000}\theta = 0.069L\theta \quad \text{kW} \tag{20.9}$$

For $\theta = 30°$,

$$L \simeq \text{heat removed (kW)}/2.1 \quad \text{litres/min} \tag{20.10}$$

20.4 Arrangement of cooling paths

Equation 20.10 only tells how much water, heated through 30°C, is needed to remove a certain amount of heat. It does not say that the heat *will* be removed. For this, the cooling paths must be arranged so that the heat transfer coefficient loss/θl is right. In practice this may not even be possible, in which case a total redesign is necessary. Even when the heat transfer is satisfactory, the pressure drop across the system may be wrong. In general, a high velocity through a small pipe is desirable, although this may produce excessive pressure drop.

All that has been said so far in this chapter has been general, and can be applied to any arrangement of water in tubes, with the exception of very short tubes. It will now be applied to the specific example where a known loss is being removed from a known length of tube, as would occur in a water-cooled coil.

Total length of tube	l m
Loss to be removed	kW
Internal diameter of tube	d m
Temperature rise of water	θ_w K
Inlet temperature of water	θ_1 K
Outlet temperature of water	θ_2 K
Tube temperature (assumed constant)	θ_t K
Parallel paths (number)	p
Water quantity	L litres/min

From eqn 20.9,

$$\text{heat removed} = 0.069L\theta_w \quad \text{kW}$$

Next,

$$\text{water quantity/path} = L/p \quad \text{litres/min}$$

From Fig. 20.2, L/p and d give loss/θl.

Temperature difference at inlet $= \theta_t - \theta_1$

Temperature difference at outlet $= \theta_t - \theta_2$

Arithmetic mean temperature difference θ $= (2\theta_t - \theta_1 - \theta_2)/2$

$= \theta_t - (\theta_1 + \theta_2)/2$

From loss curve, l and θ, calculate the heat removed (kW). Compare the calculated heat removed with that required. If the values are close, check the pressure drop (Section 20.5). If not, first try changing the number of parallel paths; if this fails, one of the main constraints will have to be relaxed.

20.5 Pressure drops

Having checked that the heat dissipation corresponds to the water quantity, the pressure drop across the water circuit must be calculated. A decision can then be made about the kind of pump needed to circulate the water. In most cooling circuits, there will be four kinds of pressure drop:

- in the length of the tubes
- at the entrance of the tubes
- at the exit from the tubes
- at each bend in the system.

This subject is dealt with in detail in books on mechanics of fluids[32] and will only be summarised here for the special case of water cooling.

Pressure drop along tube: The dissipation curves given earlier in this chapter are based on turbulent flow (Re $\geqslant$ 2100) and, for these conditions, Ede[23] gives the pressure drop as

$$P = 0.092 \frac{\gamma v^2}{d} \mathrm{Re}^{-0.2} l \qquad \mathrm{Pa} \qquad (20.11)$$

where $\mathrm{Re} = v\gamma d/\eta$ and the other symbols have the same meaning as in the rest of this chapter.

If water in the temperature range is assumed, values of the constants can be inserted, giving

$$P = \frac{5.55}{1000} \frac{l\gamma v^{1.8}}{d^{1.2}} \qquad \mathrm{Pa} \qquad (20.12)$$

A mean value of 0.24 has been used for $\eta^{0.2}$; see Table 20.1.

Pressure drop at entry: From Bernoulli's equation, the pressure drop at a sudden change of section can be obtained. The normal entry condition to a cooling pipe is from a reservoir or header tank, so the initial velocity can be taken as zero. A reasonable approximation for the pressure drop is then

$$\Delta P = \frac{1}{2}\left(\frac{v^2}{2}\right)\gamma \quad \text{Pa} \tag{20.13}$$

where v is the water velocity in the pipe.

Pressure drop at exit: The assumption is made that the pipe is discharging into a large header, so that its final velocity is zero. The pressure drop is then

$$\Delta P = \left(\frac{v^2}{2}\right)\gamma \quad \text{Pa} \tag{20.14}$$

Pressure loss at bends: A detailed discussion for bends of various radii and for various angles of bend is given in Reference 32. The loss of pressure is expressed as an additional loss, to be added to the pressure loss associated with friction. For a right-angle bend in a smooth pipe, and for a bend radius greater than 3 × pipe diameter,

$$\Delta P = 0.1\left(\frac{v^2}{2}\right)\gamma \quad \text{Pa} \tag{20.15}$$

For $R/D = 2$ use 0.15, and for $R/D = 1$ use 0.2. For detailed curve, see Reference 32.

Total pressure drop and associated head of water: The total pressure drop is

$$P = \gamma\left[\frac{5.55}{1000}\frac{lv^{1.8}}{d^{1.2}} + v^2(0.75 + 0.05 \times \text{no. of bends})\right] \text{Pa} \tag{20.16}$$

In practice this pressure drop is often expressed as a head of water in metres:

$$\text{pressure drop} = P/\gamma g \quad \text{metres} \tag{20.17}$$

Example 20.2

An electroheating device is cooled by a labyrinth of 12 mm ID water pipes comprising 14 straight sections, each 0.6 m long, joined at the ends by right-angled connectors and 0.1 m straight bridging pieces. If the labyrinth is fed from a header tank and discharges to a header tank, in both of which the velocity can be assumed to be zero, find the pressure drop (a) in pascals (b) in metres of water head, if the water velocity is 0.8 m/s.

Solution

Since the device is being used for cooling, it is reasonable to assume that the water temperature is in the 10° to 60°C range; so the equations of Section 20.5 apply.

The length of pipe between headers is $l = (14 \times 0.6) + (13 \times 0.1) = 9.70$ m, and the number of bends is 26.

(a) From eqn 20.16,

$$P = \gamma\left[\frac{5.55}{1000}\frac{lv^{1.8}}{d^{1.2}} + v^2(0.75 + 0.05 \times \text{no. of bends})\right]$$

$$= 995\left[\frac{5.55}{1000}\frac{(9.70)(0.8)^{1.8}}{(0.012)^{1.2}} + (0.8)^2(0.75 + 26 \times 0.05)\right]$$

$$= 995[7.27 + 1.31] = 8540 \quad \text{Pa}$$

(b) The head in metres is 8540/9795 = 0.87 m.
For comparison, one atmosphere is 1013×10^5 Pa, so the pressure drop is quite low.

20.6 Non-circular pipes and ducts

In electroheat work, square and rectangular pipes are used, and sometimes water is used between two plane surfaces. It has been found by experiment that the dimensions of these non-circular sections can be converted to an equivalent circular section by the use of the concept of 'equivalent diameter', given by

$$\text{equivalent diameter } d = \frac{4 \times \text{cross-section}}{\text{wetted perimeter}} \tag{20.18}$$

This should not be used if part of the perimeter is re-entrant, and should be used with care for concentric passages. Note that the results of eqn 20.18 are only 'equivalent' for the calculation of heat transfer and friction.

Example 20.3
Find the equivalent diameter for (a) a square pipe of section $a \times a$; (b) a rectangular pipe of section $a \times b$; (c) an infinitely-wide parallel gap, g.

Solution

(a) $d = \dfrac{4(a \times a)}{4a} = a$

So the equivalent diameter is equal to the diameter of the inscribed circle.

(b) $d = \dfrac{4ab}{2(a + b)} = \dfrac{2ab}{a + b}$

If $a = 2b$, for example, $d = 4b/3$.
(c) Let the width be w, where w tends to infinity.

$$d = \frac{4wg}{2(w + g)} \simeq 2g \qquad \text{for } w \gg g$$

Chapter 21

Billet with initial temperature distribution

In the literature, it is claimed as an advantage of induction billet heating that the coils can be arranged to give a 'tapered' temperature distribution, which is advantageous for extrusion. It is, therefore, interesting to see how such an axial temperature inequality evens out with time.

Carslaw and Jaeger[26] (p. 101, eqn 6) give an expression for the temperature distribution in a bar with an initial temperature distribution $f(x)$, where both ends and sides are thermally insulated:

$$\theta = \frac{1}{l}\int_0^l f(x')\,dx' + \frac{2}{l}\sum_{n=1}^{\infty} \exp\left[\frac{-n^2\pi^2}{l^2}\left(\frac{k}{c\gamma}\right)t\right]\cos\frac{n\pi x}{l}\int_0^l f(x')\cos\frac{n\pi x'}{l}\,dx' \tag{21.1}$$

The first term gives the final, steady-state temperature rise of the whole bar, being the mean value of the function. The second term is the transient. Without doing any further work, it can immediately be seen that the transients die away with time constants which are fractions of the quantity $\tau' = c\gamma l^2/\pi^2 k$ seconds; that the time constants become much shorter as n increases because of n^2; and that the time constant falls rapidly with shorter lengths.

The factor $k/c\gamma$ changes with temperature, but in this chapter can be taken as constant at the extrusion temperature. Take 500°C as the extrusion temperature of aluminium; the value of $k/c\gamma$ is then $74 \times 10^{-6}\,m^2/s$. Consider a billet 1/3 m long. The time constant τ' for aluminium is given by

$$\tau' = \frac{(10^6)}{(74)(\pi^2)}\frac{1}{(3)^2} = 152 \quad s$$

So it will take about 2.5 minutes for the temperature differential of the billet to fall to 37% of its initial value.

While eqn 21.1 will cope with more complex temperature distributions, it is interesting to consider a linear temperature distribution $\theta = Cx$. Obviously, the first integral will give a final temperature of $Cl/2$, so only the transient need be considered:

$$\int_0^l f(x') \cos \frac{n\pi x'}{l} \, dx' = C \int_0^l x \cos \frac{n\pi x}{l} \, dx$$

$$= \frac{l^2 C}{n^2 \pi^2} \left[\cos \frac{n\pi x}{l} + \frac{n\pi x}{l} \sin \frac{n\pi x}{l} \right]_0^l$$

$$= \frac{l^2 C}{n^2 \pi^2} [\cos n\pi - \cos 0]$$

The bracket gives -2 for n odd and zero for n even, so the whole term is zero for n even and the summation need only be taken for odd n, giving

$$\frac{-4lC}{\pi^2} \sum_{n=1,3,5}^{\infty} \frac{\exp(-n^2 t/\tau')}{n^2} \cos \frac{n\pi x}{l} \tag{21.2}$$

This gives the temperature differential at every point x in the bar. In the practical problem, it is the difference in temperature between the ends that is important:

$$\left.\begin{aligned} &\text{at } x = 0, \quad \cos \frac{n\pi x}{l} = 11 \\ &\text{at } x = l, \quad \cos \frac{n\pi x}{l} = -1 \end{aligned}\right\} \text{for all odd } n$$

Thus

$$\text{temperature difference} = \frac{8lC}{\pi^2} \sum_{n=1,3,5}^{\infty} \frac{\exp(-n^2 t/\tau')}{n^2} \tag{21.3}$$

The series is rapidly convergent and it is only necessary to take $n = 3$ up to $t/\tau' = 0.25$. The temperature between the ends compared with the initial temperature difference is shown as a function of normalised time on Fig. 21.1.

Example 21.1
An aluminium billet, 1 m long, is to be extruded at 500°C. If the final temperature difference between the ends is to be 100°C and the handling time is 40 s, find the initial temperature difference.

Solution
The time constant is

$$\tau' = \left(\frac{c\gamma}{k}\right) \frac{l^2}{\pi^2} = \frac{(10^6)\,(1^2)}{(74)\,(\pi^2)} = 1370\,\text{s}$$

and so

$$t/\tau' = 40/1370 = 0.03$$

From Fig. 21.1,

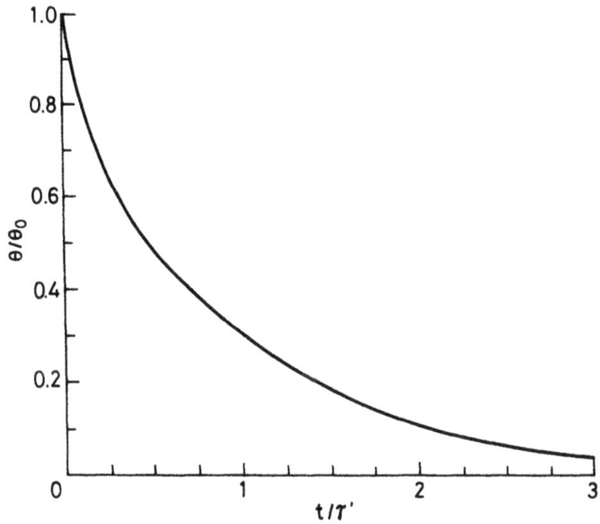

Fig. 21.1 *Variation of temperature along the length of a billet with time. θ_0 is the initial temperature difference between the ends, linearly distributed. $\tau' = (c\gamma/k)(l^2/\pi)$*

$$\theta/\theta_0 = 0.9$$

$$\theta_0 = 100/0.9 = 111°C$$

If, however, the billet is 300 mm long instead of 1 m,

$$\tau' = (0.3)^2 \times 1370 \quad \text{s}$$

$$t/\tau' = (0.03)/(0.3)^2 = 0.33$$

$$\theta/\theta_0 = 0.58$$

$$\theta_0 = 100/0.58 = 172°C$$

Part IV

Practical heating

Chapter 22

Through-heating by induction

22.1 Introduction to applications of through-heating

Induction heating is a convenient method for bulk-heating metals to a set temperature. It replaces furnaces, which tend to be large and which have the disadvantage of long start-up and shut-down times, so that their effectiveness is low. By contrast, the induction heater is relatively small in size and is immediately available for use. It is clean and relatively efficient. The power goes directly into the workpiece. The heating times are usually short – a few minutes (except when a large thermal mass is being heated) – and the process fits well into automated production methods. Although electricity costs are higher per unit of energy, this is offset by higher efficiencies. In Chapter 7 it was shown that induction heating takes place within about one skin depth (say 9 mm for iron, 10 mm for copper at 50 Hz, cold). The heat has to reach the rest of the billet or slab by conduction. Since non-ferrous metals have high heat conductivities this presents few problems, but for steel and other low-conductivity metals we have to avoid surface overheating, or perhaps even melting. Heat flow is an important part of the study of through-heating (see Part III).

Because we are not trying to confine the heating to the surface (as we do in hardening), lower frequencies can be used. Supply frequency (50/60 Hz) is obviously attractive because no conversion is involved; medium frequencies (1 to 25 kHz, say) are also extensively used and are becoming increasingly popular. Through-heating is used for forging, forming, annealing and indirect melting. Aluminium and other non-ferrous metals are heated for extrusion. These uses are power-intensive, as the throughput in tonnes per hour is usually high, so the power sources have to be substantial (normally hundreds of kilowatts, though several megawatts are not uncommon). Heating of steel presents a special problem because of the Curie point. In Examples 7.1 and 7.2, we showed that the power input rises from 0.7 to 1.7 MW/m^2 as the temperature rises and then falls suddenly to 0.4 MW/m^2 at the Curie point (760°C). It is, therefore, appropriate to use supply frequency up to the Curie point and then to change to a higher frequency (1000/3000 Hz) to continue heating up to forging temperature – in the range 1150 to 1300°C. See Table 22.3 for guidance on frequency selection.

Another feature of induction heating is repeatability. For a given material, the power input repeats accurately over the heating cycle. This allows time to be used as a measure of temperature, since the material, heated for a measured time, can be assumed to be at a given temperature with a high accuracy. This is useful where there is a constant production rate, using automatic feeding and ejection; the time cycle is chosen so that the material is at the correct temperature for the next operation. It is used in most automatic ferrous induction-forging lines, where the final temperatures are usually in the range 1150 to 1300°C and where variations of ±30°C can often be tolerated. The coils are then energised continuously, with no power switching. Several workpieces may be passing through the coil at any instant, or a few long bars may be sequentially heated.

Where the final billet temperature is critical and the billet may not be used immediately it is at temperature, a different control method is used. The temperature itself is monitored and used to control the power input; the power is pulsed to keep the exit billet at temperature until it is needed by the operator. The temperature is sensed by prod thermocouples pressing on the end of the billet.

22.1.1 Comparison with fossil-fuel-fired heating furnaces

Until relatively recently, fossil-fuel-fired furnace heating has been attractive because of the lower capital cost, together with the then lower cost of oil and gas, so it is important to compare the furnace with induction through-heaters. Strict cost comparisons are only possible for specific installations at a given time, because of the instability of fuel costs, but it is always important to look beyond capital cost and to take into account other facts.

- The time spent by the billet in the induction heater is much shorter than that spent in a radiant-heated furnace. As a result, the time available for scaling is reduced. Typical scaling with furnaces is 3 to 4%; this is reduced to 0.5% with electroheating. This scaling results in scrap and the accelerated wear of forging dies. (One company estimates an increase of up to 30% in die life because of the minimal scaling during induction heating.)
- Since there are no combustion products in the heating region, the fast heating produces clean billets, virtually free of oxidation and decarburisation (steel). Traditionally, with high-carbon, alloy, high-speed, and tool steels, up to 5% by weight of surface material has been decarburised; this has to be removed from the product, which represents both a material charge and a machining charge. The short time at high temperature, combined with the favourable heating conditions of the induction processes, mean that high-speed and tool steels meet the Class 1 decarburisation tolerance of BS4659: 971 without need for surface machining.
- The accurate temperature control, coupled with absence of scale, results in longer die life.
- Since induction heating is rapid, it fits well into automated production lines.

- The rapid start (typically of the order of minutes) allows the heater to be switched off for meal breaks, die changes and production holdups, thus saving energy. The equivalent furnace takes hours to reach working temperature and is normally left running, even during idle shifts, to avoid damage to the furnace walls and delays when production is resumed.
- The induction heater normally occupies much less shop-floor space than the furnace. Also, since the sides of the heater are not hot, the surrounding space can be used and the environment is improved for the workforce.
- Since there are no combustion products and no heat radiation, the ventilation and fume extraction is minimal.
- Health and safety are improved since there are no explosive gaseous mixtures needing elaborate precautions, no noxious fumes, and no risk of burns from furnace casings. Obviously, the electrical features will be designed to normal safety standards.
- It is possible to design induction heaters to give uneven 'taper' heating. For example, in extrusion it is often advantageous to heat the leading end of the billet to a higher temperature to reduce initial ram pressure and to compensate for the heat produced in the extrusion process. Fuel-fired billet heating would need a quenching stage to achieve this, with associated energy loss and extra installation cost, although fast tunnel gas furnaces can achieve a form of taper heating.
- With furnace heating, a change of process temperature can take up to an hour. When many changes are needed in a day, this is disadvantageous. With an induction heater, the new temperature is set and achieved within minutes.

Since scaling and decarburisation are not problems below 800°C, there is an argument (where gas or oil furnaces already exist) for preheating to 800°C in a furnace and then using electroheat over the critical 800 to 1200°C range. Again, the precise economics must be investigated.

22.1.2 Induction-heating configurations

Figure 22.1 shows the principle of the induction billet heater. The billet is surrounded by a coil, supplied with alternating current. The coil is normally water-cooled and has a refractory lining. In practice, the coil is usually much longer than a single billet and the billets are pushed through a horizontal coil, supported on water-cooled skid-rails (Fig. 22.2). The billets may be pushed into the charging end of the coil, using a hydraulic or pneumatic ram. Alternatively, the billets can be continuously-fed using a gripping mechanism or friction, which has the minor advantage of keeping the coil full. A more expensive and (electrically) less efficient method of moving the billets uses a walking-beam mechanism; for bars up to 38 mm diameter, the walking-beam arrangement may be preferable.

At the exit end, extractor rollers are normally provided; these are specially important for continuous-feed heaters to avoid progressive cooling of the nose of the billet as it emerges.

One heating coil covers a range of diameters, with some loss of effectiveness as the billet diameter decreases relative to the coil. It is easy to provide for coil changing, when the work changes, using spring-loaded electrical connections and quick-make water connectors. Changes can be made in about 20 minutes.

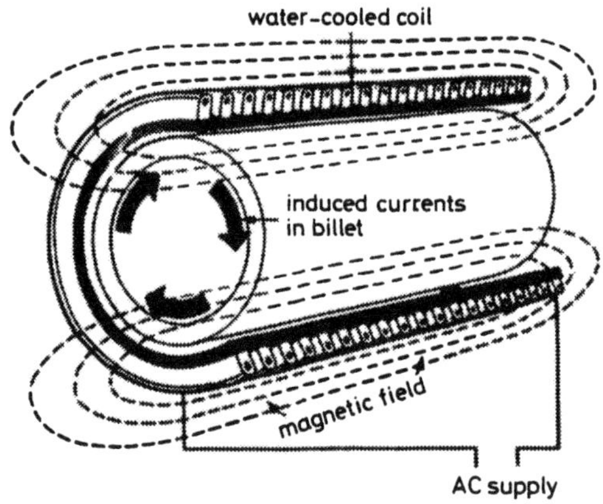

Fig. 22.1 *Principle of the induction billet heater*

Typical aluminium billet diameters are 125 to 185 mm with lengths up to 660 mm. A typical output rate is one billet every 7 seconds.

22.2 Temperature effects

22.2.1 Specific heat

A measure of the effectiveness of a heater is the energy needed to heat a given weight of material. With a perfect heater, the energy can be calculated from the

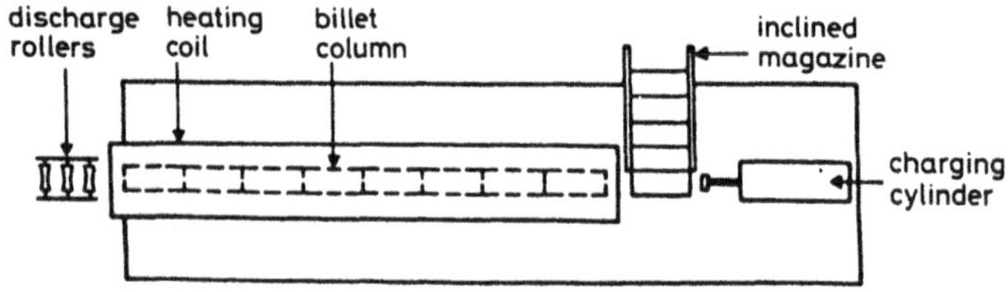

Fig. 22.2 *Pusher-feed tunnel heater*

mean specific heat over the temperature range, together with any latent heat involved in metallic transformations. Mass specific heat c is measured in W s/kg K or J/kg K, so that energy is $cm\theta$ J or $c\theta$ J/kg. In practice, it is convenient to use kilowatt-hours per tonne for commercial calculation.

Typical values of c for common metals are as follows:

	J/kg K	kWh/t K
Aluminium	900	0.25
Brass/bronze, copper	377	0.105
Iron	460	0.128
Nickel	439	0.122

Note that the values for iron and nickel are only valid up to the Curie point.

Curves of heat energy for various materials are given in Fig. 22.3. From this, it would apparently need 230 kWh/t to heat iron from 20 to 1200°C, whereas

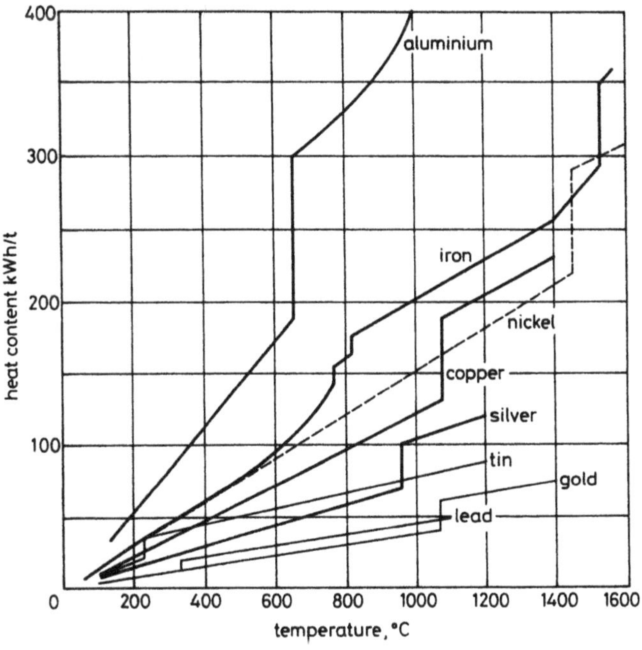

Fig. 22.3 *Heat content of metals: datum temperature 20°C (after Stansel[33])*

320 kWh/t at the coil is a good practical figure, showing a 70% heating efficiency. (In addition to this, the converter efficiency has to be included.) Actual figures depend on the size of the billet and the frequency. For forging temperatures for steel, values range from 275 to 440 kWh/t; for annealing temperatures, they vary from 155 to 275 kWh/t. Figure 22.4 indicates these values for various materials as a function of diameter. (As the diameter increases, there is less tendency to flux-cancellation in the billet, and the efficiency approaches a maximum value.) These practical curves take into account heat losses, as well as the various electrical losses.

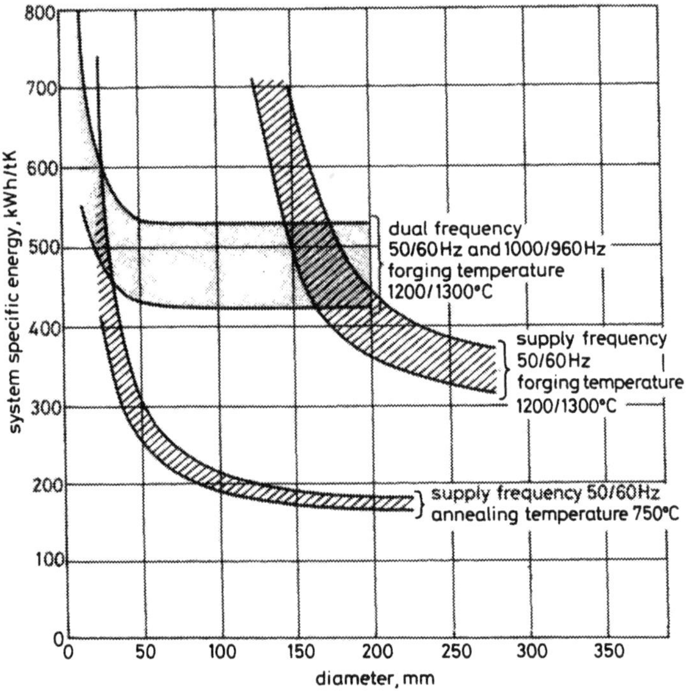

a

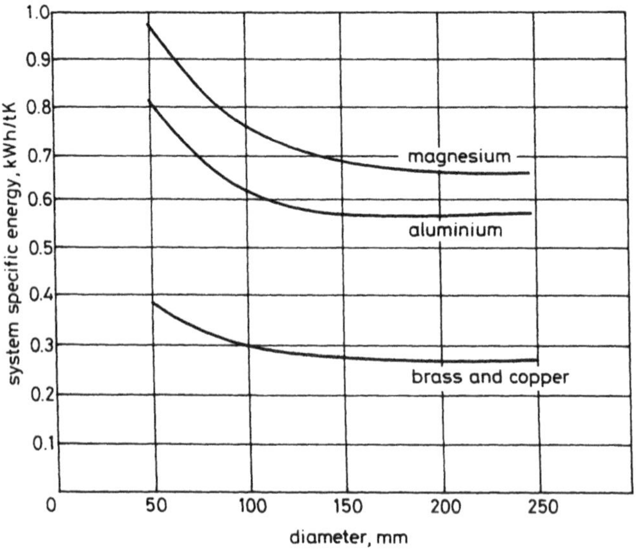

b

Fig. 22.4 *Actual heat needed, including system inefficiencies, for various billet diameters: (a) steel (b) non-ferrous metal. Multiply by rise in temperature above 20°C to obtain kWh per tonne*

Table 22.1 *Values of ρ, α, and melting point for various metals*

	α	$\rho_{20°C}$ (Ωm)	Melting point (°C)
Aluminium	4.29×10^{-3}	2.65×10^{-8}	660
Brass (65% Cu, 35% Zn)	1.6×10^{-3}	6.4×10^{-8}	931
Bronze (90% Cu, 11% Zn)	1.86×10^{-3}	3.9×10^{-8}	1046
Copper	3.93×10^{-3}	1.72×10^{-8}	1084
Pure iron	6.5×10^{-3}	9.71×10^{-8}	1538
Lead*	3.36×10^{-3}	20.65×10^{-8}	327
Silver	4.1×10^{-3}	1.59×10^{-8}	962

* Lead has a slightly nonlinear resistivity relationship.

22.2.2 Resistivity

The resistivity of metals varies with temperature. Most metals vary linearly:

$$\rho_0 = \rho_1[1 + \alpha(\theta - \theta_1)] \qquad (22.1)$$

where ρ_0 is the resistivity at any temperature θ (°C) in terms of the resistivity ρ_1

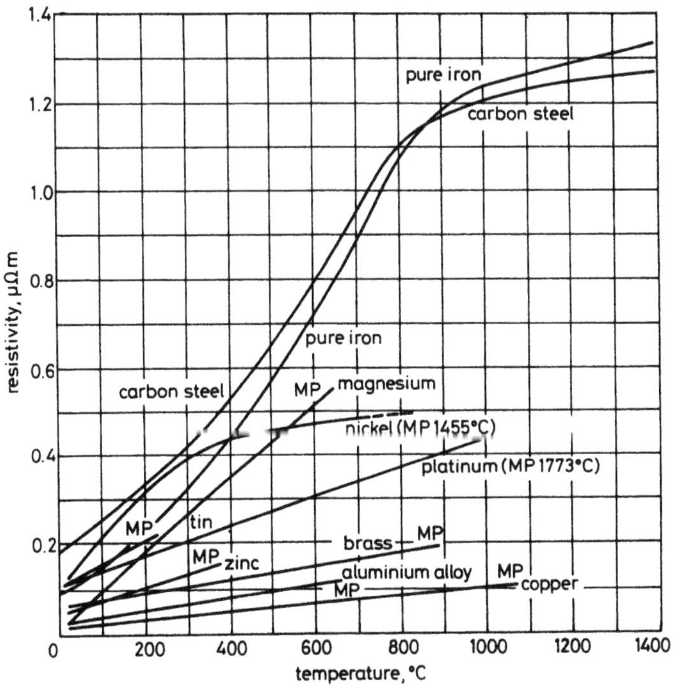

Fig. 22.5 *Variation of resistivity with temperature*

at temperature θ_1. Table 22.1 gives values of α, ρ_{20} and melting point for various metals. Figure 22.5 shows the resistivity variation of a range of metals; note the non-linearity for steel.

In Chapter 12, it was shown that loss varies as ρ/δ for a given applied field.

Since δ varies as $\sqrt{\rho}$, this means that the loss is proportional to $\sqrt{\rho}$, if other quantities remain constant, so it is important to decide which value of ρ to use in calculations which involve a range of temperatures. Since resistivity usually varies linearly with temperature over wide temperature ranges, the concept of 'integrated resistivity' ρ_m can be used. (Even for non-linear materials, such as steel, the method is reasonably accurate.)

For a linear material, eqn 22.1 can be rearranged:

$$\rho_\theta = \rho_1 + \frac{\theta - \theta_1}{\theta_2 - \theta_1}(\rho_2 - \rho_1) \tag{22.2}$$

where ρ_1 and ρ_2 are the resistivities at the starting and finishing temperatures θ_1 and θ_2.

The rate of change of temperature is proportional to input power (see Section 15.2), which has been shown above to be proportional to $\sqrt{\rho}$, other things being equal. Hence

$$\frac{d\theta}{dt} = \text{constant} \times \sqrt{\rho} \tag{22.3}$$

Rearranging and substituting from eqn 22.2,

$$\int_{\theta_1}^{\theta_2} \frac{d\theta}{\sqrt{\{\rho_1 + [(\theta - \theta_1)/(\theta_2 - \theta_1)](\rho_2 - \rho_1)\}}} = \int_0^t (\text{constant})\, dt$$

Hence, the heating time t is given by

$$t = 2\frac{(\theta_2 - \theta_1)\sqrt{\rho_2} - \sqrt{\rho_1}}{(\rho_2 - \rho_1)(\text{constant})} \tag{22.4}$$

If the resistivity were constant and equal to ρ_m,

$$\int_{\theta_1}^{\theta_2} d\theta = \int_0^t (\text{constant})\sqrt{\rho_m}\, dt$$

$$t = \frac{\theta_2 - \theta_1}{(\text{constant})\sqrt{\rho_m}} \tag{22.5}$$

Equating values of t from eqns 22.4 and 22.5,

$$\rho_m = \left[\frac{\rho_2 - \rho_1}{2(\sqrt{\rho_2} - \sqrt{\rho_1})}\right]^2 \tag{22.6}$$

$$= \frac{(\sqrt{\rho_2} + \sqrt{\rho_1})^2}{4}$$

Example 22.1

Find the integrated resistivity for mild steel heated from room temperature to 760°C (the Curie point) and from 760°C to 1250°C, given that $\rho_{20} = 0.16\,\mu\Omega\,m$, $\rho_{760} = 1.02\,\mu\Omega\,m$ and $\rho_{1250} = 1.23\,\mu\Omega\,m$.

Solution
From 20°C to 760°C,

$$\rho_m = (\sqrt{1.02} + \sqrt{0.16})^2/4 = 0.497\,\mu\Omega\,m$$

From 760°C to 1250°C,

$$\rho_m = (\sqrt{1.23} + \sqrt{1.02})^2/4 = 1.123\,\mu\Omega\,m$$

Compare these with the respective arithmetic mean values of 0.59 and 1.125 $\mu\Omega$ m (see below). Note also that the theory is not exact for steel, as the resistivity becomes non-linear as temperature rises. Fortunately, the variation is fairly linear over each of the two ranges given above.

It is easy to rearrange eqn 22.6 in terms of the arithmetic mean resistivity $\rho_a = (\rho_1 + \rho_2)/2$:

$$\rho_m = \rho_a\left[\frac{1}{2} + \frac{\sqrt{(\rho_1\rho_2)}}{\rho_1 + \rho_2}\right]$$

But $\rho_2 = \rho_1(1 + \alpha\theta)$; hence

$$\rho_m = \rho_a\left[\frac{1}{2} + \frac{\sqrt{(1 + \alpha\theta)}}{2 + \alpha\theta}\right]$$

From Table 2?.1, $\alpha\theta$ varies from about 1.5 for brass to about 8 for iron. Values of the square bracket above are given in Table 22.2.

Table 22.2 *Relation between integrated resistivity ρ_m and arithmetic mean resistivity ρ_a*

$\alpha\theta$	0	0.5	1	2	3	4	5	6	7	8	10
ρ_m/ρ_a	1	0.99	0.97	0.93	0.90	0.87	0.85	0.83	0.81	0.80	0.78
$\sqrt{(\rho_m/\rho_a)}$	1	0.99	0.98	0.96	0.95	0.93	0.92	0.91	0.90	0.89	0.88

Since the integrated resistivity is always used in its $\sqrt{\rho}$ form, the bottom line in Table 22.2 gives a measure of the error involved in using ρ_a instead of ρ_m. It is really only worthwhile when large temperature rises are involved. With a computer, the calculations are repeated at various temperatures and a curve of loss against temperature is plotted; the ρ_m simplification is then not needed.

Steel and other ferromagnetic materials have a non-linear variation of ρ with temperature. Full details have been found for many alloys.[34] Figure 22.5 shows the variation for carbon steel, which can be taken as a typical curve. At elevated temperatures, ρ is increasingly independent of carbon content.

22.3 Heat radiation and convection

Heat lost by radiation plays an important part in the slower forms of induction heating at high temperatures. Temperatures below about 850°C usually result in very little loss in most metals when totally enclosed in the solenoid type of heating coil. Above these temperatures, losses become significant, especially in the through-heating of steel to temperatures in the region of 1250 to 1300°C for forging. Most of these losses are conducted through the coil liner and removed by the coil cooling-water.

The radiation loss (in W/m^2) is given by eqn 17.2:

$$\mathscr{P}_R = 5.67 \times 10^{-8} \varepsilon (T_s^4 - T_a^4)$$

where ε is the emissivity coefficient of the surface, T_s is the absolute surface

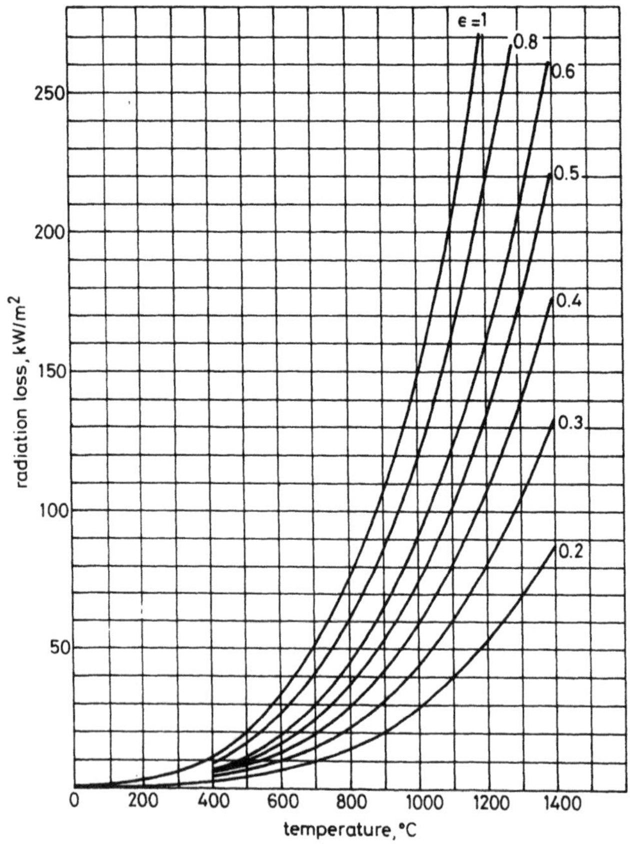

Fig. 22.6 *Variation of radiation losses with temperature and emissivity ε (for an ambient temperature of 20°C). Emissivity for various materials: steel 0.8; copper and brass 0.5; aluminium 0.3; oxidised titanium (980°C) 0.53*

temperature (K, where K = °C + 273), T_a is the absolute ambient temperature (K), and $\mathscr{P}_R$ is the radiated loss per square metre of the surface.

Figure 22.6 shows how radiation loss varies with T_s for a temperature T_a of 293 K (= 20°C), assuming various values for ε. Some typical values of ε are given, based on a normal surface for a commercial workpiece. They must be used with caution as many of the surfaces oxidise with temperature rise, increasing ε and the radiation loss. For instance, copper may vary in the value of ε from 0.03 for a polished surface to 0.70 for a heavily-oxidised surface. Also, very rarely does a billet radiate to ambient; in practice, the coil liners will be at a much higher temperature, thus reducing loss.

From these considerations, it can be seen that it is very difficult to calculate accurately the radiation losses in induction heating. Radiation is discussed in more detail in Chapter 17. For through-heating of steel and other metals up to 1300°C, power inputs are of the order of 0.3 to 1.5 MW/m^2. Most coils are well insulated with liners, and the power lost by radiation from the work surface through the coil liner, averaged between room temperature and 1300°C, is usually about 40 kW/m^2. This is not negligible and should be calculated for the whole work-surface area. The coil liners used in through-heating are described in detail in Section 22.10. They vary in thickness between 5 and 25 mm, and consist of either a metal inner shell and refractory felt, or a cast or rammed solid refractory shell with steel rails. These materials act as good muffles and keep the heat inside the coil. (The radial temperature drop across the total liner may be as high as 280 to 560 K).

Convection losses are small in through-heating, as the workpiece is contained in a shell which does not permit air movement. For workpieces in the open air, a good approximation for convection losses (in W/m^2) is

$$\mathscr{P}_{con} = 1.54(\Delta T)^{4/3} \tag{22.7}$$

where ΔT is the temperature rise above ambient in kelvins.

Example 22.2
Calculate the convection loss-density at 1000°C for an induction-hardened rod. Assume a room temperature of 20°C.

Solution
From eqn 22.7,

$$\mathscr{P}_{con} = (1.54)(980)^{4/3} = 15\,\text{kW/m}^2$$

This is negligible compared with power inputs of the order of 1 MW/m^2.

22.4 Temperature distribution in induction through-heating

Most processes that follow heating (forging, forming, or extrusion) are critically dependent on the uniformity of temperature in the workpiece, as this affects die life and metallurgical uniformity. Uniformity can always be obtained by heating slowly or by allowing the work to even out in a muffle or soaking chamber, but these are both at variance with the need for fast production rates and minimum floor area.

The designer's problem is to put in maximum power (without damaging the surface regions), whilst achieving the required low surface–centre temperature differential. A good understanding of heat flow is essential. (Another constraint is the mechanical stress set up in the billet during heating.)

The classical problem of the temperature distribution in various configurations with constant power input at the surface was solved by Carslaw and Jaeger.[26] These solutions are discussed and extended in Chapter 15, where it is shown that, after a transient temperature rise, the temperature distribution becomes parabolic and, thereafter, the temperature profile remains the same, with all parts rising at the same rate. This is shown clearly in Figs 15.3 and 15.4 by generalised curves: these are very important for the understanding of the induction-heating process, as they show that the temperature difference between the surface and the centre is fixed. Its value is given by eqn 15.5c:

$$(\theta_s - \theta_c)_{\tau \geqslant 0.25} = \mathscr{P}_s R/2k$$

where $\mathscr{P}_s$ is the surface power input density (W/m^2), R is the radius of the billet (m) and k is thermal conductivity (W/m K). Example 15.4 shows that, for practical values of $\mathscr{P}_s$, R, and k, the temperature differential is large (1250 K). It is important to realise that this differential will be maintained at a constant value for the whole time that $\mathscr{P}_s$ is applied; the temperatures θ_s and θ_c simply rise together. As a result, the billet has to soak for a time after the removal of power to equalise the temperatures. The theory of soaking is given in Chapter 16, and Figs 16.1 and 16.2 summarise the temperature variation in generalised form. A useful concept is the time taken to reduce the initial differential temperature $\theta_s - \theta_c$ to 10% of that value, which is given by $\tau = 0.154$ on Fig. 16.2. Hence

$$t_{10\%} = 0.154 \frac{(c\gamma)R^2}{k} \quad \text{seconds} \tag{22.8}$$

where c is specific heat (J/kg K), γ is density (kg/m^3), and k and R are defined above. The time given by eqn 22.8 is roughly correct. The precise soak time for a given differential temperature can be obtained from Fig. 16.2, but in practice the rule-of-thumb figure of 1.1°C per millimetre of radius is achieved in roughly the time given by eqn 22.8. (From eqn 15.5c it will be seen that $(\theta_s - \theta_c)/R$ is given by $\mathscr{P}_s/2k$, so that the final temperature difference per millimetre of radius is directly related to power input.)

Example 22.3

Calculate the soak time needed to reduce the temperature differential of a 200 mm diameter aluminium billet to 10% of its value at power turn-off.

Solution

From eqn 22.8, the required time is

$$t_{10\%} = 0.154 \frac{(c\gamma)}{k} (0.1)^2$$

$$= 0.154 \frac{(2700)(946)}{204)} (0.1)^2 = 19\,\text{s}$$

Compare this value with the time calculated for a steel billet in Example 16.1, i.e. 192 s. The 10:1 change in time is entirely due to the constants ($c\gamma$) and k: aluminium has five times the thermal conductivity and half the volume specific heat. The soak time for aluminium and other non-ferrous metals is small compared with the heating time and may simply be the time needed to transport the billet to the press, whereas steel billets need a soaking period.

22.5 Effects of current depth and radiated heat

The ideas of the last section were based on power flow at the surface of the billet, but the heat is actually induced in the metal. The theory of Chapter 15 is modified in Chapter 18 to allow for this. The revised temperature distributions are given in Figs 18.5 to 18.8. Note that these are steady curves, i.e. the transient has died away ($\tau \geqslant 0.25$). The curve for $R/\delta = \infty$ is the same as given in Chapter 15 but, for all other values of R/δ, the temperature differential is decreased. Efficient heating requires that $R/\delta > 1.5$ (see Chapter 9). As an example, a 200 mm diameter steel billet heated above the Curie point at 60 Hz corresponds to $R/\delta = 1.6$. From the curves, it will be seen that the heating and soaking times given in the last section are pessimistic.

Radiation also reduces the temperature differential by reducing the surface temperature. In Chapter 18 this is calculated, based on certain simplifying assumptions, and the results are summarised on Fig. 18.10. It must be emphasised that these curves should be used with caution, as the radiation power varies with surface temperature and so is different for each instant in the heating cycle. Radiation certainly cannot be neglected. In the later calculations, we make an approximate allowance for this, but with a computer is is easy to allow for radiation step-by-step over the whole temperature range.

We shall see later in coil design (Section 22.8) that the temperature distribution when heating steel is further complicated by the 'soak time' involved in the Curie transition, as well as the non-linearity of the heat-flow 'constants' and the actual soak coil. In modern practice, the final 'delivered' temperature differential (usually 50°C, independent of diameter, and sometimes 100°C or 25°C) is specified and the manufacturer arranges the combination of coils – below-Curie, above-Curie and soak – to achieve this. In the design, account has to be taken not only of normal running, but also of what happens if the line runs slow for production reasons or has to restart after a stop. Every firm solves these problems individually; we shall discuss neither slow-running nor restart.

22.6 Temperature distributions in a heating cycle

Based on the ideas of the previous sections, Fig. 22.7 shows the variation of temperature with time during a heating cycle for a non-ferrous billet, i.e. a billet not affected by the Curie point. The transient stage during t_t corresponds to the curves for $\tau < 0.25$ in Fig. 15.3, after which the surface and centre curves rise together until the power is turned off at t_h. The temperature difference during t_s depends on the finite-depth effect and on radiation, as previously discussed. The temperatures then equalise during a time t_e. The time allowed for t_e takes into account any cooling after leaving the heating coil and before being used. If the surface cooling is abnormally high, then the curves may take the form of curve B, shown dashed.

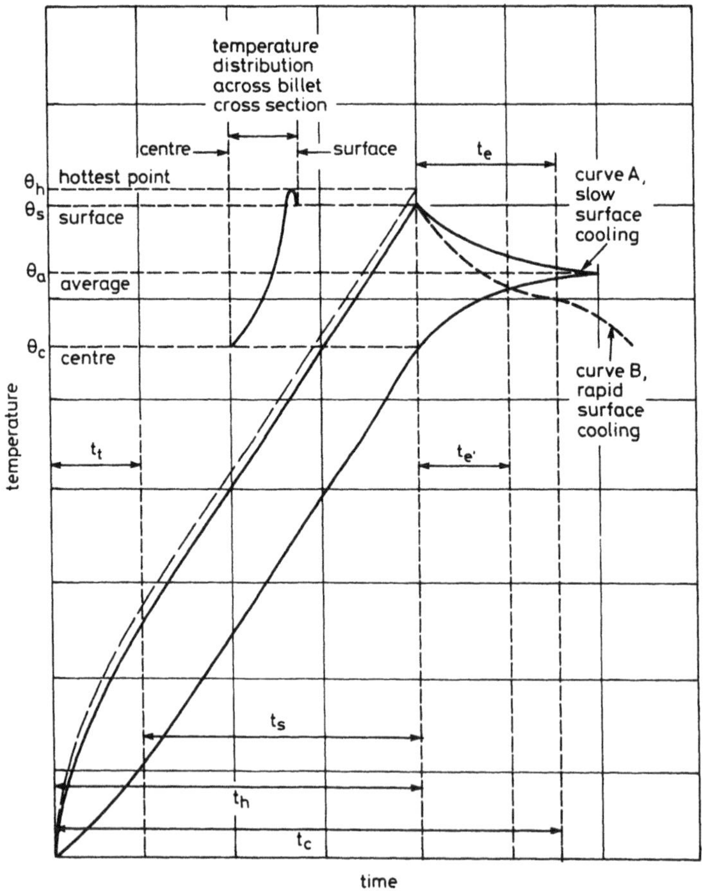

Fig. 22.7 *Typical heating cycle in non-ferrous through-heating. Times: t_t transient heat time; t_s steady-state heat time; t_e equalising time; $t_{e'}$ equalising time in crossover case; t_c total heating time per cycle*

Note that the hottest point is below the surface; in practical examples of overheated aluminium, a ring of partially-melted metal can be seen about 10 mm below the surface. In calculating power input, care must be taken to ensure that θ_h is below the melting point and that the differential temperature $\theta_s - \theta_c$ is low enough not to set up stress cracking. Although no precise relationship has been established for this, an empirical figure is 1.1°C per radial millimetre, for non-ferrous materials.

In practice, the temperature distribution, obtained experimentally by embedding chromel-alumel thermocouples, corresponds fairly closely to the calculated result. The difference between the hottest point and the surface temperature is often small, and is usually negligible shortly after the heat time has elapsed, even in metals of low thermal conductivity. It is possible to measure the temperature distribution at the end surface of a billet; this is sometimes done in the billet heaters, but the results must be viewed with caution as the method is imprecise.

Temperature distributions for steel are complicated by saturation and by the Curie point. At first, when the metal is cold, the effect of saturation is that the

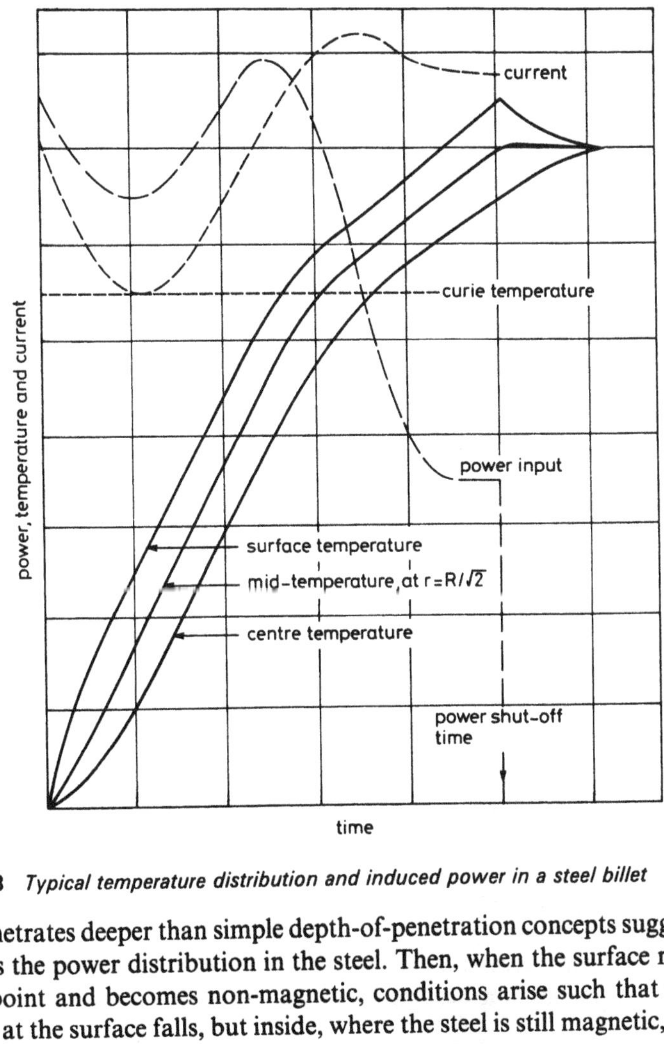

Fig. 22.8 *Typical temperature distribution and induced power in a steel billet*

flux penetrates deeper than simple depth-of-penetration concepts suggest, which changes the power distribution in the steel. Then, when the surface reaches the Curie point and becomes non-magnetic, conditions arise such that the power density at the surface falls, but inside, where the steel is still magnetic, the power input is still high. Imposed on this is the variable resistivity of the steel, which also changes the power level at every depth. These effects have been pictured as a 'magnetic wave' moving in from the surface, but this is a simplified view of a very complex situation. It is possible to get a more accurate calculation, using computers. A typical temperature–time curve for a steel billet is given in Fig. 22.8, showing temperatures at the surface, at the centre, and at $r = R/\sqrt{2}$. After the initial transient, the three points move in parallel until the surface reaches the Curie point, at which time the surface heat-rate drops. The billet becomes successively non-magnetic as each layer reaches the Curie point, until the whole billet is non-magnetic.

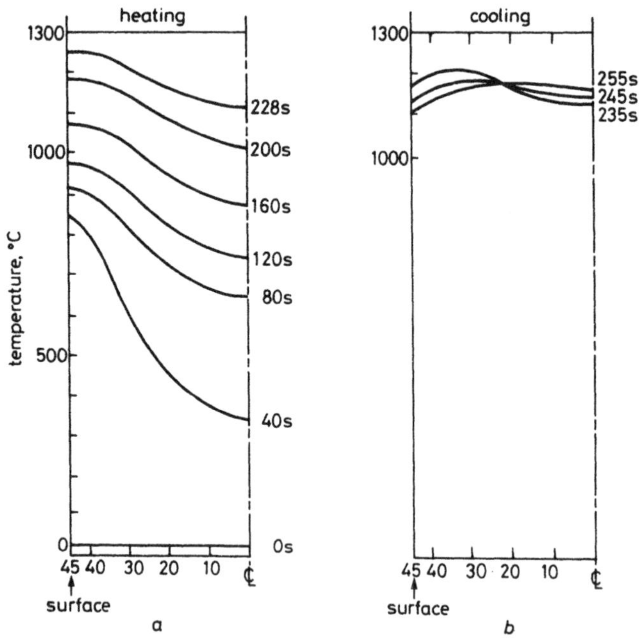

Fig. 22.9 *Temperature distribution at different stages during (a) heating and (b) cooling of a steel billet with a diameter of 90 mm, as calculated on a computer. Field strength 90 kA/m, frequency 1090 Hz (after ASEA)*

The current and power are shown on the same graph. The initial dip in the power and current curves is thought to be due to the changing resistance of the current path, which would decrease power drain. As the current depth increases with time, the cross-section of the current path increases faster than the resistivity, resulting in a net drop in resistance to current. This increases power whilst the steel is magnetic. At the Curie point, power drops, as already explained. Current rises as the path resistance drops and as more of the current is made up of the magnetising component; it settles to a steady value as the current depth and magnetising section become constant.

Figure 22.9 shows computed heating and soaking temperature-distributions through a 90 mm diameter steel billet. The field intensity was 90 kA/m at a frequency of 1090 Hz. The computer takes into account that the heat is produced in a finite depth, rather than entirely at the surface.

22.7 Selection of frequencies for induction heating

For through-heating, economic considerations are as important as optimum frequency choice. Dual frequency should be considered for steel because of the change of permeability at the Curie point, but this is unlikely to prove economic except for very large heaters. Instead, a compromise frequency is used that gives satisfactory results in terms of acceptable kg/kWh over the whole temperature range, letting the below-Curie region take care of itself as long as the specified temperature differential (usually 50°C) is achieved.

Table 22.3 lists some practical frequencies used for heating steel of various diameters. These figures obviously depend on the kg/kWh versus diameter curves for a particular manufacturer's heater range, so they should only be used as indicators at this stage. For diameters below 6 mm, radio frequency should be considered, but with care, as the coil voltages involved tend to be high.

It was shown in Chapter 9 that, for effective heating of billets, $d \geqslant 3\delta$. Using this, with a value for μ_r of unity and an integrated resistivity at 1200°C of 0.8 $\mu\Omega$ m, we can draw a graph of d against f, which will obviously have a slope of $1/\sqrt{f}$ because of δ (Fig. 22.10). For an integrated resistivity of 0.6 $\mu\Omega$ m and an assumed value of $\mu_r \simeq 2$, a second line is drawn for 800°C. These limits agree well with References 35 and 36. Figure 22.10 should be used as an indicator only, as the most economic frequency is not necessarily the same as the optimum frequency, based on $d/\delta = 3.5$.

Non-ferrous metals with high electrical resistivity (stainless steel, nickel, titanium, etc.) approximate closely to steel above the Curie point, and the same rough frequency divisions apply (Table 22.3 and Fig. 22.10). For low-resistivity, non ferrous metals (aluminium, brass, copper, magnesium, etc.), where the depth of penetration is greater, it is possible to use 50/60 Hz for most practical billets (say, between 125 mm and 300 mm diameter). In fact, 50/60 Hz aluminium heaters have been installed for bars down to 50 mm diameter. The

Table 22.3 *Steel through-heating frequencies for forging (1260°C)*

Frequency	Range of acceptable diameters (mm)	Optimum diameter (mm)
50/60 Hz	⩾140	
1 kHz	140 down to 40	90
3 kHz	100 down to 28	60
10 kHz	90 down to 20	50
25 kHz	80 down to 15	30

advantages of working with mains frequency often outweigh any ineffectiveness due to low efficiencies; nevertheless, each installation should be studied carefully to decide whether conversion equipment would give better productivity. Table 22.4 shows typical frequencies.

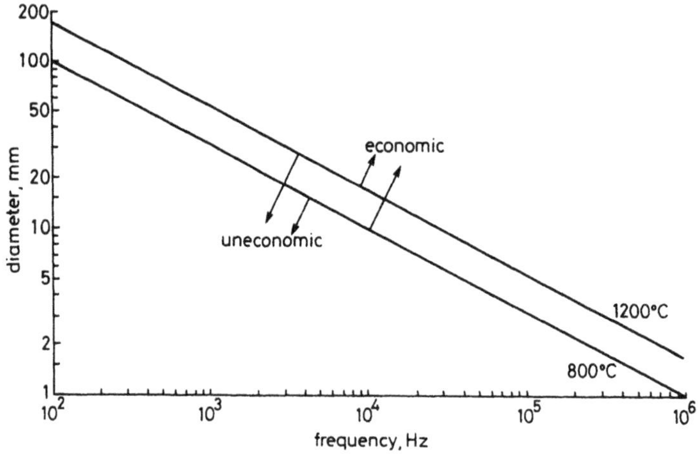

Fig. 22.10 *Economic diameter against frequency, for iron heated to 1200°C*

Since power input varies as $\sqrt{f}$, it might be expected that efficiency would rise with frequency. Whilst this is generally true, the equipment needed to produce the higher frequencies usually has lower overall efficiency. For example, at 450 kHz the coil efficiency for heating steel to the Curie temperature is 90% or better but the radio-frequency generator has an overall efficiency from line to work of about 50 to 65%. A supply-frequency coil may be only 55% efficient for non-ferrous heating but, as no frequency conversion is needed, the overall line-to-work efficiency would be about 50%. Medium frequencies give coils of 70% efficiency for steel, but the efficiency of the conversion equipment must be included, so the overall efficiency is 50 to 70%, depending on the metal.

From this, it appears that a useful overall efficiency of 55 to 70% can be expected for the induction heating of most metals (Steel below the Curie point gives better figures, say 70 to 75%), so the running costs are the same. Decisions

Table 22.4 *Typical frequencies for non-ferrous metals*

Bar diameter (mm)	Frequency
0 to 12	450 kHz
12 to 25	10 kHz and 3 kHz
25 to 75	1 kHz
Over 75	50 Hz

have to be made on capital cost, floor space occupied, and the cost of maintenance. Mains-frequency equipment has very low maintenance costs.

Since costs vary with time and place, it is difficult to be precise, so they will not be discussed further here.

22.8 Coil design for given applications

For through-heating, we need to know:

- required throughput (t/h, kg/h, or billets/h)
- dimensions of workpiece
- temperature rise
- material
- allowable temperature-differential over radius.

From these, we must specify the power density and the frequency, which allow the coil to be designed. The temperature distribution and heat losses mentioned earlier have to be calculated.

The impedance of the power source must be kept in mind and the load must be matched to this, if the maximum power available from the source is to be used.

To illustrate the derivation of the data needed for coil design, we shall use two examples (with explanatory notes) for aluminium and steel heating, respectively. Obviously the method can be extended to other shapes (slabs, etc.) and to other metals.

22.8.1 Coil design for aluminium billet heater

Metal	commercially pure aluminium
Production rate	2.25 t/h
Temperature rise (average over billet cross-section), $\Delta\theta$	490 K (20–510°C)
Maximum surface–centre temperature differential, $\theta_s - \theta_c$	40 K
Mass of each billet, M	23.9 kg
Billet length, l_w	500 mm
Billet diameter, d_w	150 mm
Metal density, γ	2700 kg/m^3
Metal resistivity, ρ_m (see note 1)	0.535 $\mu\Omega$ m
Maximum allowable temperature differential along length, θ	8 K
Metal kilowatt-hours per tonne over $\Delta\theta$ (see note 2)	145
Metal thermal conductivity over $\Delta\theta$, k (see note 3)	218 W/m K

Notes

1 This is the integrated value for the temperature range, calculated as follows:
$\rho_{20} = 0.027\,\mu\Omega\,\text{m}$, $\rho_{510} = 0.089\,\mu\Omega\,\text{m}$,
$\rho_m = (\sqrt{0.089} + \sqrt{0.027})^2/4 = 0.535\,\mu\Omega\,\text{m}$.

2 From specific heat; see Fig. 22.3.

3 There is a small change of value over the temperature range; this is the average value. See Appendix A.1.

Frequency: From Table 22.4, 50 Hz is suitable.

Thermal power P_t:

$$P_t = 2.25\,\text{t/h} \times 145\,\text{kWh/t} = 326\ \text{kW}$$

Surface power losses $\mathscr{P}_R$: The main loss is radiation. If it is assumed that the work is inside the coil for most through-heating applications, an empirical rule is that the radiation loss is reduced by 25% of its free-space value. Equation 17.14 gives the mean radiated power-density as

$$\mathscr{P}_{Rm} = \frac{5.67 \times 10^{-8}\varepsilon}{5}\left(\frac{T_2^5 - T_1^5}{T_2 - T_1}\right)$$

For this example, $T_1 = 20 + 273 = 293\,\text{K}$, $T_2 = 510 + 273 = 783\,\text{K}$, and $\varepsilon = 0.3$ (see Fig. 22.6). So

$$\mathscr{P}_{Rm} = \frac{(5.67 \times 10^{-8})(0.3)}{5}\left(\frac{783^5 - 293^5}{783 - 293}\right) = 2028\,\text{W/m}^2 \simeq 2\,\text{kW/m}^2$$

This is reduced by 25% to allow for the conditions in the heater; then $\mathscr{P}_{Rm} = 1.5\,\text{kW/m}^2$.

Power density $\mathscr{P}_s$: In Section 22.4, the power density for a given temperature difference between the surface and the centre is discussed (eqn 15.5c):

$$\theta_s - \theta_c = \mathscr{P}_s R/2k$$

In our example, this gives

$$\mathscr{P}_{s\,\text{idealised}} = \frac{(2)(218)(40)}{(75/1000)}\,\text{W/m}^2 = 0.233\,\text{MW/m}^2$$

In Chapter 18, this was modified to allow for finite current depth. Here, $R = (75/1000)\,\text{m}$ and, from eqn 2.7,

$$\delta = \sqrt{\left[\frac{2 \times (5.35 \times 10^{-8})}{(4\pi 10^{-7})(2\pi 50)}\right]} = 0.017 \quad \text{m}$$

$$R/\delta = 75/17 = 4.41$$

The finite depth is, therefore, important (see Fig. 18.7). For $R/\delta = 4.4$,

$$(\theta_s - \theta_c)/(\theta_s - \theta_c)_{idealised} = 0.72$$

The actual power density is then

$$\mathcal{P}_s = 0.233/0.72 = 0.324\,\mathrm{MW/m^2}$$

Note that the key decision in billet heating is the allowable temperature differential between surface and centre. This controls the power input and, hence, all other dimensions. If, for example, the specification had followed the recommendation given in Section 12.4 to use 1.1°C per radial millimetre, the differential would be 75 × 1.1 = 82.5°C and $\mathcal{P}_s$ would become (82.5/40)324 kW/m^2 = 0.668 MW/m^2. The reader is recommended to try for himself the effect this change would have on the remainder of this calculation.

Effect of radiation on temperature differential: The radiation loss calculated is such a small proportion of the power density that it can be ignored in the calculation of temperature differential.

Minimum work-length l_w: From the power density and the thermal power, the minimum heated surface area A_s is

$$A_s = \frac{326 \times 1000}{0.324 \times 10^6} = 1.01\,\mathrm{m^2}$$

Then the minimum length is

$$l_w = \frac{1.01}{\pi(150/1000)} = 2.14\,\mathrm{m}$$

As each billet is 500 mm long, the nearest multiple is four billets, which is very close to the calculated l_w. Four billets, each weighing 23.9 kg, will be heated together in the coil.

The production rate of 2.25 t/h gives 2250/23.9 = 94 billets/h.

Heating time per billet t_h: This is the inverse of billets/h multiplied by the number of billets being heated simultaneously:

$$t_h = 60 \times 4/94 = 2.55\,\mathrm{min}$$

Total work power P_w: The total power density is the sum of the thermal power-density and the radiation loss-density:

$$\mathcal{P}_w = \mathcal{P}_c + \mathcal{P}_R = 324 + 1.5 \quad \mathrm{kW/m^2}$$

Hence

$$P_w = \frac{(326)(\pi 150)(2)}{1000} = 307\,\mathrm{kW}$$

Estimated coil dimensions: The inner diameter or dimension of a through-heating coil usually depends on the weight and temperature of the billet and the distortional and dimensional tolerances. The method of feed-through also determines the diameter. As a general guide, Table 22.5 gives some indication of practical diameters for through-heating coils. The construction is detailed in Section 22.10. The air gap is here defined as the radial difference between coil inner dimension (diameter, width, or length) and work outer dimension.

The length of coil windings depends largely on the chosen method of work handling and the metal heated. In most steel installations, where the production rate is high, the coils are often separated. The resulting unheated areas have a negligible effect on the temperature distribution, as a sufficiently large number of billets is usually progressing through the tunnel of coils for the heat to diffuse. Therefore, the length of each coil is determined mainly by available power supplies and the maximum length allowed for each line of coils. In most non-ferrous installations, the billet dimensions are larger than those of steel billets, and each heating unit has only one coil, with no gap along its length. Therefore, by adding 25–75 mm to the minimum work-length, the coil length can be estimated closely enough for electrical calculation purposes.

From Table 22.5, the recommended air gap is 12 mm. Therefore, the inner diameter of the coil windings is

$$d_c = d_w + 24 = 174\,\text{mm}$$

The coil length is approximately 50 mm longer than the total work-length, i.e.

$$l_c = 2000 + 50 = 2050\,\text{mm}$$

Table 22.5 *Recommended air gaps (mm) for through-heating coils*

Frequency	Billet temperature (°C)	Work diameter (mm)		
		0–60	60–125	125–250
50/60 Hz	550	12	12	12
	850	12	20	40
	1250	25	25	40
1 kHz	850	25	30	38
	1250	30	38	40
3 kHz	850	25		
	1250	30		
10 kHz	850	25		
	1250	30		
450 kHz	1250	3 to 12		

22.8.2 Coil design for ferromagnetic billets

This second example is similar to that in Section 22.8.1, but introduces the problem of ferromagnetic material.

Metal alloy and composition	0.44% carbon steel
Production rate	0.85 t/h
Temperature rise, $\Delta\theta$	1230 K
Maximum temperature differential, $\theta_s - \theta_c$	83 K
Mass per billet, M	2.6 kg
Billet length, l_w	100 mm
Billet diameter, d_w	65 mm
Metal density, γ	7850 kg/m^3
Metal resistivity between 20 and 760°C, ρ_m	0.497 $\mu\Omega$ m
Metal resistivity between 760 and 1250°C, ρ_m	1.12 $\mu\Omega$ m
Metal kWh per tonne between 20 and 1250°C	235
Metal kWh per tonne between 20 and 760°C	140

For notes on resistivity and heat content (kWh/t), see Section 22.8.1.

Steel has a much higher tensile strength than aluminium, so the stress-cracking criterion is less likely to apply. It is *assumed* that this is not the constraint in this application. If the surface is overheated to the point where it becomes tacky (say, 1300°C), there will be a tendency for the billets to stick ('sticky billets'). Also, no point in the billet should exceed 1400°C. (Many operators prefer that the surface temperature does not exceed 1250°C.) Within these constraints, the billet is heated as quickly as possible by applying the maximum practical magnetic field intensity H at the surface over the heating period, taking into account the soaking condition during the Curie transition and the actual soaking coil.

Frequency f and field intensity H: From Table 22.3, for this diameter, a suitable combination of frequencies is 50 Hz below the Curie point and 1000 Hz above it. This is a typical dual-frequency application. The magnetic field intensity H is again the starting-point for the coil design (see Section 22.9). For the purpose of this example, take

$$H_{s\,20\text{–}760} = 200\,\text{kA/m (RMS)}$$

$$H_{s\,760\text{–}1250} = 110\,\text{kA/m (RMS)}$$

Thermal power P_t: The total power needed to heat the load to 1250°C is

$$P_{t\,1250} = \text{production rate} \times \text{specific energy requirement}$$

$$= 0.85 \times 235 = 200\,\text{kW}$$

Similarly,

$$P_{760} = 0.85 \times 140 = 119\,\text{kW}$$

So the power at 1000 Hz (above Curie) = 200 − 119 = 81 kW.

Power density and heating time below the Curie temperature (20 to 760°C): At $H_s = 200\,000$ A/m, $\mu_r = 9.4$ (using the RMS value of H in Figs 13.1 or 7.3), and

$$\delta = \sqrt{\left[\frac{2\rho_m}{\mu_r\mu_0\omega}\right]} = \sqrt{\left[\frac{2(49.7 \times 10^{-8})}{(9.4)(4\pi \times 10^{-7})(2\pi \times 50)}\right]} = 0.0164 \quad \text{m}$$

Hence

$$d/\delta = 0.065/0.0164 = 3.97$$

From Fig. 9.4, $p = 0.36$ for $d/\delta = 4$ and, from eqn 9.18,

$$P_w = 2\mu\pi f H_s^2(l_w A_w)p$$

$$= 2(9.4)(4\pi \times 10^{-7})(\pi)(50)(\sqrt{2} \times 2 \times 10^5)^2 l_w(\pi 0.065^2/4)0.36$$

$$= 0.18 l_w \quad \text{MW}$$

Therefore

$$\mathscr{P} = 0.18/\pi d_w = 0.88\,\text{MW/m}^2$$

It is recommended in Chapter 13 that this figure be multiplied by 1.47 to allow for the non-linear theory, giving 1.29 MW/m².

The surface power loss by radiation, $\mathscr{P}_{Rm}$ is obtained from eqn 17.14:

$$\mathscr{P}_{Rm} = \frac{5.67 \times 10^{-8}\varepsilon}{5}\left(\frac{T_2^5 - T_1^5}{T_2 - T_1}\right)$$

Up to the Curie temperature, $T_2 = 760 + 273 = 1033$ K, $T_1 = 20 + 273 = 293$ K and $\varepsilon = 0.8$ (see Fig. 22.6). Hence

$$\mathscr{P}_{Rm\,20\text{–}760} = \frac{(5.67 \times 10^{-8})(0.8)}{5}\left(\frac{1033^5 - 293^5}{1033 - 293}\right)$$

$$= 14\,400\,\text{W/m}^2 = 14.4\,\text{kW/m}^2$$

Since well-insulated coils are used for heating steel, this can be halved in practice to 7 kW/m², which is negligible compared with 1290 kW/m².

From eqn 15.2c, the heating time (up to the Curie temperature) is

$$\frac{(c\gamma)R}{2\mathscr{P}}\theta = \frac{(4.9 \times 10^6)(0.065/2)}{(2)(1.28 \times 10^6)}(740) = 46.0\,\text{s}$$

Power density and transition time, at Curie point (760°C): At the Curie point, the relative permeability falls to unity, and further heating is done at 1 kHz.

Energy has to be put into the steel without change of temperature.

Given that $H_s = 110\,\text{kA/m}$, $\mu_r = 1$, and $\rho_{760} = 112 \times 10^{-8}\,\Omega\,\text{m}$,

$$\delta = \sqrt{\left[\frac{(2)(112 \times 10^{-8})}{(4\pi \times 10^{-7})(2\pi \times 1000)}\right)} = 0.017 \quad \text{m}$$

$$d/\delta = 0.065/0.017 = 3.89$$

$$p = 0.37$$

$$P_w = 2(4\pi \times 10^{-7})(\pi)(1000)(1.1 \times 10^5)^2 l_w(\pi 0.065^2/4)\,0.37\,\text{W}$$

$$= 0.117 l_w \quad \text{MW}$$

$$\mathscr{P} = 0.117/\pi d_w = 0.57\,\text{MW/m}^2$$

From Fig. 22.3, the energy needed to go through the Curie point is 17.7 kWh/t or 500 MW s/m^3. Hence

$$\text{transition time} = \frac{(\text{transition energy})R}{2\mathscr{P}} = \frac{(500 \times 10^6)(0.065/2)}{(2)(570\,000)}$$

$$= 14.3\,\text{s}$$

This is quite long compared with the time taken for the billet to get to Curie temperature, and can be regarded as a soaking period for the levelling out of temperature. The actual process of passing through the Curie point is more complicated than this, remembering that eqn 15.2c is an averaged equation. The actual temperature–time profile could be worked out by taking the profile just as the Curie point was reached and applying the theory of Chapter 16 to a soaking period. Superimposed on this would be the input of power at the Curie temperature level, which would create its own temperature–time profile, and the sum of the two would give the resultant (except that, when the surface reaches the Curie temperature, not all the billet is non-magnetic). It really needs a computer to deal with this very complex problem and programs exist for this at the Electricity Council Research Centre, Capenhurst.[47]

Transition time, allowing for radiation: Assume that the billet is radiating to 660°C, i.e. 100°C lower than Curie temperature. Then from eqn 17.2,

$$\mathscr{P}_R = 5.67 \times 10^{-8}\varepsilon(T_s^4 - T_a^4)$$

$$= (5.67 \times 10^{-8})(0.8)(1033^4 - 933^4) = 17\,\text{kW/m}^2$$

This is roughly 3% of the induced power, so that the transition time will be increased to 14.7 s.

Power density and heating time above Curie temperature (760 to 1250°C): Here, the power input will vary slightly with temperature due to the small change in ρ, but the value of p is relatively insensitive to a small change in d/δ, so it is reasonable to assume that the power input over the higher temperature range stays constant at 0.57 MW/m^2.

Radiation changes over the temperature range, and the method of Section 17.1 must be used. From eqn 17.5,

$$\mathscr{P} = CT_{\mathscr{P}}^4$$

$$T_{\mathscr{P}} = \sqrt[4]{\left[\frac{(0.57 \times 10^6)}{(5.67 \times 10^{-8})(0.8)}\right]} = 1883\,\text{K}$$

From eqn 17.8,

$$\Theta_1 = \frac{T_1}{T_{\mathscr{P}}} = \frac{760 + 273}{1883} = 0.55$$

$$\Theta_2 = \frac{T_2}{T_{\mathscr{P}}} = \frac{1250 + 273}{1883} = 0.81$$

From eqn 17.7,

$$f(\Theta_1) = 0.58 \qquad f(\Theta_2) = 0.9$$

Then by eqn 17.9,

$$t = \frac{(c\gamma)R}{2CT_{\mathscr{P}}^3}[f(\Theta_2) - f(\Theta_1)]$$

$$= \frac{(5.6 \times 10^6)(0.065/2)}{2(5.67 \times 10^{-8})(0.8)(1883)^3}(0.9 - 0.58) = 96.2\,\text{s}$$

Total heating time T:

$$T = 46.0 + 14.7 + 96.2 \simeq 157\,\text{s}$$

Coil length l_c and coil diameter d_c: The linear speed of the billet through the coil is fixed by the production rate (in kilograms per second):

$$v = \frac{\text{production rate}}{\pi R^2 \gamma}$$

$$= \left(\frac{0.85 \times 1000}{3600}\right)\frac{1}{(\pi)(0.065/2)^2(7850)} = 0.009 \quad \text{m/s}$$

Active length of 50 Hz coil $= vt_1 = 0.009 \times 46 = 0.41\,\text{m}$
Active length of 1 kHz coil $= v(t_2 + t_3) = 0.009(15 + 96) = 1.0\,\text{m}$

The specified billet length (0.1 m) is small compared with these, so they need not be adjusted for exact multiples of billet length. In practice, these lengths would be increased by a distance roughly equal to the diametral clearance. From Table 22.5, these are 40 and 76 mm, respectively, below and above the Curie point.

Overall length of 50 Hz coil	$l_{c1} = 0.41 + 0.04 = 0.45\,\text{m}$
Overall length of 1 kHz coil	$l_{c2} = 1.0 + 0.08 = 1.08\,\text{m}$
Internal diameter of 50 Hz coil	$d_{c1} = 65 + 40 = 105\,\text{mm}$
Internal diameter of 1 Hz coil	$d_{c2} = 65 + 76 = 141\,\text{mm}$

Work power P_w: This is given by the appropriate power density $\mathscr{P}$, multiplied by the surface area of the billets in the active part of the coil:

Work power of 50 Hz coil P_{w1}	$= (1.29 \times 10^3)\,(0.41)(\pi \times 0.065)$	$= 109\,\text{kW}$
Work power of 1 kHz coil P_{w2}	$= (0.57 \times 10^3)\,(1.0)(\pi \times 0.065)$	$= 116\,\text{kW}$
Total work power		$= 225\,\text{kW}$

This gives (225/0.85) = 265 kWh/t; cf. 235 kWh/t from Fig. 22.3, which does not allow for radiation loss.

This coil design method gives good results in practice. With a computer, it is easy to set up the expressions to explore the effects of changes in the specification and/or applied field. The results of one such study are shown in Fig. 22.11, where changing the applied field is explored; the curves show clearly that, for diameters > 150 mm, there is little benefit in using $H > 1\,\text{MA/m}$. In fact, half this value is adequate for most applications; the small penalty in greater heating time, or longer coils, is offset by simpler coils and better control. We shall return to this figure in Section 22.8.3.

22.8.3 Semi-empirical approximations

From eqn 15.2c,

$$\mathscr{P}_s = \frac{(c\gamma)R}{2t_h}\,\theta_m$$

which gives the heating time

$$t_h = (c\gamma)R\theta_m/2\mathscr{P}_s$$

From eqn 18.33,

$$\theta_s - \theta_c = \frac{\mathscr{P}_s R}{2k}\,F(k_2)$$

For optimum heating, $d/\delta \simeq 3.5$, for which $F(k_2) = 0.52$, say 0.5.

From eqn 18.33, $\mathscr{P}_s$ can be obtained for a given $\theta_s - \theta_c$ and substituted in eqn 15.2c, giving a heating time t_h (in seconds) of

$$t_h = \left(\frac{c\gamma}{k}\right)\frac{R^2}{8}\left(\frac{\theta_m}{\theta_s - \theta_c}\right)$$

$$= \frac{1}{32}\left(\frac{c\gamma}{k}\right)\left(\frac{\theta_m}{\theta_s - \theta_c}\right)d^2 \qquad (22.9)$$

This suggests that the heating time t_h is proportional to d^2, if the rest of eqn 22.9 is constant. The dependence on d^2 is verified in practice but is obviously more complex than this, since the analysis of Section 22.8.2 shows that the solution

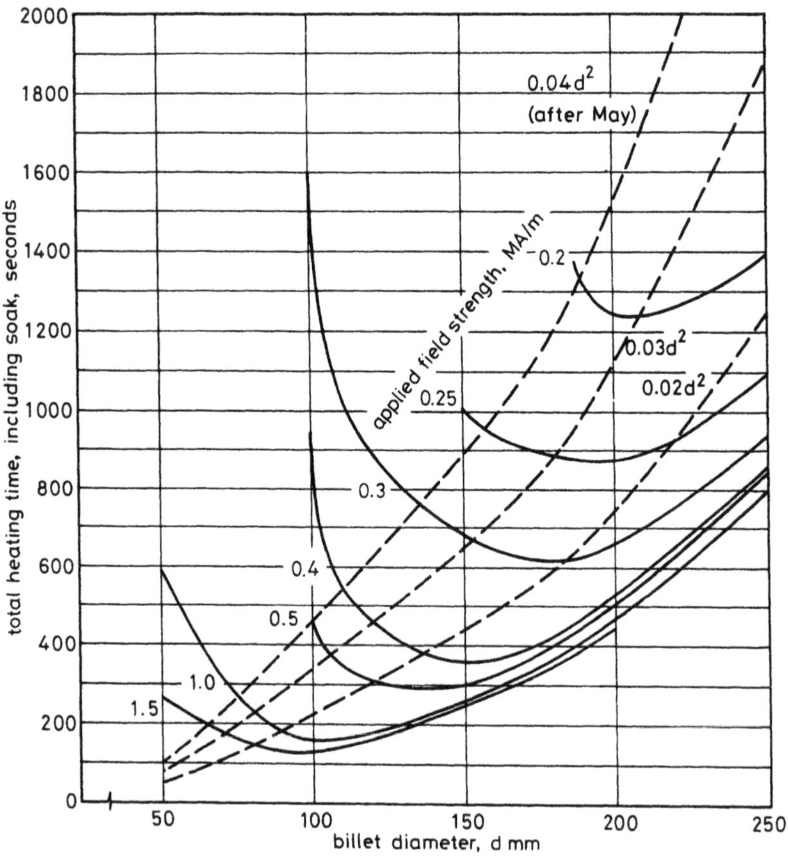

Fig. 22.11 *Computed total time to heat EN3 steel billets of various diameters to a mean temperature of 1250°C at 50 Hz for various field strengths in MA/m. The final surface-to-centre temperature difference is specified as <50 K. The dotted curves are discussed in Section 22.8.3 (Private communication)*

has to include complications imposed by the Curie transition and multiple power densities. For a given material, θ_m is known but $\theta_s - \theta_c$ is open to choice and the 'constant' $(c\gamma/k)$ is variable. So it is not profitable to substitute such values into eqn 22.9. Instead, we shall quote practical values for the constant part of the expression, but retain the d^2 part. May[38] gives a practical value of the constant of proportionality for steel for billets in the expression

$$t_h = d^2/25 \tag{22.10}$$

where d is the diameter of billet (mm). May[38] gives a similar expression for square bars as

$$t_h = b^2/20 \tag{22.11}$$

where b is the length of side (mm). (This implies that the RMS value of the distance from the centre is $1.13b$.)

Figure 22.11, which is based on a full analysis, shows May's $0.04d^2$ assumption, together with $0.03d^2$ and $0.02d^2$, both of which fit better over parts of the curves.

May[38] also gives an expression for power input per unit length:

$$P = 126 + 0.25d \tag{22.12}$$

in kilowatts per metre of billet length. This expression allows for radiation. (It is not clear what assumptions May made about temperature range, but it is believed that steel to 1250°C is implied.)

Example 22.4
Use May's approximations for the coil derived in Section 22.8.2.

Solution
From eqn 22.10,

$$t_h = (65)^2/25 = 169\,\text{s} \qquad \text{(cf. 158)}$$

From eqn 22.12,

$$P = 126 + (0.25 \times 65) = 142\ \text{kW/m}$$

The active heating coils are 1.63 m long in total, giving $(1.63 \times 142) = 231$ kW, cf. 228 kW calculated. These are seen to be useful 'empirical' equations.

Table 22.6 gives similar approximate heating times for other materials. It should be emphasised that the multipliers in Table 22.6 must be used with skill and modified in the light of experience. For example, Sundberg[30] gives measured

Table 22.6 *Approximate billet heating times*

Material	t_h (s)
Aluminium	$5.4 \times 10^{-3} d^2$
Copper	$6.2 \times 10^{-3} d^2$
Brass	$7.8 \times 10^{-3} d^2$
Steel	$40 \times 10^{-3} d^2$

For square sections use $d = 1.13b$, where b is the length of side in millimetres. d is given in millimetres.

Table 22.7 *Relation between coil length and throughput*

Material	Coil length (m)
Aluminium	$0.71T$
Copper	$0.25T$
Brass	$0.33T$
Steel	$1.8T$

T = production rate (t/h).
This table is empirical only because it uses the value of t_h from Table 22.6.

curves for the heating of steel over a range of diameters from 20 to 80 mm which suggest that t_h lies between $d^2/33$ and $d^2/40$, for 1 and 4 kHz, with a maximum temperature of 1250/1270°C and a centre temperature > 1060°C.

Using the heating times given above, a relationship can be found between coil length and throughput, since both heating time and billet weight are proportional to d^2. In section 22.8.2 it was shown that

$$\text{active coil length} = t_h \frac{\text{production rate (kg/s)}}{\pi R^2 \gamma}$$

$$= (\text{constant})\, d^2\, 10^{16} \left(\frac{T \times 1000}{3600} \right) \frac{4}{\pi d^2 \gamma}$$

$$= \left(\frac{0.35 \text{ constant}}{\gamma} 10^6 \right) T \quad \text{metres}$$

where T is the production rate (t/h) and γ is the density (kg/m^3), and the constant is obtained from Table 22.6. For example, for steel, $\gamma = 7850$ kg/m^3 and the constant $= 40/10^3$; so the coil length $= 1.80T$.

Table 22.7 shows this and other values for the relation between coil length and production rate.

22.9 Equivalent circuit coil-design method

As there is a close analogy between induction heating and the transformer, the coil can be designed by obtaining the values of the resistances and reactances and solving the equivalent circuit. The coil is usually a single-layer winding, water-cooled to remove the I^2R losses and the heat losses arriving from the workpiece.

Starting from the magnetic field strength H_s needed at the work surface to produce the heat, there are three components of flux, as shown in Fig. 22.12a:

Φ_w flux linking the workpiece

Φ_g flux in the air gap

Φ_c flux passing through the coil surface

Workpiece flux Φ_w: In Chapter 9, the power loss (in watts) and flux in the workpiece (in webers) are related to the RMS value of the applied surface H_s by the p and q expressions in eqns 9.18 and 9.8:

$$P_w = 2\mu\pi f(\text{billet volume})H_s^2 p$$

$$\Phi_w = \mu H_s A_w(q - jp)$$

where P_w is the workpiece power (W), H_s is the RMS surface magnetic field strength (A/m), Φ_w is the RMS value of the total flux inside the workpiece (Wb), A_w is the cross-section (in m²) of the workpiece, and p and q are numerics from Fig. 9.4.

Using these equations, the flux needed for a given P_w can be calculated.

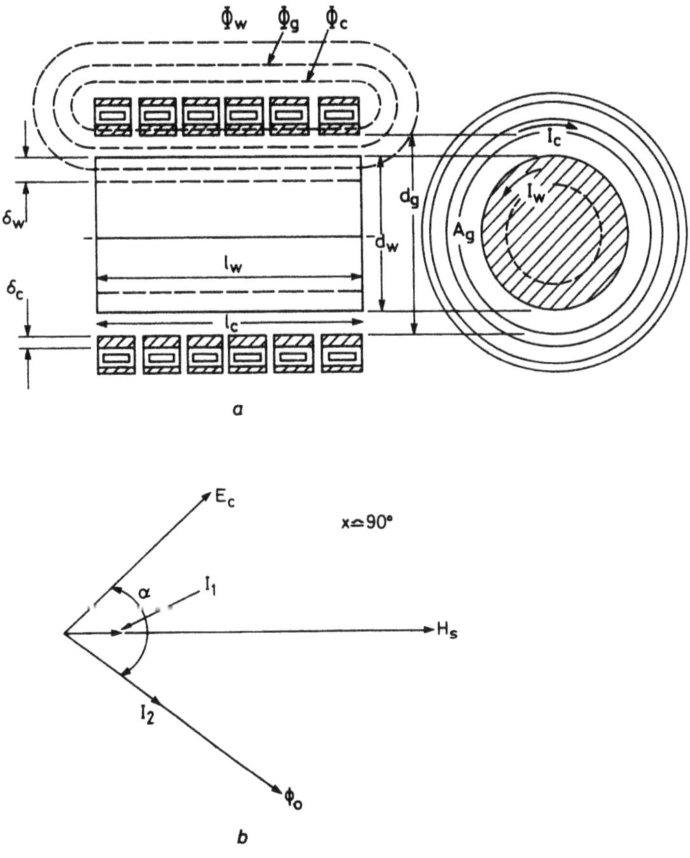

Fig. 22.12 *(a) Flux paths and (b) phasor diagram for a solenoid coil*

Air-gap flux Φ_g: The air gap in induction-heating coils is normally large to allow for the thermal insulation and mechanical clearance (Table 22.5), so this flux is substantial. Since it is reactive, it lowers the coil power-factor.

$$\Phi_g = \mu_0 H_s A_g \tag{22.13}$$

Coil flux Φ_c*:* Skin-effect operates on the coil, just as it does on the workpiece, forcing the coil current to flow on the inner surface of the copper. Since the coil is non-ferrous and normally electrically thick, there is a 45° phase shift. It is easy to rearrange eqn 7.10 in the form

$$\Phi = \frac{\mu_0 H_s}{\alpha\sqrt{2}} \frac{1 - j}{\sqrt{2}} = \frac{\mu_0 H_s \delta_c}{2}(1 - j)$$

The 'width' by which this must be multiplied is the inner periphery of the coil πd_c, giving

$$\Phi_c = k_r \frac{\mu_0 \delta_c (\pi d_c)}{2} H_s (1 - j) \tag{22.14}$$

where k_r is a correction factor, allowing for the spacing between the turns, etc. (k_r lies between 1 and 1.5, with 1.15 as a typical value).

The total RMS flux is the sum of these components:

$$\Phi_0 = \Phi_w + \Phi_g + \Phi_c$$

Note that this is a phasor addition; see Fig. 22.12b. Collecting terms,

$$\Phi_0 = \mu_0 H_s \left[\left(A_g + \mu_r q A_w + k_r \frac{\pi d_c \delta_c}{2}\right) - j\left(\mu_r p A_w + k_r \frac{\pi d_c \delta_c}{2}\right)\right] \tag{22.15}$$

If the coil is short, and has no external magnetic circuit to provide an easy path for the flux, then some magnetomotive force (MMF) is needed to drive this flux through the external reluctance. Assume that the coil is either long, or has a ferromagnetic external circuit, and that all the coil MMF is used to make Φ_0. Therefore

$$H_s = I_c N_c / l_c \tag{22.16}$$

and the RMS coil voltage is

$$E_c = j\,4.44 f N_c \Phi_{0m} = j\frac{2\pi}{\sqrt{2}} f N_c (\sqrt{2}) \Phi_0 \tag{22.17}$$

where Φ_{0m} is the peak value of the flux. Substituting eqns 22.15 and 22.16 in 22.17,

$$E_c = \frac{2\pi f \mu_0 N_c^2}{l_c} I_c \left[\left(\mu_r p A_w + \frac{k_r \pi d_c \delta_c}{2}\right) + j\left(A_g + \mu_r q A_w + \frac{k_r \pi d_c \delta_c}{2}\right)\right] \tag{22.18}$$

It has been common practice, following Baker,[13] to represent this by a series electrical equivalent circuit. Baker represented the theory above by an equivalent circuit which closely resembled that normally used for the transformer, where the resistance of the workpiece and the reactances of the workpiece, air gap and coil (R_w, X_w, X_g and X_c respectively) are all in series, and connected in parallel with an ill-defined quantity X_e, representing the impedance of the end-effects. This is manifestly wrong. He then simplified this circuit to the simple series circuit.

The assumption used in deriving the theory above is that all paths share a common magnetic field intensity H_s. This is a debatable assumption, as it implies that there is close coupling of the work with the coil, which may not be true in practice, and that all the flux defined as Φ_c links all the coil turns; but it is a reasonable engineering assumption, giving good results in practice.

Equation 22.18 can be represented by the parallel magnetic equivalent circuit of Fig. 22.13a; it can also, in an artificial way, be represented by the simple series

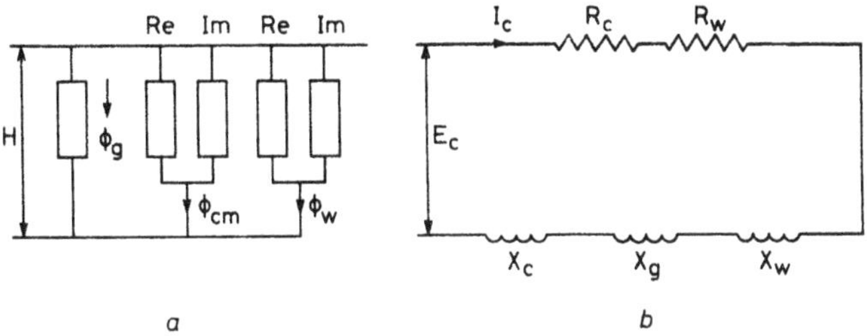

Fig. 22.13 *(a) Parallel magnetic and (b) series equivalent circuits*

electrical equivalent circuit of Fig. 22.13b, but this must be interpreted carefully. The coil-design method of this section implies acceptance of Baker's series circuit and should be used with caution.

The impedance corresponding to eqn 22.18 is given by

$$Z = (R_w + R_c) + j(X_g + X_w + X_c) \qquad (22.19)$$

where

$$\text{work resistance } R_w = K(\mu_r p A_w) \quad \text{ohms} \qquad (22.20)$$

$$\text{coil resistance } R_c = K\left(\frac{k_r \pi d_c \delta_c}{2}\right) \quad \text{ohms} \qquad (22.21)$$

$$\text{gap reactance } X_g = K(A_g) \quad \text{ohms} \qquad (22.22)$$

$$\text{work reactance } X_w = K(\mu_r q A_w) \quad \text{ohms} \qquad (22.23)$$

$$\text{coil reactance } X_c = K\left(\frac{k_r \pi d_c \delta_c}{2}\right) \quad \text{ohms} \qquad (22.24)$$

$$K = 2\pi f \mu_0 N_c^2 / l_c \quad \text{ohms/m}^2 \qquad (22.25)$$

These values can now be used to calculate the major coil properties, i.e.

efficiency, the coil power-factor and kVA, the volts per turn and the number of ampere-turns:

$$\text{coil efficiency } \eta = \frac{R_w}{R_c + R_w} \tag{22.26}$$

$$\text{coil power factor } \cos\phi = \frac{R_w + R_c}{Z}$$

where

$$Z^2 = (R_w + R_c)^2 + (X_g + X_w + X_c)^2 \tag{22.28}$$

$$\text{coil power } P = P_w/\eta \tag{22.29}$$

$$\text{coil VA} = P/\cos\phi = I_c^2 Z \tag{22.30}$$

$$\text{coil volts per turn } \frac{E_c}{N_c} = \frac{\text{coil VA}}{\text{total ampere-turns}} = \frac{\text{coil VA}}{H_0 l_c} \tag{22.31}$$

$$\text{coil ampere-turns } I_c N_c = H_0 l_c \tag{22.32}$$

I_c and E_c are RMS quantities. The design of the coil is perhaps best understood by an example.

22.9.1 Coil design for ferromagnetic billet: equivalent circuit method

Taking the data of Section 22.8.2, for the coil used to heat above the Curie point:

Total work power, P_w	116 kW
Temperature rise, $\Delta\theta$	760 to 1250°C
Work diameter, d_w	65 mm
Work resistivity over $\Delta\theta$, ρ_w	1.17 $\mu\Omega$ m
Coil length, l_c	1.10 m
Coil inner diameter, d_c	140 mm
Frequency, f	1000 Hz
Preferred coil voltages, E_c	380 or 760 V
Work relative permeability, μ_w	1.0
Coil correction factor, k_r (assumed)	1.5
Coil resistivity, ρ_c	0.02 $\mu\Omega$ m

p and q factors: From Section 22.8.2, $d_w/\delta = 3.9$, for which $q = 0.56$, $p = 0.37$ from Fig. 9.4.

Equivalent circuit components (eqns 22.20 to 22.25):

$$R_w = K\left(1 \times 0.37 \times \frac{\pi}{4}\frac{65^2}{10^6}\right) = (1.23 \times 10^{-3})K$$

$$R_c = K\left(\frac{1.5}{2}\pi\frac{140}{10^3}\frac{2.25}{10^3}\right) = (7.42 \times 10^{-4})K$$

$$X_g = K\frac{\pi}{4}\left[\left(\frac{140}{10^3}\right)^2 - \left(\frac{65}{10^3}\right)^2\right] = (1.21 \times 10^{-2})K$$

$$X_w = K\left(1 \times 0.56 \times \frac{\pi}{4}\frac{65^2}{10^6}\right) = (1.86 \times 10^{-3})K$$

$$X_c = R_c = (7.42 \times 10^{-4})K$$

$$K = \frac{2\pi 1000}{1.10}\left(\frac{4\pi}{10^7}\right)N_c^2 = (7.18 \times 10^{-3})N_c^2$$

In the above,

$$\delta_c = 0.017\sqrt{(2/117)} = 2.25 \times 10^{-3}$$

These component values can only be evaluated (where necessary) later, when N_c has been decided.

Coil efficiency, power factor and coil VA (eqns 22.26 to 22.30):

$$\text{coil efficiency } \eta = \frac{1.23}{1.97} = 0.62\ (62\%)$$

$$\text{coil power factor } \cos\phi = \frac{(1.97 \times 10^{-3})}{\sqrt{[(1.97 \times 10^{-3})^2 + (14.70 \times 10^{-3})^2]}} = 0.13$$

$$\text{coil VA} = \frac{119}{0.62 \times 0.13} = 1476\,\text{kVA}$$

Coil volts per turn and ampere-turns (eqns 22.31 and 22.31): First,

$$\frac{Z}{N_c^2} = \frac{7.18}{10^3}\sqrt{\left[\left(\frac{1.97}{10^3}\right)^2 + \left(\frac{14.70}{10^3}\right)^2\right]} = \frac{1.07}{10^4}$$

Then

$$\text{coil volts per turn} = \frac{E_c}{N_c} = \sqrt{\left[\text{VA} \times \frac{Z}{N_c^2}\right]}$$

$$= \sqrt{\left[(1476 \times 10^3)\frac{(1.07)}{(10^4)}\right]} = 12.53$$

$$\text{ampere-turns} = I_c N_c = \sqrt{\left[\frac{\text{VA}}{(Z/N_c^2)}\right]}$$

$$= \sqrt{\left[\frac{(1476 \times 10^3)}{1.07/10^4}\right]} = 117\,400$$

This agrees well with the value taken for H in Section 22.8.2.

Coil turns, current, and volts: Using a coil voltage of 380 V (obtained from 400 V at the generator, less 5% for transmission drop),

$$N_c = \frac{\text{volts}}{\text{volts per turn}} = \frac{380}{12.53} = 29.2 \simeq 30 \text{ turns}$$

With a coil length of 1.1 m, the pitch is 37 mm. In medium-frequency coils, it is usual to allow space for air-insulation; therefore use 30 mm tubing, with 7 mm spacing.

The coil current is

$$I_c = \frac{117\,400}{30} = 3915\,\text{A}$$

As the depth of penetration in copper at 1 kHz is only 2.3 mm, there is no point in using solid copper conductors. It can be shown that a wall thickness of approximately 2.5 mm can be used, giving an effective cooling tube.

Coil loss and cooling water: The heat to be removed from the coil copper is obtained by subtracting the useful work power from the total power input. This assumes that all the radiated power is absorbed by the coil. Thus

$$\text{coil loss} = \frac{\text{power into work}}{\eta} - \text{thermal power}$$

$$= \frac{116}{0.62} - 81 = 187 - 81 = 106 \quad \text{kW}$$

That is, more power has to be removed from the coil than appears as useful heat because of radiation (see Section 22.8.2), and the overall efficiency is 43%. The cooling-water quantity $Q_{\text{water}} = 0.008 \times 106 = 0.85$ litre/s, which assumes that the water temperature rise is 30°C. (In practice, a more serious cooling calculation is obviously needed later; see Chapter 20.)

It is interesting to examine the constituent parts of this example:

- R_w and R_c are of the same order of magnitude. This means that nearly half the useful power is dissipated in the coil copper loss; hence the 62% efficiency.
- By far the largest component of reactance is that of the air gap between the

coil and the workpiece. This shows how important it is to keep this clearance to the minimum, consistent with heat insulation and mechanical clearance. It also shows the detrimental effect of using an inefficient coil, instead of changing the coil when the work size changes.

- One effect of the large gap reactance is to reduce the power factor. The natural power factor of coil and load under non-magnetic conditions is 0.6 to 0.7. It is solely the gap reactance that reduces it to 0.13.
- The other major effect of gap reactance is to increase the number of volts per turn. In this example, if the power factor were increased to 0.5, the coil kVA would be reduced by 4, so that the volts per turn would be reduced by $\sqrt{4}$ to about 6.5.

22.10 Mechanical construction of coils

The basic purpose of a through-heating coil is to couple closely with the workpiece, consistent with loading and unloading clearances, adequate thermal and electrical insulation, and mechanical rigidity. The coil must be carefully engineered, as it works in a hostile environment and yet has to be highly reliable. Most coils are designed to be integrated with the work-handling equipment and matching units, such as transformers, capacitors, etc. The coil layout, after the main parameters have been decided, must take into account the water system, vibration and the effect of stray fields.

High-conductivity copper is always used as the conductor material and the coils are always water-cooled, using hollow tubes. Small heaters can be made with the copper coil coated with nylon or PVC in a fluidised bed and then cast in a refractory cement. Water-cooled stainless-steel rails, faced with a hard stellite coating, run axially through the air gap to prevent wear on the concrete. Although simple, cheap, and robust, this construction does not have optimum thermal or electrical insulation. The concrete is subject to cracking due to thermal shock and, when the coil is due to be rebuilt, it is difficult to recover the coil without damage.

At supply frequencies, the depth of conductor is normally 12 mm, remembering that the skin depth of cold copper at 50/60 Hz is 10 mm. The coil is made of tube, or may be of square section with an integrally-extruded water tube (D-section). In the past, compound conductors have been fabricated by joining a square conductor and the cooling tube by brazing with silver-solder, but this is expensive and less reliable than integral tubing. The coils are wound on a mandrel, using a large lathe. After forming, the coils are cleaned preparatory to receiving the insulation. Figure 22.14 shows a coil for a billet heater.

Even though the coil is water-cooled, and hence constrained to a low temperature, the outer surface may be exposed to a much higher temperature, so Class H materials are desirable, although not strictly necessary. Modern high-temperature insulating materials are of high quality and the designer is faced with a wide choice. In the past, braided glass sleeving 1.5 mm thick was fitted over the conductor and then vacuum-impregnated with a high-temperature organic varnish, but this is no longer necessary. Silicone-impregnated glass tape can be used, either wound half-lap around the copper, or used as sleeving.

Fig. 22.14 *A coil for a DFR 100 billet heater, nylon-coated prior to being bound with asbestos tape. The coil is mounted in the coil box, the water and electrical connections are made, and the whole is set in a refractory compound (Newelco (Uskside) Ltd)*

Kapton taping can be used. One manufacturer sprays the coil with a high-temperature ceramic insulating material that combines adequate electrical insulation with excellent temperature qualities. During the manufacture of the coils, taps can be provided to allow various heat patterns to be obtained in the load. At supply frequency, the number of volts per turn is low, so the insulation between turns can be minimal. Also, the voltages used for through-heaters seldom exceed 1200 V and are usually much lower, so that the insulation level does not need to be high.

Nearly all through-heating coils use thermal insulation between the coil and the workpiece. This protects the winding and its insulation from the radiant heat and acts as a muffle around the work. Figure 22.15 shows some typical insulation methods used in through-heating coil construction for 50/60 Hz and 1, 3, and 10 kHz. Stainless-steel liners can only be used at supply frequency, as medium frequency would heat the liner; these liners are split lengthways to prevent circulating currents.

The refractory can be asbestos millboard up to 760°C. Above this temperature, however, high-temperature refractory felt must be used. Modern alumino-silicate ceramic materials,[39] such as Carborundum Fiberfrax, are suitable and are available in many forms. Replaceable skid plates of stainless steel can be used to reduce wear on the liner.

At higher frequencies and temperatures, only refractory materials can be

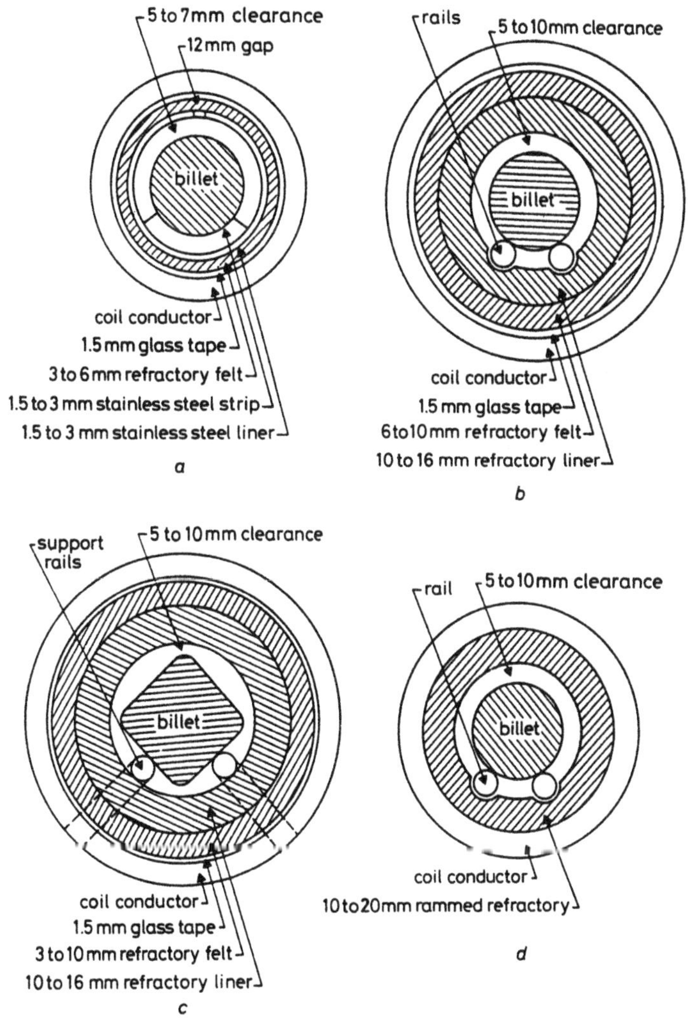

Fig. 22.15 *Through-heating coil construction: (a) 50/60 Hz coil insulation (875°C maximum) (b) high-temperature coil insulation (1250°C), all frequencies (c) heavy billet and round-cornered square billet coil insulation (1250°C) (d) high-temperature rammed refractory coil insulation (1250°C)*

used. The liners are either extruded to the required shape, as in Figs 22.15b and c, or rammed or cast, as in Fig. 22.15d. Skid rails, usually made of water-cooled stainless-steel tubing, are built into the liners, as shown; these protect the liners mechanically. They are normally set into recessed grooves in the liners and are often coated with a hard material at the wear-surface. Extruded or prefabricated liners must be limited in thickness, because of potential thermal and mechanical shock, so refractory felt is usually needed to achieve the necessary insulation thickness and to act as a shock absorber betwen the liner and the coil, which may be vibrating. An alternative method uses short semi-circular or circular firebrick sections (say, 100 mm long), grooved to take the rails. When these sections are coated on the inside with refractory cement, they appear to combine the advantages of rammed and extruded liners, since they are not as fragile as extruded liners and can have thicker walls. Silicon carbide can also be used instead of firebrick. No rails are then needed, because of the wear properties of silicon carbide, but it is more brittle than firebrick and does not stand excessive mechanical shock.

Cast or rammed refractory coils can be constructed using a mixture of mica and resin. The coil is then moulded directly into its housing; the method holds the coil rigidly and protects it from industrial atmospheres. A 3 mm wear plate backed by 0.5 mm of aluminium can be used to protect the refractory surface.

Figure 22.15c shows a type of support suitable for round-cornered square (RCS) or heavy billets. The rails, which are independent of the refractory liner, are supported by radial struts through the coil turns and the liners. If the coil is only to be used for low-frequency heating of square billets, the cross-section can be square, with a water-cooled stainless-steel liner.

The coil and windings are usually clamped between end-plates. It is important that the end-plates do not link with the main flux.

22.11 Slab reheating

In modern hot-strip rolling, continuously-cast slabs tend to be used in place of the previous ingot castings.[40] In 1965, McLouth Steel Corporation (USA) decided to use this method in their new mill. These slabs have to be reheated before passing to the rolling mill; they can arrive at the reheater cold, warm (200 to 500°C) or hot (straight from the continuous caster 500 to 700°C). They considered pusher-type reheat furnaces, but found that 2500 mm was the thickest slab that could be reheated, and walking-beam furnaces, but felt that these could not cope with the above mix. A roller-hearth furnace was rejected on grounds of space, high operating cost and the difficulty of mixing loads.

To meet this need by induction heating, a new concept was developed, by Ross[41] at Ajax Magnathermic, using a wide rectangular coil. The slabs are heated in a vertical position, the mechanical handling being arranged to lift the slabs up into the stationary coil. The concept is shown in Fig. 22.16a, where a wide rectangular coil is arranged around a slab; the actual heater is shown in section in Fig. 22.16b. Since slabs can be bowed, a substantial clearance must be allowed between the slab and the lining. This reduces the power factor.

The production criteria were:

- approximately 2.2 million tonnes per year
- 300 mm slab thickness
- slab widths of 0.9 m, 1.1 m, 1.3 m, and 1.5 m.

The following were also specified:

- an average production of 275 t/h for all slab widths and 500 t/h for the 1.5 m slab width
- the ability to mix cold slabs, warm slabs and hot slabs
- maximum flexibility to meet the performance of the continuous caster
- ease of computerisation
- maximum quality
- minimum space requirements, because of limited production area
- minimum operator needs.

The McLouth installation has been very successful and the requirements listed above have been met. There are six heating lines, each rated at 100 t/h when

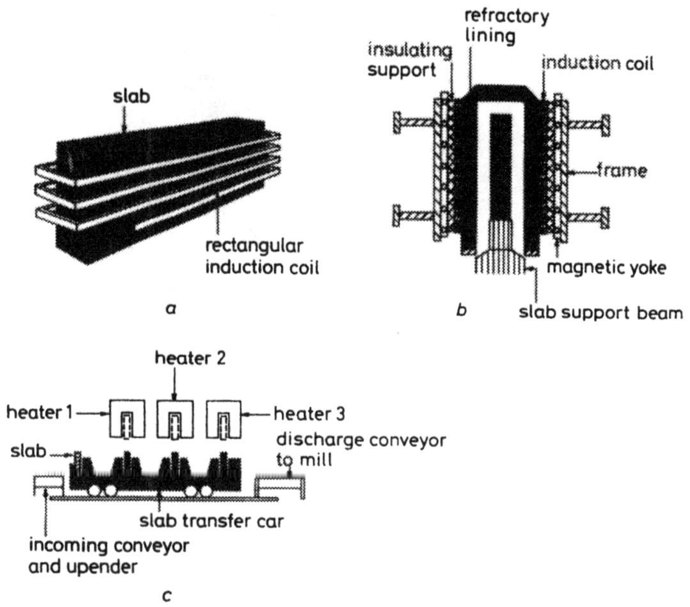

Fig. 22.16 *Single-shot vertical slab-heating mechanism (after Ajax Magnathermic Ltd)*

heating the largest slabs (300 mm thick × 1.5 m wide × 8 m long). There are three heaters per line, rated at 20, 10, and 5 MW, giving a total of 35 MW/line and 210 MW overall. Each of these figures is the maximum rating of the heaters, and it would be exceptional to have them all using maximum power at the same time.

Published data suggest that power consumption has been less than 358 kWh/t, regardless of slab size. Scale formation is less than 0.25% and is very loose. Temperature uniformity is good.

To accommodate coils of varying width, the coils are tapped. The power is switched by specially-designed SCR switches which turn the power on and off at the zero of the sine wave.

Figure 22.17 shows another kind of slab-heating coil.

Fig. 22.17 *Example of slab-heating coil construction suitable for ratings up to 5 MW. Slabs can be taken hot from a continuous casting machine or heated directly from cold (Inductoheat Banyard Ltd)*

22.12 Travelling-wave induction heating

In all the applications described so far, single-phase coils have been used. For higher powers, it is often possible to connect various coils to a three-phase supply in an attempt to balance the loading, but this is not a true three-phase load. Also, we have seen that, if coils are wound in several layers, the outer layers tend to heat the inner layers and the power factor of the external layers is bad because of the increasing air gap between the layers and the workpiece. Although punchings can be added outside the coils to channel the external flux, these do not really reduce the effective air gap.

To overcome these disadvantages, the present author proposed the use of travelling-wave induction heating[19, 42, 43] in many places where pulsating fields are now used.

The basic principles are best understood by reference to a traditional three-phase electrical machine. Windings in slots are arranged so that, when connected to a three-phase supply, they produce a sinusoidal distribution of MMF in space which rotates at a speed given by $n = f/p$, where n is the speed in rev/s, f is the supply frequency (Hz) and p is the number of pole pairs. It is normal to put a rotating member inside this stator, separated by a small mechanical gap, to produce mechanical power. If there is a 'slip' speed between the speed of the MMF wave and the mechanical speed of the rotor, then energy is 'lost' in the rotor. (For synchronous machines, $s = 0$ and there is no energy dissipation in the rotor; for induction motors, there is always a speed difference, so there is always a loss, which shows as an inefficiency of the motor.)

In the travelling-wave induction heater, use is made of this 'loss'. If the 'rotor' is held stationary and is metallic, the loss will heat the rotor; this can now be regarded as the workpiece, and the mechanism is induction heating.

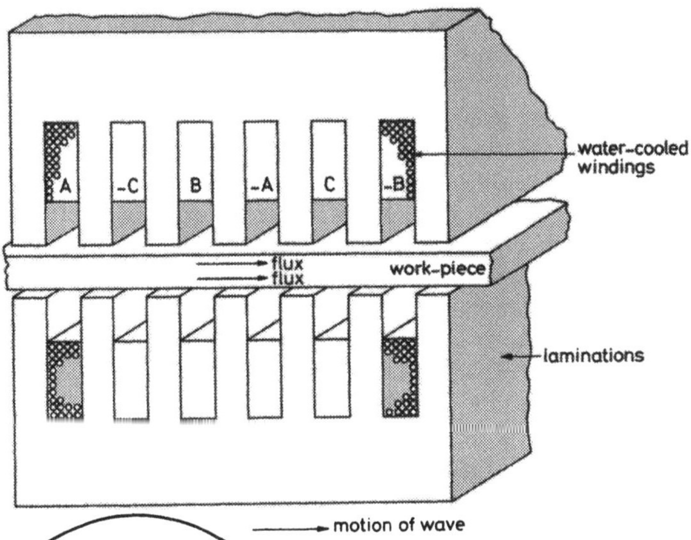

Fig. 22.18 *Travelling-wave induction heater*

In its cylindrical form, the method has uses in, for example, vessel heating (see Chapter 25) and billet heating, but it is most likely to find use in its plane form. It is then analogous to the linear motor. A two-pole travelling-wave induction heater is shown in Fig. 22.18. The air gap has a sinusoidal space distribution of flux which moves across the pole face at the speed given by $v = f\lambda$ (m/s), where

λ is the wavelength (in metres). As drawn, Fig. 22.18 shows the two sides of the slab being heated independently by useful fluxes which are predominantly along the surface of the slab.

It must be emphasised that the concept of travelling-wave induction heating is totally independent of 'transverse flux' or 'longitudinal flux'. If the material is electromagnetically 'thick', all the heating is by the longitudinal flux mechanism. If the material is 'thin', the transverse flux concept can be used by putting travelling-wave induction heating on both sides of the work and arranging that the MMFs of the opposite poles are aiding, so that the flux traverses the two gaps and the workpiece.

One major difference between pulsating-field and travelling-wave induction heaters is in 'end-effects'. These are discussed in detail in References 19, 42 and 43.

22.12.1 Advantages of travelling-wave induction heating

- A balanced polyphase load is presented to the supply. (If only one pole pair is used with a flat heater, this is not strictly true, because of the broken magnetic circuit.)
- The coils only lie in their own leakage flux, so the problem of multi-layer coils is much less. The same rules as apply in electrical machines have to be obeyed.
- The flux is led to the air gap by the laminated teeth, so that the air gap is naturally small, and the power factor is mainly controlled by the load.
- In the pulsating-field heater, the coil has to be placed near to the work to avoid excessive leakage flux and to couple the exciting coil closely with the work. As a result, it gets hot from the work. In the travelling-wave induction heater, the coils can be placed in the slots at a distance from the work surface without detrimental effect on the *main* flux, although the *leakage* flux is increased. It is easy to keep the coils and iron cool, using water.
- The construction provides a flux-carrying 'core' which gives mechanical strength and protection, as well as making an easy flux path for the main flux and preventing large leakage fluxes.
- The travelling-wave induction heater has a greater power density, because more copper can be put into the device and the iron circuit allows the MMF to be applied to the work surface.

22.12.2 Comparison of travelling-wave induction heating with the Ross coil

In Section 22.11, a brief description was given of slab reheating using the Ross coil, which is widely accepted as a valid method for this purpose. The travelling-wave induction heater described above will also do the same job, and so it is important to compare them.

Figure 22.19 shows diagrammatically how a large slab would be heated using travelling-wave induction heaters. The slab, which can be horizontal or vertical, would be held and the inductor elements, each of which would have at least four poles and preferably six poles, would be mechanically thrust against the slab. Only thermal insulation would separate the inductor from the slab surface, so that the air gap is minimal, for the best power factor. As many elements as are needed for the length are pushed forward and there should be provision for

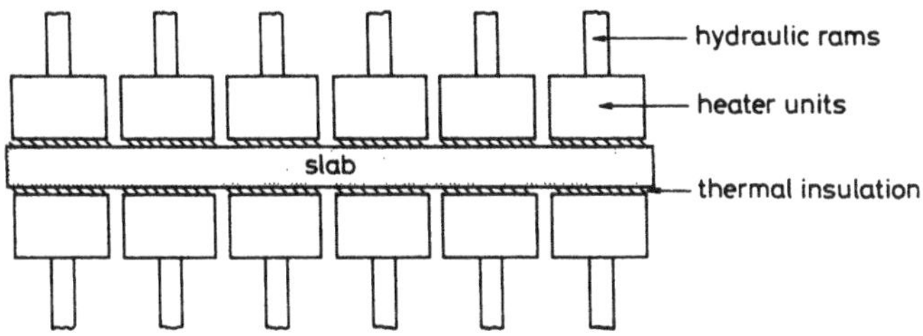

Fig. 22.19 *Slab heating using travelling-wave induction heaters*

distributing the uncovered area equitably between the elements: for example, if the slab is 8 m long and the inductors are 550 mm wide, then there will be fourteen full inductors per side and roughly 300 mm over, which could be distributed in 15 × 20 mm spaces along the length. In practice, the spaces could and should be fewer and coarser.

With relatively short inductors, bowed slabs would easily be accommodated, with little extra air gap. The slab would be heated from both sides.

Since travelling-wave induction heaters can have greater power density than Ross coils, because of the extra copper, they should do the same job better, i.e. reach temperature quicker with a better power factor and efficiency. Naturally, they can do nothing about the soak time needed to achieve uniform temperature through the work, as this depends only on thermal conductivity.

The Ross coil is more accommodating for variable width (changed by tapping); the travelling-wave heater is better for variable length, as the number of modules can easily be changed. The power factor of the latter will be better because of the shorter air gap.

Another practical advantage of the module system is that the reliability is improved. If one module out of the 28 calculated above fails, it is easy to fit a spare, compared with the single Ross coil.

22.12.3 Implementation of travelling-wave heaters

The author had a 1 MW prototype built under the auspices of BTG (NRDC). This was tested under university laboratory conditions, with scant provision for handling red-hot steel. The model worked, up to a power density of 1 MW/m^2, but the heating was not uniform, since the device had only two poles. EdF built a bigger prototype which was widely reported.[44] The reader is directed to this reference for further information but, as far as the author knows, the idea has not been taken further. Perhaps economic conditions have not been right for such developments; after all, the Ross coil has not been exploited elsewhere. The travelling-wave heater is not limited to large installations, however, and it is being used for the preheating of moulds. This, and other uses, is dealt with in Section 24.11.

22.13 Multi-layer coils

Using normal hollow tubular coils, it is not desirable to use multiple layers, as the outer layers, i.e. those further away from the workpiece, heat the inner layers as well as the work. Multiple layers are possible if the conductors are thin, compared with the skin depth, but this does not allow the use of inner-cooled conductors.

In Section 22.9, the equivalent resistance of the coil was derived with the implicit assumption that the conductors are electrically thick ($t \gg \delta_{copper}$). In fact, the theoretical minimum resistance, and hence loss, occurs in a single-layer winding when the winding thickness t equals $0.5\pi\delta_c$ and is about 9% less than the value given by eqn 22.21. Apart from supercooling the winding, which is manifestly impractical, the only way to reduce coil losses further is to use multiple layers of electrically-thin conductors ($t < \delta_c$).

The Electricity Council Research Centre (ECRC)[45,46] at Capenhurst has promoted this new concept of multi-layer windings for induction-heating coils, as a means of increasing the efficiency of the induction-heating process. Their original design was disk-wound with flat copper strip, insulated with polyimide film taping incorporating a fluorocarbon-resin sealant. The complete winding was contained in a sealed vessel through which demineralised water circulated to cool the winding at the edges of the strip.

The gain in efficiency, compared with a conventional single-layer coil, depends on the application. Economic designs, using 12 to 14 layers, can halve normal coil losses: when heating aluminium or copper alloy billets at 50 Hz for extrusion, the saving in energy usage is some 25–30% of the consumption of a single-layer coil.[48]

The ECRC concept has been adapted for commercial exploitation, using a modified construction.[102] To avoid the problems and expense of the cooling method originally proposed, they use hollow-section conductors. By careful choice of conductor section, very little efficiency is sacrificed, whilst both manufacture and maintenance are simplified.

This concept is a major contribution to billet-heating coil design and we can expect further improvements.

22.14 Tapered heating for extrusion

Induction heating is extensively used to prepare billets for extrusion, and there is no difference in principle between this and the heating described in the earlier part of this chapter. It is also possible to heat 'logs', i.e. very long billets, and to hold these at temperature until needed, when the required amount is cut by a hot shear.

The process of extrusion needs a soft 'nose' on the billet to reduce the starting force. Thereafter, energy is being added to the billet mechanically by the heat generated at the die. It is, therefore, desirable to 'taper-heat' the billet, so that the front is 100° to 120°C hotter than the rear. If this is not done, there is a risk of dimensional differences and metallurgical inconsistency between the ends of the finished product. (Note that it is also possible to use uniformly-heated billets and to vary the ram speed during the extrusion cycle, but this is tedious.)

The design of coils to produce taper will not be discussed in detail here. Since temperature rise is linearly proportional to power density and, for non-magnetic materials, power density is proportional to H^2, it is clear that the magnetic field has to vary as $\sqrt{(\theta_{front}/\theta_{rear})}$, where the θs are temperature rises. For an extrusion temperature of 500°C and a taper of 100°C, this represents roughly $\sqrt{(480/380)} = 1.1$. A more exact calculation must include the variation of the 'constants' with temperature.

In Chapter 21, it is shown that a billet with non-uniform axial temperature distribution evens out that difference with a time constant of $c\gamma l^2/\pi^2 k$, giving 2.5 min for an aluminium billet 0.33 m long at 500°C. Example 21.1 shows that, for a soak time of 40 s after heating, the initial temperature difference between the ends has to be 172°C to obtain 100°C difference at the extruder. Figure 21.1 allows the reader to fit his own conditions to the calculations. It is clear that a long 'soak' period is incompatible with taper heating, although the time constant above is long compared with the time taken to obtain a reasonably uniform temperature in the cross-section.

22.15 Scale-model analogue methods

Experimental coil design has been proposed[49] in which the work and coil dimensions are scaled down and the frequency is raised in proportion. This allows large and costly coils, drawing very high powers at low frequencies, to be designed from tests using smaller and more economical coils at higher frequencies.

The changes are as follows:

- All linear dimensions of the coil and load are reduced by a factor k.
- The frequency is increased by $1/k^2$.
- The number of coil turns is maintained.
- The applied voltage is maintained.

The results of these changes are:

- Power factor and efficiency are retained.
- Coil and work power input and coil loss are reduced by k.

- Heating pattern is retained.
- Surface power-density is increased by $1/k$.
- Heating time to a given temperature is reduced in the ratio k^2.

As the chosen variable is frequency, the ratio k is determined from this. For example, increasing the frequency from 50 Hz to 3000 Hz gives $1/k^2 = 3000/50$, i.e. $k \simeq 1/8$.

A 200 mm diameter and 0.8 m long coil at 50 Hz can be modelled by a coil of 25 mm diameter and 100 mm length using 3000 Hz. The coil turns, voltage, power input, etc. are experimentally determined to obtain the required heat pattern and $1/k$ (i.e. eight) times the required surface power-density. The full-scale coil can then be accurately designed, using the same voltage and turns as the model.

22.16 Heating of powdered-metal preforms for sintering

There have been considerable developments in the use of forgings made from powdered metals, especially for automotive parts. The author and colleagues at Aston University were involved in a cooperative project with industry to investigate the principles involved in sintering low-alloy steel powders. There was a lack of physical-property data for sintered steels of different compact densities, so this had to be tackled first.[50] Simple relationships were found linking percentage compaction with the important properties, such as resistivity, needed in the application of the theory of induction heating.

The theory of Parts II and III of this book was applied to simple cylindrical-shaped preforms of various densities and compared with solid steel; the results proved that, with the correct constants, the theory is valid.[51] The interested reader is referred to these papers for details.

Chapter 23

Surface heating by induction

In the last chapter, induction was used for heating large masses of metal, where the aim was a uniform temperature throughout. There, the possible high intensity of induction heating was limited by the rate at which the heat could be conducted away from the surface. Surface hardening makes use of this 'disadvantage'. This application of induction heating is very important. Whilst the basic principles are obviously the same, the use is very different. Figure 23.1 shows examples of parts suitable for induction hardening.

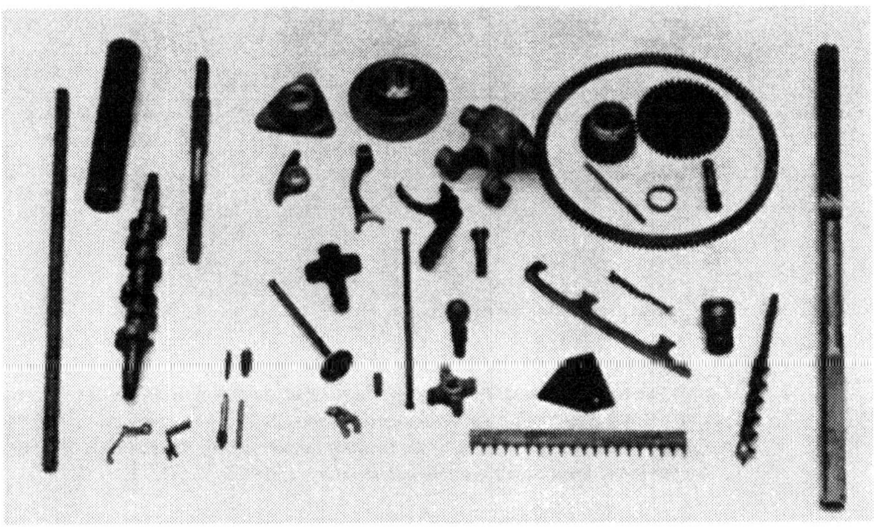

Fig. 23.1 *Samples of components suitable for induction hardening: only the areas needing to be hardened are treated (Radyne Ltd)*

By using higher frequencies to obtain high power densities, heat can be concentrated quickly in a localised region. This region can have its metallurgical nature changed without affecting the rest of the material; the result is a shallow hardened region backed by a less-brittle core – combining the virtues of surface wear resistance with general strength, as well as virtually no distortion during the hardening. For example, the torsional strength of a surface-hardened shaft may be three times that of a 'through-hardened' shaft of the same diameter. Figure 23.2 is an example of a typical shaft-hardening equipment.

Fig. 23.2 *Shaft hardening equipment. The shaft is rotated as the coil and quench travel upwards. The generator power and speed of travel can be programmed on the PLC (program logic controller) to vary the heating pattern along the shaft; for example, for hardening the lobes on a camshaft (Radyne Ltd)*

One alternative to induction hardening is to heat the workpiece in a furnace, using the carburising process. This is a batch process, taking a long time and using skilled labour. Because of the metallurgical transformation, carburising often needs a special alloy steel, which is more costly than the simple carbon steel or cheaper alloys frequently used in induction heating. By its nature, the 'through-hardening' process uses large amounts of energy in heating the whole part to temperature – energy which is wasted in the quench – and it is difficult to control the regions which have to be hardened. By contrast, induction hardening heats a localised region to a high temperature without affecting the rest of the material, saving energy and allowing selective hardening.

It is easy to automate most induction-hardening equipment, and automatic or semi-automatic arrangements can be incorporated in a continuous production line. Once set up, the induction method reproduces the same conditions accurately for each workpiece, giving high-quality products.

The 'through-hardening' method exposes the work to combustion products, with a risk of contamination, and may produce oxidised surfaces which need subsequent cleaning. Both of these disadvantages are minimised by induction heating.

Because of these advantages over through-hardening, induction hardening is increasingly used, so it is important to have a basic understanding of the process.

23.1 Metallurgical principles of heat treatment

Figure 23.3 shows a simplified equilibrium diagram for low-carbon steel. At temperatures below 724°C, steel with 0.3 to 0.83% carbon is in its normal state

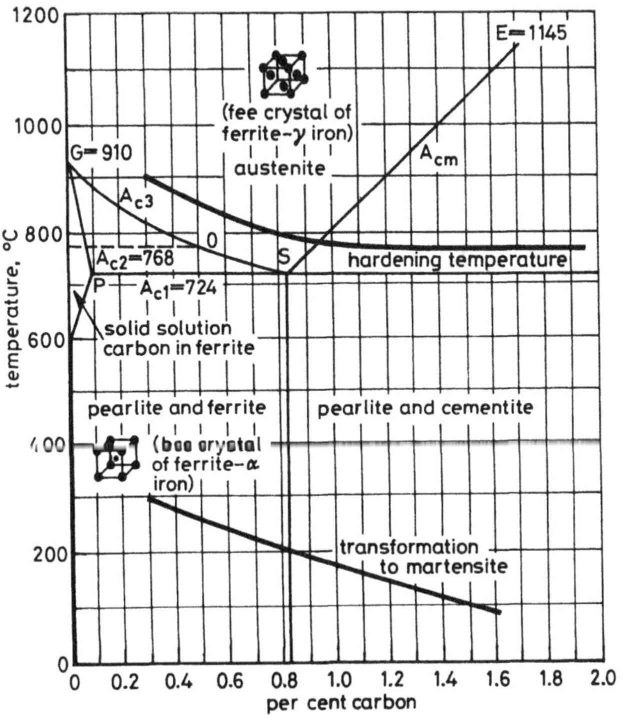

Fig. 23.3 *Hardening temperatures for plain carbon steel (after Dreyfus[12] and ASM[62])*

with the pure iron (ferrite) accompanied by pearlite (a mixture of ferrite and cementite, cementite being ferrite and carbon in layers). If the temperature is raised above 724°C (the so-called A_{c1} line), the cementite starts to go into solution and, when the A_{c2} line is reached, all the material has changed to austenite – a ferrite with a completely different crystal structure from the original ferrite. If the surface is now cooled *rapidly* to 350/200°C, it can miss the reformation of the original ferrite and pearlite and pass directly to the martensite stage, when the carbon comes rapidly out of solution, leaving martensite, a thin hard crystalline structure in which the crystals interlock to form a very hard steel. This rapid cooling is called 'quenching' (see Section 23.2).

With steels of higher carbon content (0.8 to 1.7% carbon), the number of martensite crystals increases and the steel becomes harder. The steel is not taken all the way to the A_{cm} line but to a lower temperature, shown as the hardening temperature, and cooled from there. The required hardening temperatures and quench temperatures are shown on Fig. 23.3.

If the quench is not done rapidly, the material will revert to its original composition. The rate of cooling must not be below a critical value. For 0.3% carbon steels, this rate is only just achievable with intensive water sprays: as the carbon content rises, the critical value falls. Alloying ingredients, such as molybdenum, chromium, manganese and nickel, lower the critical cooling-rate.

Two facts emerge from this simplified metallurgy.

- The heating rate should be rapid, so that the hardening temperature is reached quickly, to avoid grain growth in the overheated surface.
- The quench should be rapid and effective, as the deeper layers will not form martensite unless they, too, are cooled at the critical rate.

There is a relationship between power density, hardened depth, heating, time and frequency, which will be discussed later (Section 23.3). There is also a choice between very rapid heating, which just allows time to heat to the hardened depth required, and a 'heat-conduction' cycle, which allows heat flow to about three times the depth required. The latter produces a more gentle transition between the hardened surface and the soft core. Power densities can be lower and heating times longer, allowing greater accuracy in time control. Obviously, the detailed hardening is specified by the metallurgist.

The grade of steel, the rapidity of cooling, the previous metallurgical treatment of the metal, and the coil design all affect the depth of hardness. A 0.45 carbon steel has good hardening properties.

Two other points should be noted from Fig. 23.3. At the 0.83% carbon line, the steel is in a eutectoid state. This means that when the point S is reached, the A_{c3} and A_{c1} lines must meet and the transformation from pearlite to austenite is instantaneous, with no intermediate or partial solution state. It also implies that the optimum combination of ferrite and cementite, as pearlite, is at 0.83% carbon. Above this percentage, the excess carbon forms extra cementite; below it, the lack of carbon results in pure iron or ferrite.

As one of the main advantages of induction heating is the high rate of power input and temperature rise, steels which have rapid solution-rates are most suitable. These include nearly all heat-treatment steels, except 'high-speed' steels containing high percentages of chromium and tungsten. Another advantage of this rapid heating-rate is that considerable overheating of the surface above the critical temperature can be tolerated without deterioration, because grain growth is small during the short times involved. Surface temperatures as high as 220°C above the critical temperature are not uncommon.[53] This often leads to greater surface hardness. Figures 23.4 to 23.8 show examples of surface-hardening operations. Times of 1 s or less are not unusual in some applications, and plain 0.45% carbon steel has been heated to hardening temperature in 0.3 to 0.5 s.

Fig. 23.4 *Indexing machine for hardening rockers. After mounting in foreground, they are levelled by a guideway before being heated in pairs (1 and 2 in photograph). After heating, using 30 kW, 450 kHz, the quench system quenches and then cools them to handling temperature. The inset shows hardened rockers for car engines (Radyne Ltd)*

23.1.1 Typical induction-hardening steels

The basis of selection of induction-hardening steels is similar, but not identical, to that of choosing steels to be hardened by other methods. Increased carbon content in plain carbon steels results in increased hardenability. For instance, 0.2% carbon steel, or 1020 steel, hardens only to Rockwell 48C, but 0.45%, or 1045, steel results in Rockwell 60C. Additions of alloying agents such as chromium, nickel, molybdenum, tungsten, etc., by moving the noses of the equilibrium curve to the right, result in greater hardenability.

For example, in through-hardening, a plain carbon steel requires a rapid rate of cooling or quenching to avoid the noses of the equilibrium curve and,

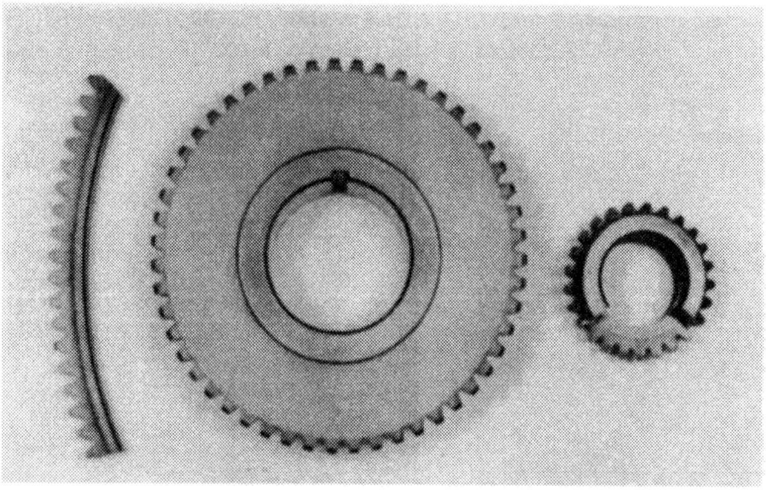

a

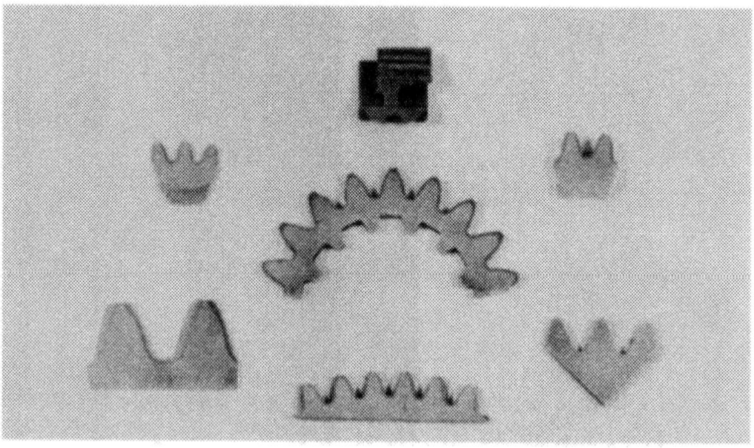

b

Fig. 23.5 *(a) Gear wheels hardened by the single-shot method. Etched teeth show the hardening pattern, which is not as good as tooth-by-tooth hardening but is adequate for normal industrial applications. (b) Heat patterns obtained by underwater induction hardening, tooth-by-tooth. Used when gear must withstand high forces. Note heat pattern absent from tip, which is said to reduce gear noise (Radyne Ltd)*

therefore, to avoid softening. Layers below the quenched surface, which are cooled more slowly because of delays in heat conduction, are likely to soften owing to incomplete martensite transformation. Even in surface hardening, heat conducted to the surface from the centre of the piece often tends to slow down the cooling. This produces the same effect, with the formation of pearlite causing a softer surface. If a more drastic quench than oil is used, distortion is often excessive. Therefore, by adding only small percentages of these alloying agents, oil quenching can be used to minimise distortion. At the same time, the required hardness can be achieved.

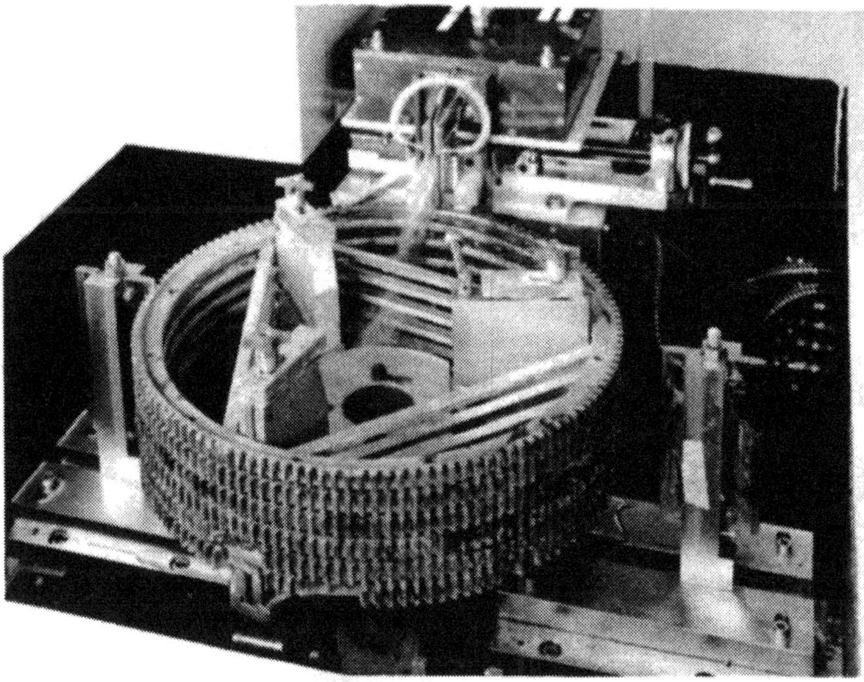

Fig. 23.6 *Detail of ring-gear hardening machine. Up to six starter rings are hardened in one set-up. Coil with integral quench heats sides of two teeth on each ring. Table then indexes around and repeats (Radyne Ltd)*

Another reason for adding alloys is to improve the tensile, yield, impact and fatigue properties of the steel. This improved strength enables the desired hardness to be achieved.

Surface heating by induction considerably influences the choice of steel. The problem of incomplete transformation below the surface does not occur, because only the surface layers have been heated. Also, the fast rate of heating of the surface often means that a fast quench can be used without risk of distortion due to internal stresses. The centre of the piece therefore retains its initial strength, as it has not been heated or hardened at all. These two effects mean that a plain carbon steel can often be substituted for a more expensive alloy steel previously hardened by other heating methods.

Another reason for preferring plain carbon steels for induction hardening is that, in general, they go into solution faster and at lower temperatures than alloy steels. This means that the heating time can be shorter, with less loss of heat to the centre and with increased production rates. A further reason is that a water quench, which is more economical than an oil quench is often adequate. Alloy

Fig. 23.7 *Hardening of bearing surface of automotive suspension socket using 20 kW, 300 kHz supply. Components are manually loaded and unloaded. Equipment is fitted with four positions (two shown) and two automatic change-over switches, each feeding two stations. Component rotates during heating and is quenched from below. Heating time 13 s; quench time 4 s; production rate 500 components/hour with four positions (Stanelco Ltd)*

steel is only used when a hardened piece with an unusually high core strength is needed.

The amounts of carbon and alloy content suitable for induction surface hardening are given below in order of preference:

C = 0.35 to 0.60%		
C = 0.30 to 0.40%,	Mn = 0.6 to 1.9%	
C = 0.30 to 0.45%,	Ni = 3.0 to 5.0%	
C = 0.35 to 0.45%,	Ni = 0.46 to 0.7%,	Cr = 0.3 to 0.6%
C = 0.35 to 0.45%,	Ni = 1.5 to 3.5%,	Mo = 0.25%
C = 0.35 to 0.45%,	Cr = 1.0%,	Mn = 0.6 to 1.2%

Examples are SAE steels 1040, 1050, 1340, 2345, 3140, and 4640.

Previous heat-treatment of the steels is also of great importance, primarily because of the short heating times used in most induction-heating processes.

These times are not long enough to allow coarse- or heterogeneous-grained steel to diffuse evenly. Free ferrite usually remains after treatment, causing localised soft areas. Annealed, spheroidised, and similar pre-treatments should be avoided, where possible. The most desirable pre-treatment is normalising, which results in an evenly-distributed fine pearlite structure.

Fig. 23.8 *Simultaneous hardening of the exterior of two 12.7 mm diameter carburettor plungers on the twin-track horizontal machine. Using 40 kW radio-frequency power, 1080 plungers/hour are hardened to Rockwell 66C for a depth of 1 mm (Cheltenham Induction Heating Ltd)*

Annealing before hardening results in deep hardening, but less hardness (Rockwell 52C). Normalising before hardening results in medium hardening (Rockwell 53C). Hardening and tempering, before hardening, results in a shallow layer (0.75 mm) of very hard metal (Rockwell 62C). In practice, normalising before hardening results in a good metallurgical structure for deeper hardening.

Some steels need a short delay between power removal and quenching, to complete the carbide solution process. Other quenching steels (SAE 1000 and 1300 series) have been developed to eliminate this delay.

23.2 Quenches and quenching

Obviously, the quench is a vital part of the induction-hardening process. Natural cooling is rare because it is too slow; most hardening is carried out with forced cooling, using water, oil, brine, or forced air; water is most common because of its excellent heat-transfer properties. *Single-shot hardening* heats the whole surface to be hardened at the same time, and then either plunges the workpiece into coolant or sprays the whole heated surface. *Scanning quenches*

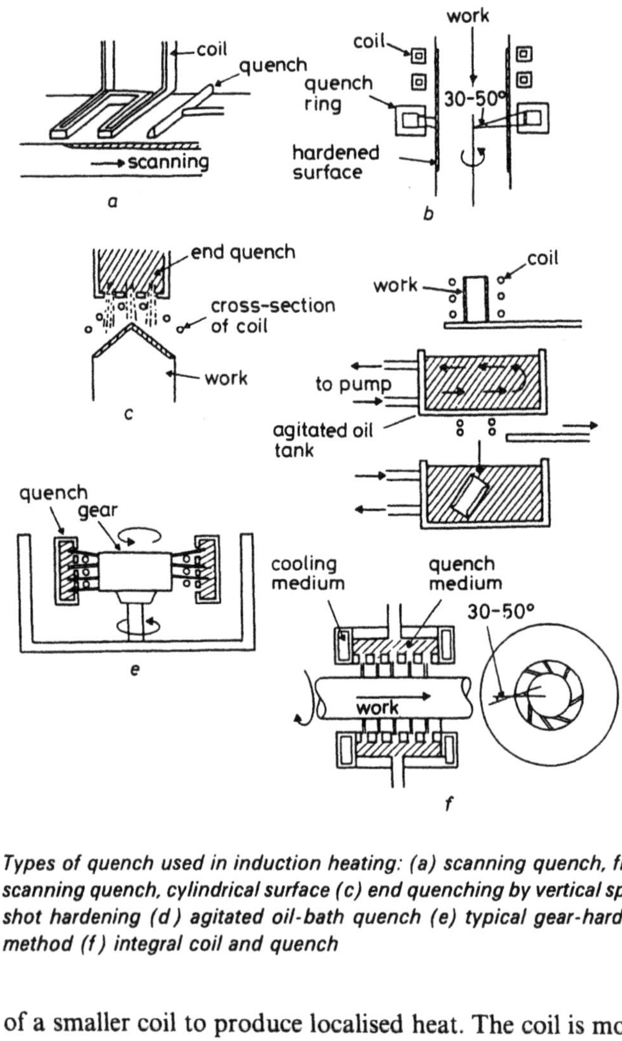

Fig. 23.9 *Types of quench used in induction heating: (a) scanning quench, flat surface (b) scanning quench, cylindrical surface (c) end quenching by vertical spray for single-shot hardening (d) agitated oil-bath quench (e) typical gear-hardening quench method (f) integral coil and quench*

make use of a smaller coil to produce localised heat. The coil is moved relative to the heated area and the quench applied almost immediately after heating. In this way, a small coil and relatively low power can be used to heat a large surface, at the expense of cycle time.

Figure 23.9 illustrates the basic types of quench used in induction heating. Figures 23.9a and b show scanning quenches of the flat-surface type[54,55] and the circular type.[56] The coil, the quench, and the work are arranged to scan to give the best heating/cooling cycle. The angle of spray incidence is important; up to

50° must be used, which allows uniform cooling and an even hardness depth with no local hot spots. Note the gap between the heating coil and the quench to allow the controlled delay mentioned above. Nearly all shaft hardeners use the circular type, with the shaft rotating to give greater uniformity and the coil travelling axially. Figure 23.9c shows end-hardening by spraying through the coil.[57] Figure 23.9d is a typical agitated-bath, single-shot quench, where the part is heated and then dropped into the bath.[56,57] Figure 23.9e is a static quench, where the part remains in the coil after heating and the quench is sprayed between the coil turns. This method is common in gear hardening and the work is rotated during both heating and cooling for uniformity. (The spray must again be at 30° to 50° to the work.) Figure 23.9f shows a combined coil and quench, where the coolant is forced out through spray holes in the coil face. This is particularly useful for minimum delay time, as the heat can be turned off and the quench applied almost instantaneously.

Water is used for low-alloy and low-carbon steels, and for cast iron with simple shapes. Oils are used for alloy and high-carbon steels and high-hardness surfaces, and for non-uniform sections. (Fire safety precautions must always be taken with oil.) Brine is sometimes used for low-hardenability steels. Air can be used for stainless steels and compressed air for small areas of alloy steels.[58]

If an oil quench is too slow or is likely to create hazards associated with fire or fumes, and a water quench would produce cracking, then a polymer quenchant can be used.[59] These are organic chemicals of high molecular weight (polyalkalene glycols) which dissolve in water at room temperature, giving a true solution. Polymer quenchant can be recirculated like water and is non-flammable; it produces no objectionable fumes, nor does it irritate the skin. Its use is widespread in induction-hardening applications, where oil is beginning to be unacceptable for health and safety reasons. Polymer solutions are likely to take over, eliminating the risk of fire, fumes, and smoke. One such concentrate is Aquaquench 251, produced by Edgar Vaughan & Co.

The cooling rate, and hence the quench speed, can be varied by changing the concentration of the solution. At low concentrations (2 to 3%), Aquaquench 251 solution wets the component surface more easily than water and, hence, improves quench efficiency, which may permit the treatment of low-hardenability material which would not respond satisfactorily to a water quench. Because a thin film of polymer is deposited on the workpiece during quenching, the uniformity of quench is improved, in many cases eliminating problems associated with steam-pocketing and soft spots.

Higher concentrations of 8 to 10% Aquaquench 251 are used with components of high hardenability to replace oils, where water quenching could not be considered.

Table 23.1 shows the range of cooling characteristics obtainable with polymer quenchants; the quench times vary from quicker-than-water to that of oil.

It is recommended that polymer quenchants should be used in the 25 to 45°C range to ensure quenching uniformity and should not be used above 60°C, otherwise patchy hardening may result.

Table 23.1 *Time to cool a component from 850 to 300°C when quenched in Aquaquench 251 solutions of different concentrations, together with other quenching media, for comparison*

Aquaquench 251 concentration (%)	Time (s)	Quench	Time (s)
2	11.5	Water	12.0
5	12.5	10% soluble oil	13.0
10	14.0	An additive-treated fast quenching oil	20.5
15	19.0		
20	24.5	A high-grade mineral quenching oil	26.0

Quenching cracks are minimised by:

- eliminating flawed pieces prior to hardening
- uniform heating
- quenching at lowest cooling rate and temperature, compatible with required hardness
- uniform quenching and cooling
- elimination of sharp angles, uneven cross-sections, and other stress-raisers.[60]

Tempering after quenching helps to reduce cracking; so does decreasing the carbon content of the steel, while at the same time increasing the manganese content. By design and experiment, it is often possible to allow sufficient heat to be retained in the workpiece to provide some tempering.

The core of a large component can act as a heat sink and may provide sufficiently rapid cooling to give 'self-quenching', particularly for alloy steels.

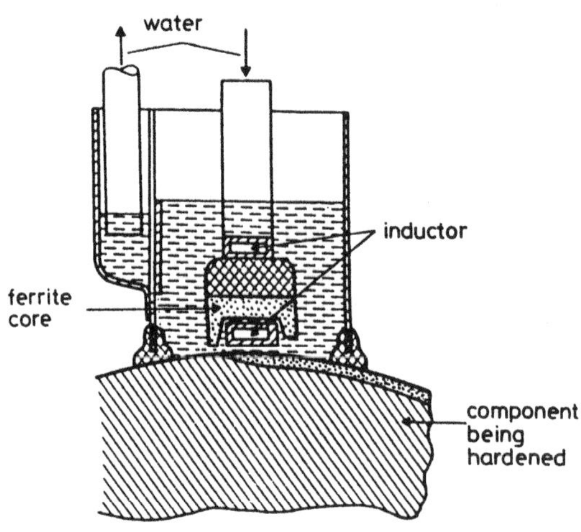

Fig. 23.10 *Sectional view of a large cylindrical article being hardened under water (R. B. Tennent Ltd)*

Another method of quenching is to perform the whole heating operation under water. Figure 23.10 shows a large cylindrical article being hardened; the coil, concentrator (see Section 23.6), and workpiece are all cooled by the water inside the seals.

Mill rolls for cold rolling of metals have a 20 mm thick hardened layer which must have a precise hardness profile. This is done by lowering the roll slowly through two coils rated at 1100 kVA (50 Hz) and 700 kVA (250 Hz). Water quenching is followed by cryogenic treatment to − 80°C. Finally, the roll is tempered in an electrically-heated oil bath.

Coil/quench arrangements are usually fairly empirical and are modified on the test bed. However, once modified, they give uniform results consistently, provided that the conditions are carefully maintained.

23.3 Selection of frequency for induction heat-treatment

Many factors enter into the choice of frequency for hardening. Higher frequencies (3 to 50 kHz, 400 to 500 kHz) are always used to obtain the shallow

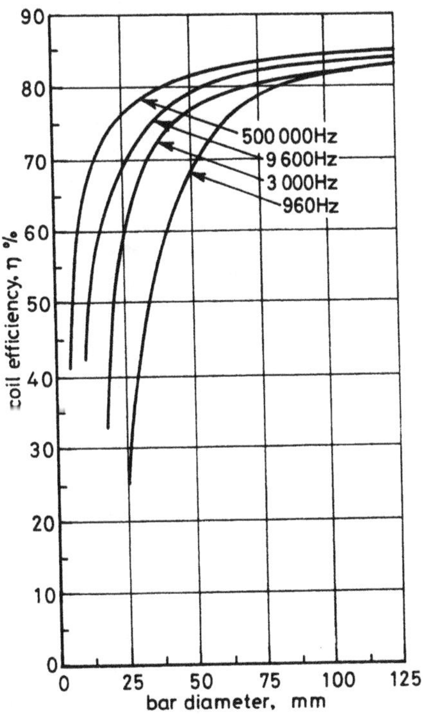

Fig. 23.11 *Coil efficiencies for bar hardening at various frequencies. Metal is 0·45% carbon steel; temperature is 1100°C; difference between coil inner diameter and bar outer diameter is 3 mm*

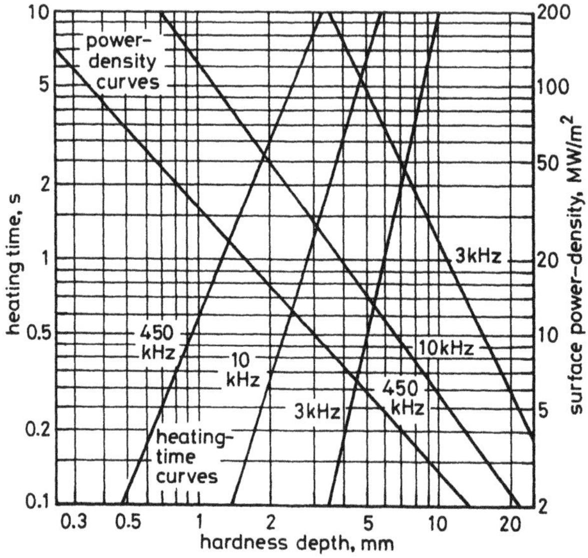

Fig. 23.12 *Relationship between heating time, surface power-density and hardened depth for various frequencies (after Lozinskii[55])*

Fig. 23.13 *Lorry axle flange showing affected zone by etching. The induction coil is programmed to pause at the start of the heat cycle for heat to penetrate the flange, which obviates the stress point at the corner (Radyne Ltd)*

skin-depths required. The power sources must be precisely controllable (preferably automatically); accuracy and repeatability are vital. A specific comparison for four frequencies used for simple hardening of steel bars is given in Fig. 23.11. These curves give the coil efficiency; the overall conversion efficiency must be considered as well. When this is included, it may be possible to use several frequencies for the same job, except for very small diameters.

Fig. 23.14 *Etched sample of camshaft showing heat-affected zones (Radyne Ltd)*

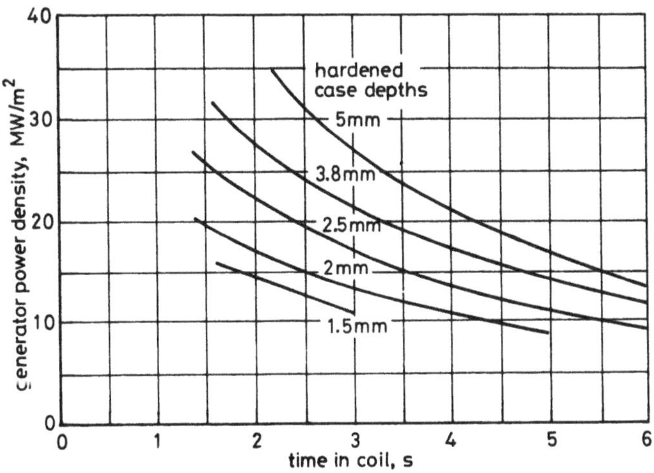

Fig. 23.15 *Shaft hardening by scanning: 3 kHz. Minimum scanning rate 6.5 mm/s; minimum shaft diameter 25 mm*

Lozinskii[54,55,61] did much work on the relationship between power input, hardened depth, frequency, and heating time. Figure 23.12 is based on his work, which was mainly carried out for 'single-shot' heating; it has been simplified to normally-available frequencies, based on hardening temperatures between 850 and 900°C. Most hardening depths lie between 0.25 and 5 mm; only 450 kHz covers this range and, hence, this is the most widely-used frequency. Medium

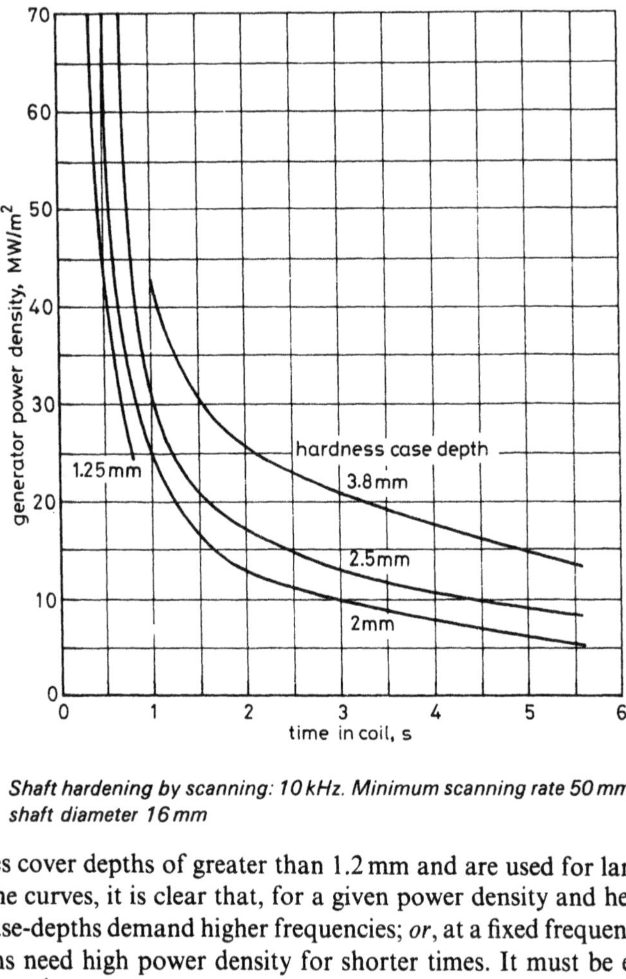

Fig. 23.16 *Shaft hardening by scanning: 10 kHz. Minimum scanning rate 50 mm/s; minimum shaft diameter 16 mm*

frequencies cover depths of greater than 1.2 mm and are used for large powers.

From the curves, it is clear that, for a given power density and heating time, shallow case-depths demand higher frequencies; *or*, at a fixed frequency, shallow case depths need high power density for shorter times. It must be emphasised that Fig. 23.12 is approximate, as there are many factors not allowed for. It is interesting to note that the straight lines have different slopes, showing that the behaviour with diameter is greatly different at the various frequencies. Single-shot heating can only be used when the piece is small enough to achieve the hardness depth with a given generator in one shot. Larger pieces have to be scanned, usually by being fed through the coil automatically. Figures 23.14 and 23.14 show examples of work using programmed heating operations.

For scanning methods, Figs. 23.15 to 23.17 give approximate relationships between generator power and case-hardened depth, with time. Coil losses are included in the curves. The 'depth' is the distance from the surface to the point where the hardness starts to fall rapidly; this varies from 50 to 80% of the actual heated depth, depending on the metal. The curves are plotted from experimental

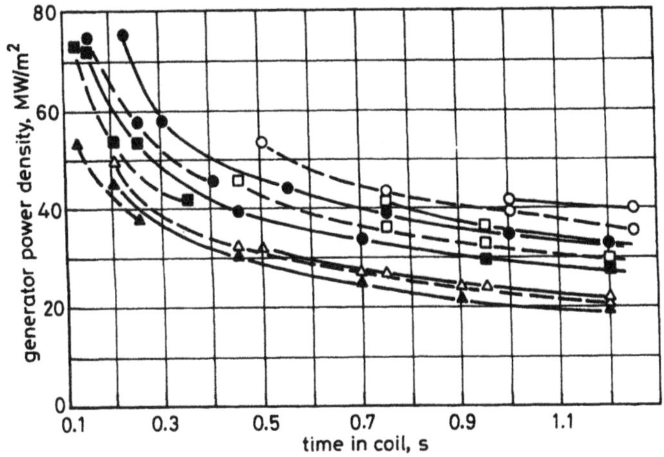

Fig. 23.17 *Shaft hardening by scanning: 450 kHz. Maximum shaft diameters: 355 mm per MW/m² generator power density at 10 kW generator power; 835 mm per MW/m² at 25 kW; 1670 mm per MW/m² at 50 kW. Metal is SAE 1045 normalised; coil length is 10 mm; two turns; difference between internal coil diameter and outer shaft diameter is 3.5 mm; case depth is 50% martensite*

	Generator power (kW)	Case depth (mm)
○——○	50	1.5
○- - -○	50	1.25
●——●	50	1.0
●- - -●	50	0.75
□——□	25	1.5
□- - -□	25	1.25
■——■	25	1.0
■- - -■	25	0.75
△——△	10	1.5
△- - -△	10	1.25
▲——▲	10	1.0
▲- - -▲	10	0.75

data for 0.45% carbon steel. Minimum speeds and diameters are given; the results were obtained with a single-turn work coil and a current transformer. A minimum power-density of 30 MW/m² is recommended at 10 kHz for shafts between 16 and 20 mm diameter. The most efficient part of the curves, with the steepest slope, should be used wherever possible. These correspond closely to the single-shot curves of Fig. 23.12.

Table 23.2 shows frequencies for induction heating.[58] The table should be used only as an indicator, as equally good results may be obtained with the lower of the frequencies given, by using a higher power-density. The choice of frequency is mainly economic, except where shallow hardening depths demand radio frequency.

Table 23.2 *Selection of frequency for induction hardening*

Hardened depth (mm)	Diameter of bar (mm)	Frequencies used in practice (Hz)
0.4–1.25	5–25	450 000
1.25–2.5	10–16	450 000 (10 000)
	16–25	10 000; 450 000
	25–50	10 000 (3000; 450 000)
	Over 50	10 000 (3000)
2.5–5	20–50	10 000; 3000
	50–100	3000 (1000; 3000)
	Over 100	1000 (3000)

Example 23.1
Select a suitable frequency for a 25 mm diameter shaft to be hardened to 2.5 mm depth using a 50 kW generator and a 25 mm long coil.

Solution
The power density is

$$(50\,000)/(25\pi)(25)(10^{-6}) = 25.5\,\mathrm{MW/m^2}$$

For 3000 Hz, from Fig. 23.15, this gives 1.6 s in the coil. For 10 kHz, from Fig. 23.16, this gives 1.2 s in the coil. The corresponding scanning rates are:

At 3 kHz: $25/1.6 = 15.6\,\mathrm{mm/s}$

At 10 kHz: $25/1.2 = 20.8\,\mathrm{mm/s}$

Other things being equal, the higher scanning rate results in higher production and so is preferred. It will also be seen that, if the generator size and the coil length are both doubled, the time stays the same, but the scanning rate and the production rate will double.

In practice, energy inputs for case depths between 2 and 4 mm lie between 30 and 60 MW s/m^2 and the relation between minimum case depth and frequency is as follows:

f (kHz)	*Minimum case depth* (mm)
3	1.5
10	0.75
450	0.38

The same hardening depth can obviously be achieved at more than one frequency, as above.

When small diameters are being heat-treated, the frequency has to be raised. Reference 8 has an interesting chapter on heating fine wires, knitting-machine needles, etc., in which frequencies of 30 MHz are discussed. The effects of saturation level on permeability should also be considered.

While throughput is important, controllability must also be considered. It may be preferable in Example 23.1 to use a longer heating time to increase the accuracy of control. For a heating time of 1 s, at 10 kHz, 30.5 MW/m^2 are needed, i.e. a 60 kW generator. This time–power combination is very difficult to control accurately. It may be preferable to increase the time to 2 s, using a 33 kW generator.

23.4 Temperature distribution in induction surface-hardening

In surface hardening, a high power-density is applied at the surface of the workpiece and the aim is to heat the surface layer without affecting the rest of the workpiece. Since the heated layer is shallow compared with the radius or thickness, it is always justifiable to treat the problem in cartesian coordinates, as if it were a semi-infinite slab. Various assumptions can be made, as follows: (a) constant surface temperature; (b) constant power input; (c) variable surface temperature; (d) variable power input. Assumptions (c) and (d) are extensions of (a) and (b). All four solutions are given in detail in Chapter 19. For present purposes, we need an indication of the temperature distribution at any point in the workpiece for any time after the start, and we will take the two simple cases (a) and (b).

23.4.1 Constant surface temperature

In Section 19.1 and Fig. 19.1, the ratio of the temperature θ_y at a point distant y from the surface to the temperature θ_s at the surface is given as a function of $y/2\sqrt{(kt/c\gamma)}$, where t is time (s) and k, c, and γ are the physical properties of the metal, assuming that θ_s is held constant. An example shows how this can be used to find the approximate heating time.

Example 23.2

Given the following data, find the approximate heating time and the surface power-density:

Metal	steel
Hardened depth, y	0.75 mm
Critical temperature at y, θ_y	820°C
Assumed surface temperature, θ_s	930°C
Density, γ	7850 kg/m^3
Thermal conductivity at θ_y, k	26 W/m K
Specific heat at θ_y, c	857 J/kg K

Solution

To find the time t seconds to raise the temperature at a depth y millimetres to θ_y°C, first assume that the surface is brought suddenly from 20°C to a temperature θ_s°C and maintained at that temperature. Thus

$$\frac{\theta_y}{\theta_s} = \frac{820 - 20}{930 - 20} = \frac{800}{910} = 0.88$$

Fig. 19.1 curve A gives a value of the abscissa of 0.1, so that

$$\frac{y}{2}\sqrt{\left(\frac{c\gamma}{kt}\right)} = \frac{1}{2}\left(\frac{0.75}{1000}\right)\sqrt{\left[\frac{(857)(7850)}{(26)t}\right]} = 0.1$$

$$t = \left(\frac{7.5}{1000}\right)^2 \frac{(857)(7850)}{(4)(26)} = 3.64 \quad \text{s}$$

In this calculation, the values of k and c at 800°C have been used. c is too large and k is small, so the calculation gives a conservative answer. If arithmetic mean values between 20 and 800°C are used (39 and 661, respectively), the time becomes 1.87 s.

This example assumes that the surface can be instantaneously brought to 930°C and held there. It is more typical of heat treatment in a furnace than induction heating.

The surface power-density (W/m^2) that would be needed in this example is given by eqn 19.5:

$$\mathscr{P}_s = \frac{\theta_s}{\sqrt{t}}\sqrt{\left(\frac{kc\gamma}{\pi}\right)}$$

Hence

$$\mathscr{P}_s = (910)\sqrt{\left[\frac{(26)(857)(7850)}{\pi}\right]}\frac{1}{\sqrt{t}} = 6.8/\sqrt{t} \qquad \text{MW/m}^2$$

That is, the power density would vary from infinity at zero time to 3.6 MW/m^2 at the end of the heating time.

23.4.2 Constant power input

Although the power input will vary with resistivity in induction heating, constant power input is probably a better assumption than constant surface temperature. In any event, Section 19.4 shows how to handle variable power.

Example 23.3
Find the surface rise time when all details are as in Example 23.2, except that the surface power density throughout is 20 MW/m^2. This is not an unreasonable density, from the previous sections.

Solution
For $\theta_y/\theta_s = 0.88$, the constant power curve (Fig. 19.1 curve B) gives an abscissa of 0.07. Hence

$$t = (3.64)(0.1/0.07)^2 = 7.43\,\text{s}$$

For small values of y, eqn 19.20 simplifies to eqn 19.21:

$$\theta_y/\theta_s = 1 - y\sqrt{\pi} + y^2 \ldots$$

So, for $y = 0.000\,75\,\text{m}$,

$$\theta_y/\theta_s = 1 - 0.0013 = 0.9987$$

$$\theta_s = 1.0013\theta_y = (1.0013)(820 - 20) = 801$$

Note that this method does not allow independent specification of θ_s and θ_y, since they are always linked by eqn 19.20.

From eqn 19.19,

$$\theta_s = \frac{2\mathscr{P}_s}{\sqrt{(\pi c \gamma k)}}\sqrt{t}$$

$$= \frac{(2)(20 \times 10^6)}{\sqrt{[(\pi)(857)(7850)(26)]}}\sqrt{t} = 1706\sqrt{t}$$

The time for the surface to rise 801°C is, therefore, $(801/1706)^2 = 0.22\,\text{s}$.

Examples 23.2 and 23.3 are not strictly comparable, as the power levels are different. The results show that the heating times are short. The two examples show the usefulness of Fig. 19.1 for simple surface-heating calculations.

23.5 Application specifications for coil design

In the previous sections, the choice of frequency, types of quench, heating times, etc. have been discussed. In this section, two examples illustrate these ideas in practice.

Example 23.4: single shot
Given the following data for a surface heating application by the single-shot

method, select frequency, power and coil dimensions.

Metal	steel
Production rate	3600 pieces/h
Case depth	1 mm
Workpiece diameter	12.5 mm
Workpiece length	25 mm
Handling time	0.5 s

Solution
The cylindrical surface is to be hardened. Pieces this size are almost always hardened by single-shot methods.

Frequency The production time is 1 s and the handling time is 0.5 s, so the heating time is 0.5 s. From Fig. 23.12, only 450 kHz will harden in this time. From Table 23.2, 450 kHz is also indicated.
Heated area $A_w = (\pi)(12.5)(25) = 982\,\text{mm}^2$.
Power From Fig. 23.12, the power density at 450 kHz for 1 mm hardened depth is $31\,\text{MW/m}^2 \equiv 31\,\text{W/mm}^2$. So power needed $= (31)(982)\,\text{W} = 30.4\,\text{kW}$.
Coil dimensions Single-shot hardening coils are coupled as closely as possible to the work surface being hardened. The turns are kept to a minimum and the area of the coil corresponds closely to that of the area to be hardened. Air gaps between windings and work vary from 0.75 mm upwards, depending on the accuracy of the locating fixtures and the condition of the work surface. The coil length is the same as the heated length, 25 mm, since the coil is close-coupled. In this example, we shall take the coil inner diameter as the workpiece diameter, plus 2 mm clearance, i.e. $12.5 + 2 = 14.5$ mm.

Example 23.5: scanning
Given the following data for a surface heating application by the scanning method, select frequency, power and coil dimensions.

Metal	steel
Production rate	1 shaft/min
Mass of shaft	11 kg
Case depth needed	2.5 mm
Shaft diameter	38 ± 0.25 mm
Shaft length	1.2 m
Area to be hardened	full 1.2 m length

Solution
Frequency and power From Figs 23.15 to 23.17, a case depth of 2.5 mm is possible with 3 and 10 kHz, but not with 450 kHz. The choice of frequency depends on heating time. If 12 s are allowed for loading and unloading, the heating time for each shaft is 48 s; so the scanning rate is $1200/48 = 25$ mm/s. *Assume* a 25 mm long coil initially, so that the heating time in the coil is 1 s.

a

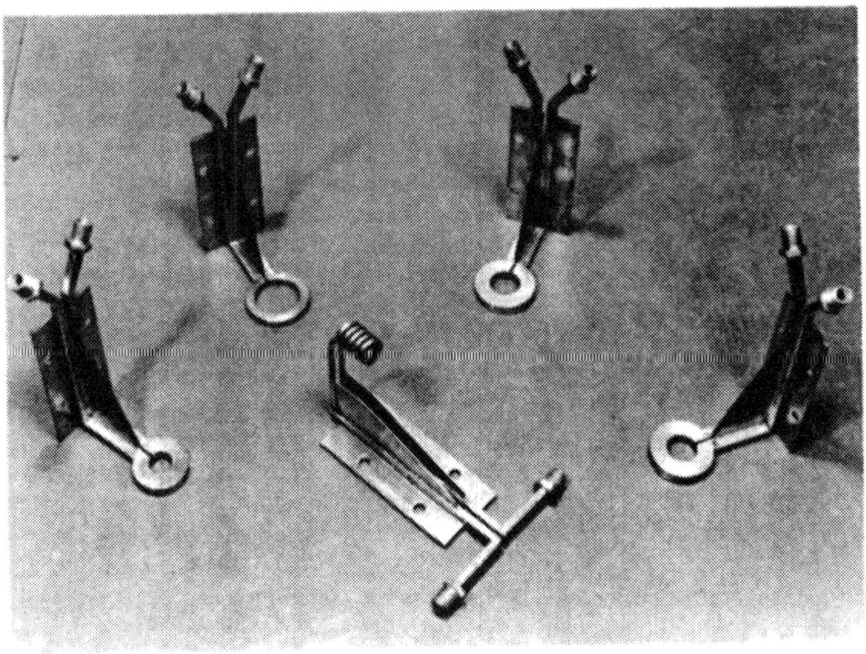

b

Fig. 23.18 *(a) Typical induction coils, with quench integral with coil. Coil on right used for single-shot gear-wheel hardening. (b) Shaft-hardening coils and internal heating coil (Radyne Ltd)*

At 10 kHz, from Fig. 23.16, we get 31 MW/m^2 for the power density associated with 2.5 mm. The heated area $A_w = (\pi)(38)(25) = 2985\,mm^2$, so the generator power $P_g = (31)(2985)\,W = 93\,kW$. A 100 kW, 10 kHz generator would therefore be suitable.

At 3 kHz, from Fig. 23.15, 1 s is too short. Suppose a 50 mm coil is used instead, so that the heating time becomes 2 s; then the power density becomes 22.2 MW/m^2. Since the heated area is now doubled, the total power is $(2)(2985)(22.2) = 133\,kW$.

So far, the choice is therefore between 100 kW at 10 kHz or 150 kW at 3 kHz. Similar calculations are needed to check for 25 and 50 kHz; then an economic decision can be made.

Coil dimensions Scanning coils usually have to allow a somewhat larger air gap because of work distortions and tolerances. Practical gaps range from 1 to 3 mm. The inner diameter must take the tolerance into account, which could be 0.5 mm, so a practical gap would be 1 mm, giving a coil inner diameter of $38 + 2 = 40$ mm. A single-turn or two-turn coil, 25 mm long, would be suitable.

23.6 Mechanical construction of coils

The coils must be close-coupled to the work, but allowing adequate mechanical and electrical clearance. Often the coils must be designed integrally with work-handling equipment and with matching units, such as capacitors, transformers, etc. Stray field-heating, vibration, water, and oil effects must be taken into account. After the electrical design, the coil must be well engineered.

The conductor is normally made from high-conductivity copper, water-cooled. Most coils are made of tubing carrying a separate water supply, but occasionally oil or 'plastic' quench media serve the dual purpose of cooling the inductor and quenching the work. In some single-shot applications having a short heating-cycle, the inductor may have no separate cooling, as the spray quench does both jobs. Silver plating is often used to improve surface conductivity, especially at the contact surfaces. Figure 23.18 shows typical induction coils.

Most scanning coils have a single turn, carrying very large currents, which may be as high as 20 kA. Single-turn coils, supplied from a matching transformer, are desirable, as the low voltage eliminates the risk of voltage breakdown in the presence of the quenching medium. The inductor is robust and stiff, ensuring a long life and dimensional stability. Another advantage of the low voltage is that close-coupling with the workpiece is possible, increasing heating efficiency and allowing closely-delineated hardened areas.

Because the frequency is high (3 to 450 kHz), the currents concentrate at the inner surface. All joints must be silver soldered. High-temperature brazing compounds, such as Silfos, should not be used as the resistance of the joint is too high; also, the joints are brittle and tend to crack with thermal expansion.

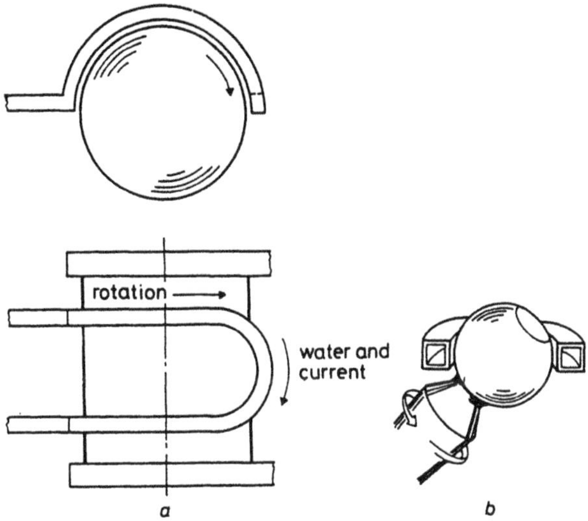

Fig. 23.19 *Special coil configurations: (a) hairpin coil (b) heating of ball joints (after ASEA)*

Coils must be mechanically rigid and held firmly, as there are substantial electromagnetic forces on the coil.

The quench may be built into the coil, or be made separately.

Although circular or cylindrical coils are obviously desirable, they are not necessary and it is sometimes convenient to use other shapes, such as hairpins (either straight or shaped). A curved hairpin coil is shown in Fig. 23.19a; this would be used where it is impossible to encompass the cylindrical surface, because of the presence of other parts, but where it is possible to rotate the cylindrical surface under the coil.

It is also possible to heat spherical parts, such as ball joints, by rotating the part about its axis at an angle to a circular coil, as in Fig. 23.19b. This heats the whole useful surface of the ball.

To increase the coupling of the coil with the workpiece, ferromagnetic 'concentrators' or 'intensifiers' are often used. These provide a low-reluctance path for the flux instead of allowing it free flow through the normal leakage paths. Skill and ingenuity are involved in the devising of suitable concentrators to optimise the heating performance, without interference with the mechanical arrangements. At 450 kHz, ferrites are used. At lower frequencies, thin laminations of electrical steel are preferred, because of their higher flux-carrying capacity. These must be insulated with high-temperature insulation (typically a phosphate coating) and should be as thin as practicable (typically 0.1 to 0.5 mm). Four per cent silicon steel is desirable, but low-loss transformer sheet is possible. Concentrators can make a big difference to heating performance;

one hardening cycle was reduced by 80%, using an inductor designed with a ferrite intensifier. Such an intensifier is shown in Fig. 23.10.

Another way to concentrate the power is to use the skin-effect property of higher-frequency currents. This is illustrated in Figs 23.20a and b.[38] The concentrators form the secondary circuits of a work-head transformer; the primary is not shown on the figure, but is closely coupled to the outside surface of the concentrator. The secondary voltage is induced around a path which includes the slot and inner bore in Fig. 23.20a and the work slot in Fig. 23.20b. The external parts run at low current densities because of the large area, but the total current is channelled into the active region by the skin-effect.

The concentrator can be made hollow and internally water-cooled. The primary coils will also be made of water-cooled copper tubing, electrically insulated and wound closely on the cylindrical surface.

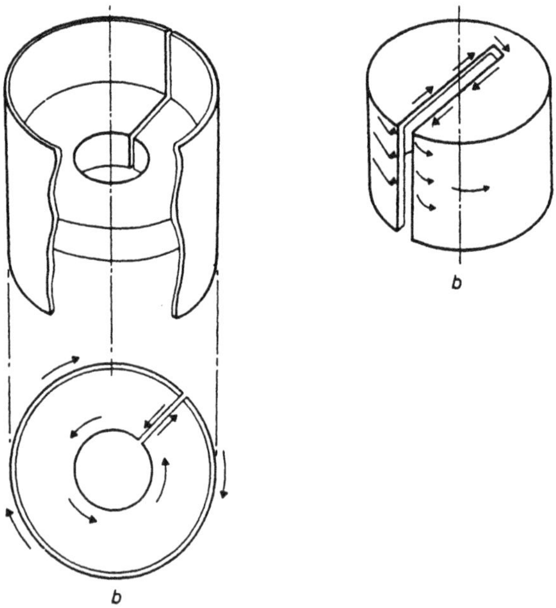

Fig. 23.20 *Current concentrators: (a) circular (b) slot (after May[38])*

Chapter 24

Other applications of induction heating

24.1 Soldering and brazing

Brazing, soft soldering, and welding need concentrated and accurately-controlled heating, as in heating for hardening. Because of the very fast heating-rates in induction heating, oxidation and distortion are minimal and the costs are less than those of most conventional methods. In many ways, the applications resemble the techniques used for hardening. The power must be capable of being repeatedly switched in a closely-controlled manner, as heat times for many joining applications are short. The techniques of both through-heating and surface heating are used, depending on the work size and the location of the joint. The frequencies most used are 3, 10, 25, 50 and 450 kHz. Figure 24.1 shows the many shapes of coil used in induction heating, and Fig. 24.2 illustrates various applications.

24.1.1 Basic joining techniques

For every filler metal, there is an optimum range of joint clearance. Curves for the proposed filler, having the form of Fig. 24.3, should be consulted and the joints should be designed with this in mind.[62] The simple basic joints are shown in Fig. 24.4; variations are shown in Fig. 24.5.

Good practice requires that the joints should be designed for long life by reducing stress concentrations and fatigue problems. Some of the techniques are shown in Fig. 24.6. In some production processes, it is more economic to use rod material applied by hand when the material is at brazing temperature; the joints should then be designed to be self-jigging, as shown in Fig. 24.7.

In many applications, the process is automatic, or semi-automatic, and preplaced brazing materials are used, as shown in Fig. 24.8. Preformed and preplaced materials are very economical in practice, since the exact amount of material is applied at the exact place. Although substantial clearances are shown in the diagrams, these are small in practice, and the flow is usually by capillary action.

The melting points of brazing alloys vary greatly, and those of solders vary

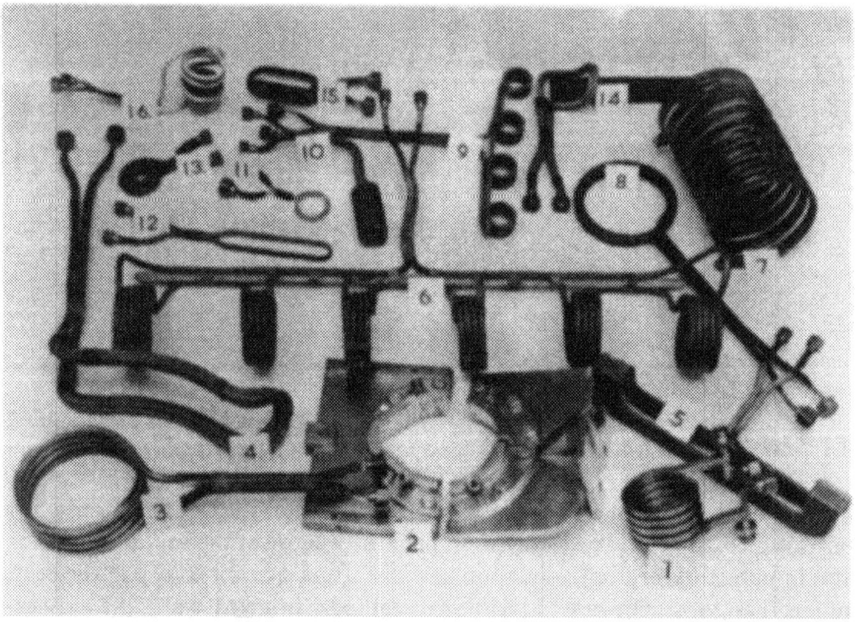

Fig. 24.1 *Typical heating coils: (1) vacuum fusion analysis coil with provision for air cooling the silicon tube (2) split-plate type coil to ease introduction of workpiece (3) multi-turn square-section coil for heating far section (4) rotary-skid coil used in conjunction with rotary indexing table (5) straight skid coil (6) horizontal zone-refining coils (7) multi-turn coil using rectangular tubing to provide good coupling (8) double two-turn coil ideal for brazing non-ferrous components (9) multi-position coil (10) internal solenoid coil (11) insulated two-turn pancake coil (12) hairpin coil (13) pancake coil for surface heating (14) tool-tipping coil (15) split coil for slot heating (16) levitation melting coil (Radyne Ltd)*

from 317 to 70°C, depending on composition. The flow-point temperature of the brazing alloys varies between 0 and 145°C above the melting point. Obviously, the joint must be brought to just above the flow-point temperature. Tables 24.1 and 24.2 show some soldering and brazing alloys, with their melting points.

The joints must always be fluxed with the appropriate flux for the material. Figures 24.9 to 24.11 show coils and components suitable for brazing.

24.1.2 Choice of frequency

Most joining applications used radio frequency (450 kHz) because of the high rate of power input and the localised heat at that frequency. Usually, non-ferrous soldering and brazing are only possible at radio frequencies, because of the high Q needed at the terminal load.

Larger ferrous brazing applications are carried out at 3, 10, 25 and 50 kHz, particularly where high production rates are needed. One advantage of medium-frequency soldering and brazing is that the volts per turn of the coil and the volts

Fig. 24.2 *Typical induction heating applications (Stanelco Ltd)*

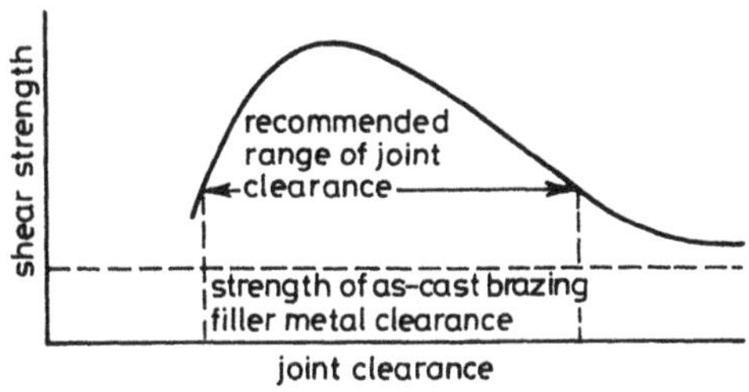

Fig. 24.3 *Optimum joint-clearance curve for filler brazing metal*

from coil to work are much smaller than in radio-frequency joining because of the lower frequency (see Chapter 22). This is important where volatile flux is used, as it often tends to short-out turns or to cause an arc between coil and work at radio frequencies. With large workpieces, it may be preferable to heat the part in a furnace, rather than use induction heating; this must be considered on its merits.

24.1.3 Design calculations

Although the same principles are involved in heating for joining as in previous chapters, the steps in designing a coil cannot be as logical; the power to be used

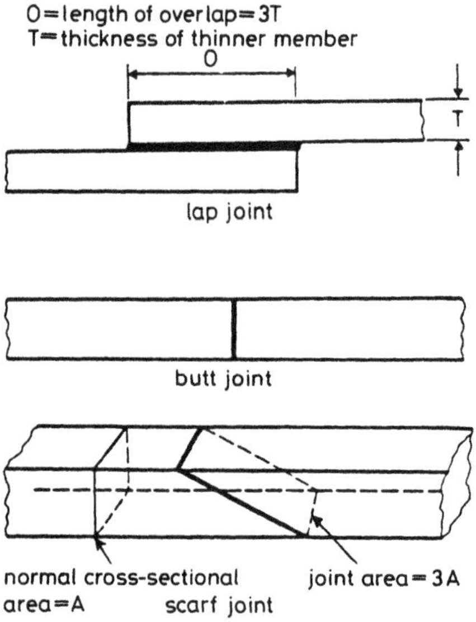

Fig. 24.4 *Basic brazing joints: lap, butt and scarf*

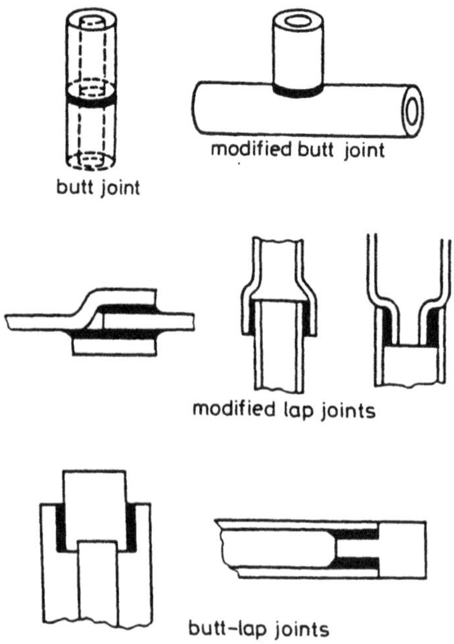

Fig. 24.5 *Modified brazing joints*

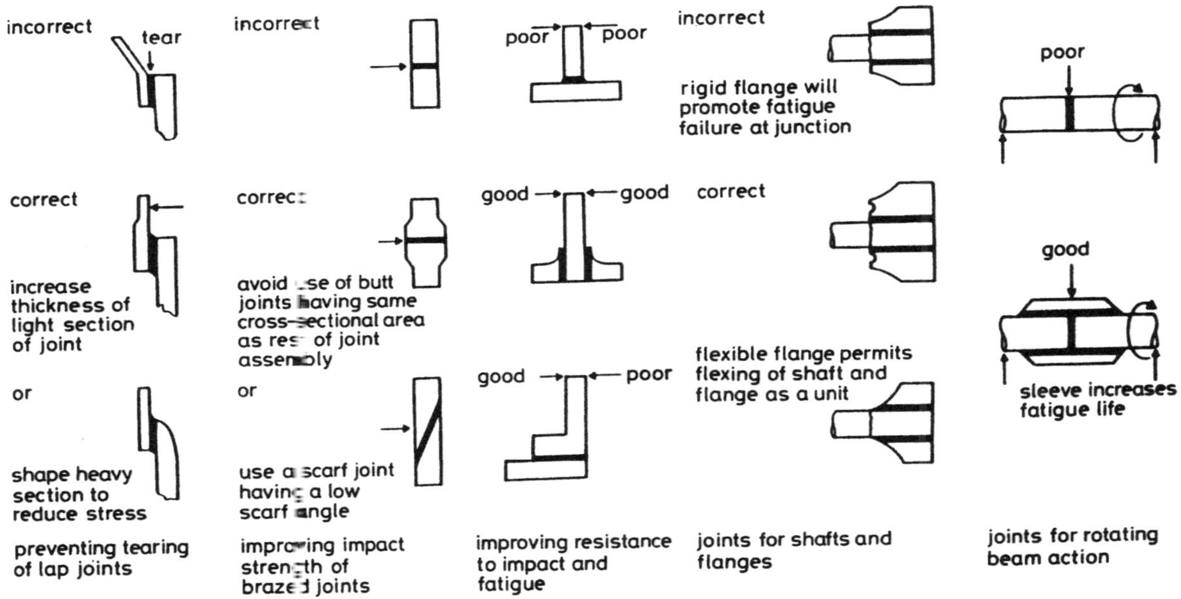

Fig. 24.6 *Correct design of joints reduces stress concentrations and fatigue failures*

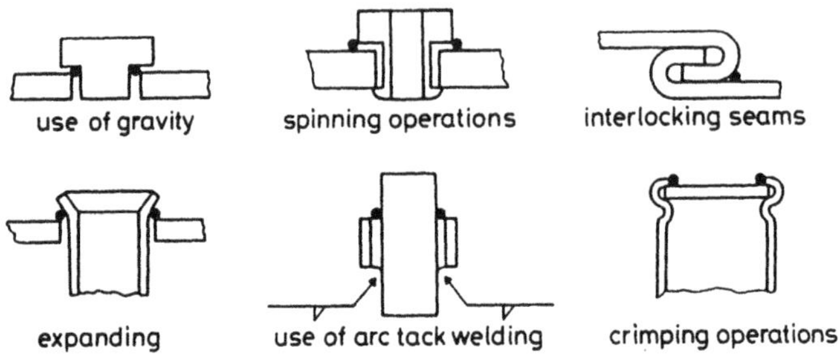

Fig. 24.7 *Self-jigging brazed joints*

is often obtained from experiment and the final equipment based on those results. Nevertheless, some rough calculations are possible and should be done.

Determine the mass of metal to be heated: The minimum amount of the workpiece should be heated consistent with the area of the joint, but heat losses play a large part in joining. If one cross-section is heavier than the other, more power should be applied to the heavier section to prevent overheating of the lighter part. If one part is steel and the other non-ferrous, the coil should be coupled closer to the non-ferrous part, as power input is greater for steel. If both parts are steel, a smaller area should be heated, as power input is greater and the temperature of the joint will reach the solder or brazing alloy flow temperature quicker than for non-ferrous parts. Steel-heating coils can, therefore, have a smaller number of turns. To avoid softening, areas which have previously been hardened should not be heated. If the work has sharp edges, heating must be slower and more metal must be heated, to prevent burning of the edges.

Table 24.1 *Chemical composition and melting range of alloys commonly used in induction soldering (after Lipsch and Capolongo[63])*

Metal (%)			Melting range (°C)	
Tin	Lead	Silver	Liquidus	Solidus
99.8	–	–	232	232
62	38	–	183	183
60	40	–	188	183
50	50	–	216	183
40	60	–	238	183
30	70	–	260	183
–	97.5	2.5	310	310
–	96.5	3.5	317	310

Other alloys are available.

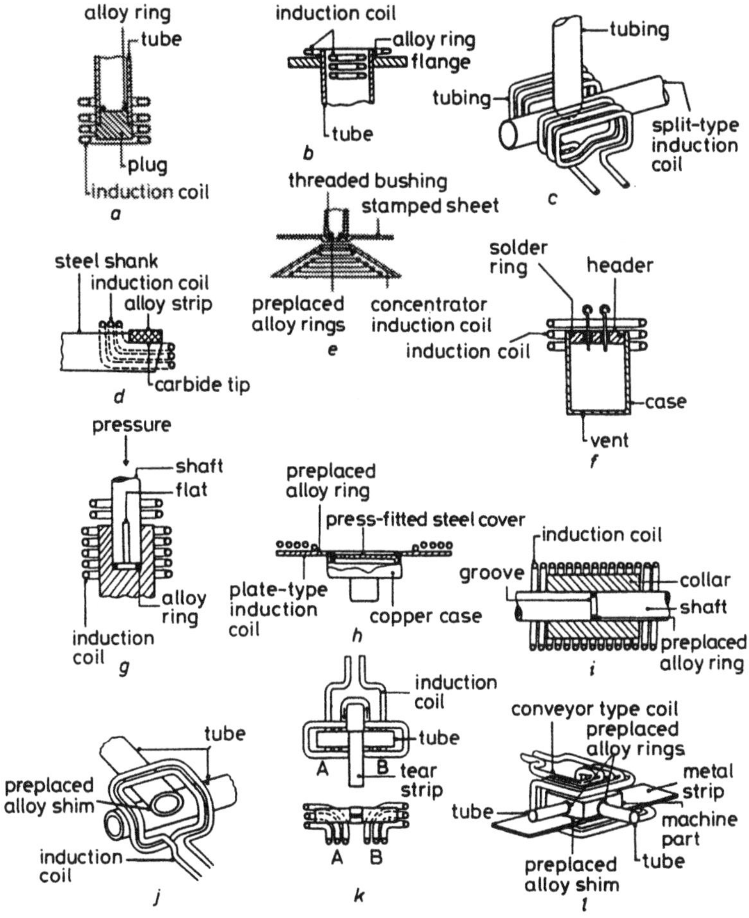

Fig. 24.8 *Typical joint and coil designs for induction soldering and brazing (after Lipsch and Capolongo[63]). (a) Plug-to-tube joint. Turns of solenoid coil may be spaced for suitable heat pattern depending upon the materials being joined. Tight fit desirable for maximum heat transfer. (b) Flange-to-tube joint. Internal–external coil. Flange chamfered to hold preplaced alloy. (c) Tube-to-tube T joint. Split solenoid coil. Alloy ring preplaced internally to provide uniform fillet without excess joining alloy along the sides of the tube. (d) Joining carbide tips to shanks. Open-end coil. Alloy strip preplaced as shown. (e) Threaded bushing joined to metal stamping. Concentrator coil localises heat. Alloy preplaced at base of bushing. (f) Header to case seal. Simple solenoid coil for uniform heating. Preplaced alloy recessed. Vent permits escape of expanded air to avoid leaky joint. (g) Shaft-to-fitting joint. Solenoid coil with varying diameter. Alloy preplaced internally at base of recess. Pressure used to seat shaft. Small flat permits escape of gas. (h) Assembly of formed metal shapes. Plate coil for localised heating. Alloy preform partially shielded from electromagnetic field by case. (i) Shaft-to-collar joint. Solenoid coil. Alloy ring preplaced in machined groove. (j) Tube-to-tube joint. L-shaped pancake coil. (k) Tear strip to tube assembly. Bucking-type coil, i.e. reversal of coil current between sections A and B prevents overheating tear strip (0.25 mm). Alloy preplaced beneath strip. (l) Assembly of components. Conveyor-type coil. Alloy shim preplaced between machine part and metal strips. Alloy rings preplaced on tubes.*

Table 24.2 *Chemical composition, melting range, and colour of alloys commonly used in induction brazing (after Lipsch and Capolongo[63])*

Alloy composition (%)					Melting range (°C)		
Silver	Copper	Zinc	Cadmium	Other	Solidus	Liquidus	Colour
72	28	–	–	–	779	779	White
15	80	–	–	5 P	641	785	Grey
50	15.5	16.5	18	–	627	635	Yellow
45	15	16	24	–	607	618	Yellow
35	26	21	18	–	607	702	Yellow
50	15.5	15.5	16	3 Ni	632	688	Yellow
54	40	5	–	1 Ni	727	857	Off white
65	20	15	–	–	668	718	White
10	52	38	–	–	766	852	Yellow
–	99.9 min	–	–	–	1083	1083	Copper
71.8	28	–	–	0.2 Li	–	–	White
85	–	–	–	15 Mn	960	971	–
–	62.5	–	–	37.5 Al	966	996	–
Ni 65/75, Cr 13/20, B 3/5, Fe + Si + C 10 max.					1010	1066	–
Si 11/13, remainder Al					577	582	–

Other alloys are available.

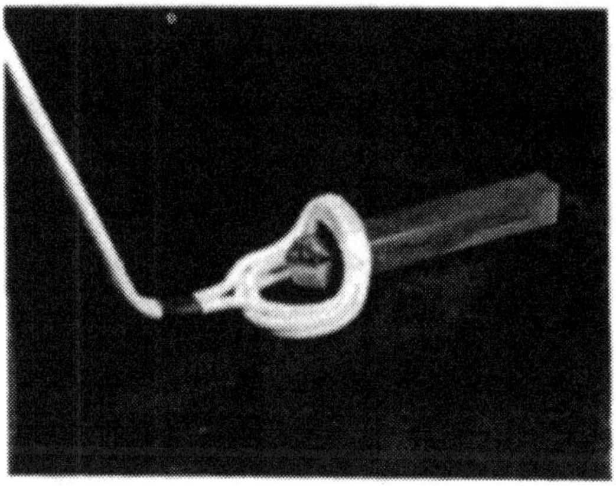

Fig. 24.9 *Tool-tip brazing coil (Radyne Ltd)*

Fig. 24.10 *Components suitable for induction brazing (Radyne Ltd)*

Current depth may sometimes determine the depth of metal heated in large pieces. Usually, however, heat conduction inward is the major factor for depth. Figure 24.12 shows the equivalent additional distance that is heated, for a given heating time. The additional volume can be calculated using the curves, when the approximate heating time is known, and the extra mass added to the design mass. (These curves are also useful in calculating the extra power needed to heat a piece for forging, allowing for heat flow.)

From these considerations, some idea of the mass to be heated can be found.

Fig. 24.11 *Saw-tip brazing fixture and coil. Tip and brazing alloy placed on anvil; saw brought forward and down on tip (Radyne Ltd)*

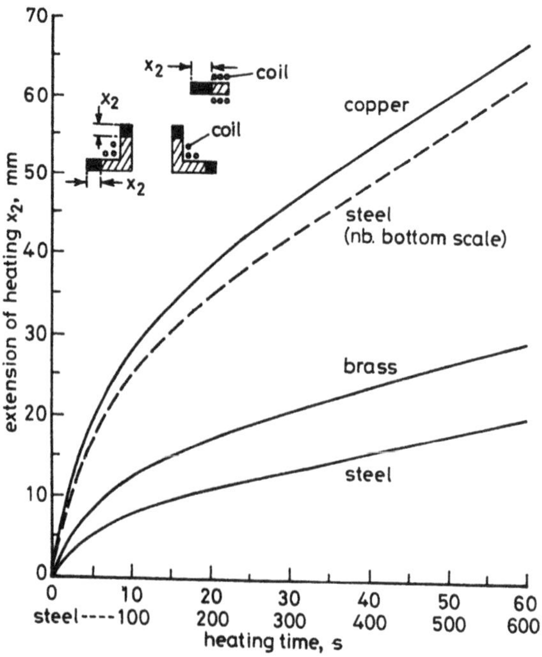

Fig. 24.12 *Extension of heating for different materials*

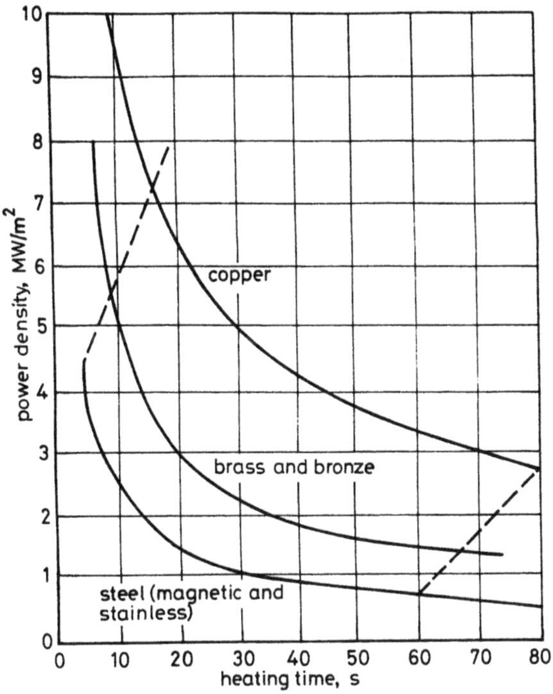

Fig. 24.13 *Radio-frequency generator rating (power density) for soft soldering at 188°C: recommended operating range between dotted lines*

Power input P: From the mass m kg of metal to be heated, the temperature rise $\Delta\theta$ °C and the heating time t s, the power input (in watts) can be calculated:

$$P = \frac{mc\Delta\theta}{t} \tag{24.1}$$

where c is the specific heat (in J/kg K), integrated over $\Delta\theta$. This value is approximate, because of the indeterminacy of the component values, but it gives some indication of power needed. Figures 24.13 and 24.14 illustrate typical radio-frequency power densities.

Radiation losses: For soldering applications, these are negligible. For brazing and welding, add 10% of P as a very pessimistic approximation.

Number of coils or pieces per coil: It is usually possible to improve the production rate by increasing the number of coils, or by heating more than one piece in the same coil, using a larger power source.

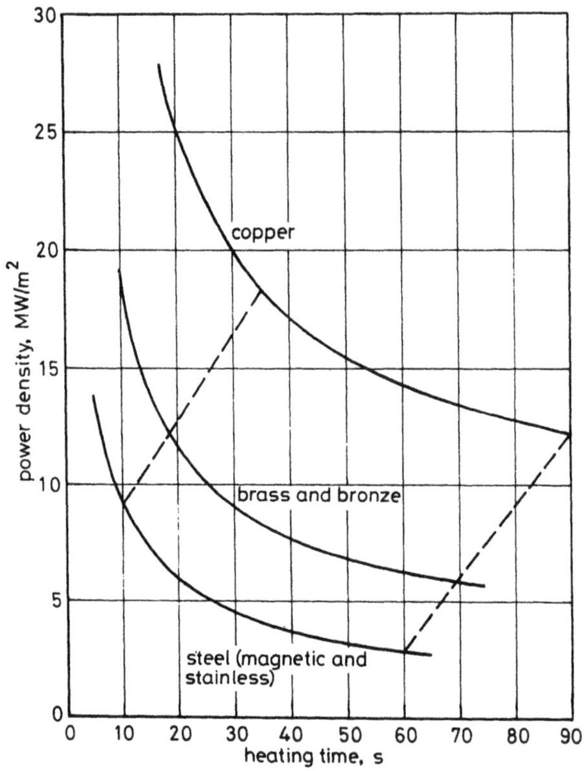

Fig. 24.14 *Radio-frequency generator rating (power density) for silver soldering at 700°C: recommended operating range between dotted lines*

Brazing or soldering alloy: If not specified for the application, a suitable alloy has to be chosen. A 50/15.5/15.5/16/3 silver/copper/zinc/cadmium/nickel alloy has the low flow-temperature of 635°C, and is suitable for induction brazing. Tin/lead/cadmium solders are suitable for soft soldering.

Configuration: See Section 24.1.1.

Coil dimensions: From the above data, the coil shape can be estimated. Clearances of 1.5 to 6 mm are practical for most radio-frequency brazing and soldering coils. Figure 24.15 shows some typical configurations; there are so many possible shapes that clearance is a matter of experience.

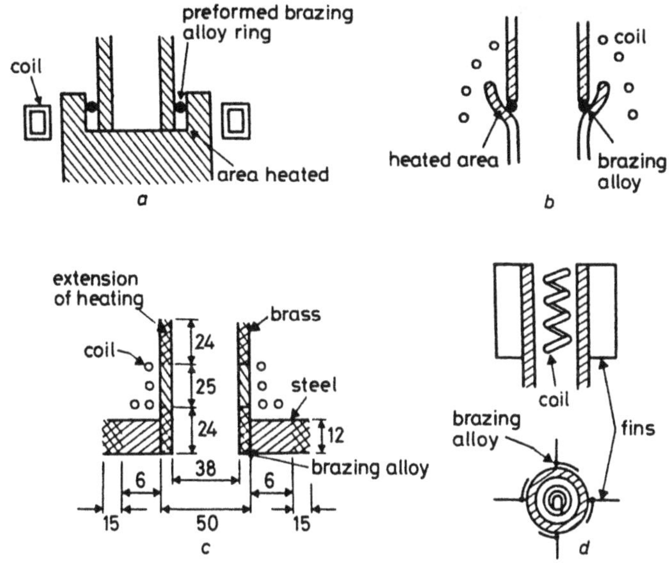

Fig. 24.15 *Typical induction brazing and soldering coil-shapes: (a) single-turn coil for steel brazing (b) multi-turn coil for non-ferrous brazing (c) steel-to-non-ferrous brazing coil (dimensions in millimetres) (d) internal heating coil for brazing fins to a shell*

Example 24.1
Specify the coil design and operating conditions for induction heating in the following brazing application:

Metals to be joined	steel and brass
Dimensions and shape of work	see Fig. 24.15c
Joining alloy	silver solder
Flow temperature	635°C (approximate)
Production rate	90 pieces/h
Heat time (100% duty cycle)	40 s
Density	7850 kg/m^3 steel
	8320 kg/m^3 brass
Specific heat over 667°C	607 J/kg K steel
	437 J/kg K brass

The silver solder is a silver/copper/zinc/cadmium/nickel alloy.

Solution
Power-supply frequency Because of the non-ferrous component, 400 to 500 kHz should be used.
Mass of metal heated Assume that the coil dimensions are as shown in Fig. 24.15c. From Fig. 24.12, the extensions of heating for a 40 s heat-time are 24 mm

(brass) and 15 mm (steel). Therefore, the total masses of metal to be heated are:

$$m\ (\text{brass}) = \pi(25^2 - 19^2)(25 + 48)(8320)/10^9 = 0.50 \quad \text{kg}$$

$$m\ (\text{steel}) = \pi[(31 + 15)^2 - 25^2]12(7350)/10^9 = 0.44 \quad \text{kg}$$

Power input needed Using eqn 24.1,

$$P(\text{brass}) = \frac{(0.50)(437)(635)}{(40)(1000)} = 3.47 \quad \text{kW}$$

$$P(\text{steel}) = \frac{(0.44)(607)(635)}{(40)(1000)} = 4.24 \quad \text{kW}$$

Therefore $P(\text{total}) = 7.71$ kW.
Add 10% for radiation loss = 0.77 kW, giving $P(\text{total}) = 8.5$ kW.
Check calculation, using Fig. 24.14
Cross-sectional area of brass workpiece = 829 mm^2
Cross-sectional area of steel workpiece = 1885 mm^2
(The steel section is taken at the joint itself.)

From Fig. 24.14, for a 40 second heating time,

$$\mathscr{P}(\text{brass}) = 7.75 \quad \text{MW/m}^2$$

$$\mathscr{P}(\text{steel}) = 3.41 \quad \text{MW/m}^2$$

Hence

$$P(\text{brass}) = (7.75)(829) = 6.42 \quad \text{kW}$$

$$P(\text{steel}) = (3.41)(1885) = 6.43 \quad \text{kW}$$

$$P(\text{total}) = 12.85 \quad \text{kW}$$

This must be reduced in the ratio 635:700, as Fig. 24.14 is prepared on the basis of 700°C, not 635°C as needed. This gives $P = 11.7$ kW; as the curves have been proven in practice, this figure should be used in preference to the 8.5 kW calculated above. However, the cross-sections of the workpiece are fairly arbitrary, so 11.7 kW should be regarded as a conservative first estimate.
Coil dimensions If we accept 11.7 kW as the power needed, the preliminary coil dimensions should be increased in the ratio 11.7:8.5, i.e. 1.37:1. The coil coupled to the brass could be 35 mm long and that to the steel about 8 mm. The actual coils would then be proven on an experimental set-up, using a variable ratio transformer.

As in induction hardening, concentrators can be incorporated to improve flux-concentration.

24.2 Tube welding

Induction heating, combined with mechanical pressure, is extensively used in the manufacture and joining of tubes.[64,65] The frequencies used are similar to those of induction heating for brazing and soldering.

The ends of tubes can be butt-welded in a very simple machine in which the tubes are separately gripped and held in line. When they are brought together inside a cylindrical inductor, they are heated to welding temperature and axial pressure forms the weld.

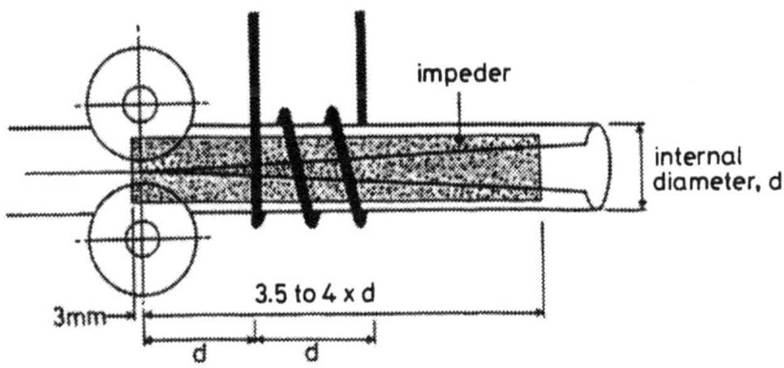

Fig. 24.16 *Position of coils and impeder relative to squeeze rolls (Thermatool Europe Ltd)*

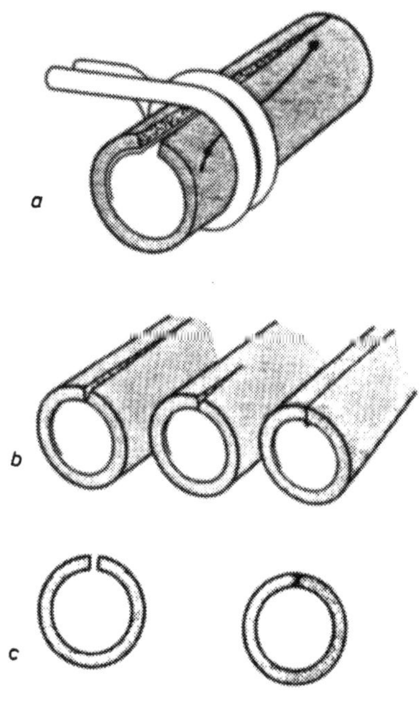

Fig. 24.17 *Induction heating for tube welding: (a) basic layout (b) undesirable conditions (c) ideal weld arrangements (Thermatool Europe Ltd)*

Much tubing is made from strip by rolling and welding. The flat strip is formed by rollers into a tube, with a small gap between the edges to be joined. An iron-cored inductor heats the edges to welding temperature and the weld is formed by pressure from another set of rollers (the squeeze rolls) which forces the edges together. The tube is then cooled by water and cut to length by flying shears. The inductor can be adjusted to control the heating so that the bead of welding metal forms on the outside of the tube (e.g. smooth-bore electrical conduit) or inside the tube for a smooth exterior (e.g. tubular-steel furniture).

24.2.1 Work coil

The layout of the work coil and impeder is shown in Fig. 24.16 relative to the position of the squeeze rolls. It induces voltage in the tube and current flows in the surface layers, establishing a complete circuit by flowing along the edges of the open V up to the point at which metal-to-metal contact is first made (Fig. 24.17a). It is very important that this contact is made prior to the entry of the tube into the squeeze rolls. High-frequency current passes more readily through the area of lowest impedance, which occurs at the apex of the closing strip-edges.

If, undesirably, the inside edges of the strip meet first, then the major portion of the high-frequency current flows through this point (Fig. 24.17b), giving excessive temperatures, which results in an 'as-cast' structure, oxide inclusions, a tendency to porosity, and (finally) a large and uneven internal weld-bead. There is a considerably-increased high-frequency power in the tube, resulting in a wide heat-affected zone and reduced welding speed.

The best results from the high-frequency technique are obtained when the mating edges of the strip are parallel with one another at the point of first contact. The current is then substantially evenly distributed across the section of the material and, under these conditions, a true 'forge weld' is obtained (Fig. 24.17c). The resulting structure gives a constant weld with maximum strength and minimum internal weld bead. Parallel mating of the strip edges is not easy to obtain, as the tube is not, at this point, contained within rolls and its natural tendency is to open out due to the residual tension in the outer surface of the strip.

The most successful forming technique so far employed to produce the parallel mating faces, prior to entering the squeeze rolls, has been to provide a tube profile that is essentially circular but has a slightly flattened top. This shape is produced by overforming of the strip edges in the breakdown passes of the forming mill.

Figure 24.18 shows a practical welder.

The work coil should be arranged as follows (see Fig. 24.16):

- The coil internal diameter must have working clearance from the tube.
- The coil length should, ideally, be equal to the internal diameter.
- The coil should be positioned such that the distance between the weld point and the nearest part of the coil is equal to the internal diameter.
- The number of turns on the coil should be sufficient to enable maximum anode (plate) current to be achieved at 100% setting of the power control.

a

b

Fig. 24.18 *Continuous seam welding of automobile transmission tube, conduit, irrigation, or gas pipe formed from strip material. Using 150 kW radio-frequency power, 50 mm 14-gauge ferrous tube can be welded at 1.5 m/s. (a) Installed machine (b) Welding point (Thermatool Europe Ltd)*

- The coil should be insulated to provide maximum protection against flash-over and give the longest work-coil life.

With the work coil set up in this manner, the highest weld efficiency will be achieved, giving the narrowest heat-affected zone and maximum weld speed, for a given input power. There is a certain overlapping of the effects of the work coil and impeder to obtain this optimum efficiency, but they will be considered separately at this stage.

The position and dimensions of the work coil can substantially alter the width of the heat-affected zone and (coupled with the impeder) its profile throughout the wall thickness of the tube. If, for example, it is necessary to increase the width of heat-affected material, it is a simple matter of pulling the work coil away from the weld point. This increases the preheat in the strip edges and, thus, the width of heated material. More power is then needed to produce the same welding speed, as a greater mass of metal along the strip edges has been raised in temperature.

The aspect ratio of the work coil has a considerable effect on the efficiency of transfer of energy between the work coil and the tube. In addition, the number of turns on the coil affects its impedance and, in turn, affects the transfer of energy from the welder to the tube. More energy is transferred if the reactive component of the work coil is reduced; reducing its diameter would do this, but there are practical limitations to getting the coil close to the tube. An alternative solution would be to reduce the number of turns on the work coil – but this would demand an even higher current from the welder. The only satisfactory answer from a tube manufacturer's point of view is to obtain a welder with an adequate Q to give sufficient coil-to-tube clearances and, at the same time, to transfer the full output power into the work.

The intense electromagnetic field around the work coil attracts metallic particles from the incoming strip and these can be locally heated, causing premature failure of the coil insulation. It is, therefore, important that the coil be kept as clean as possible, to minimise mill stoppages due to work coil failure.

24.2.2 Impeder

The impeder is a flux-concentrator held inside the tube to improve heating, as shown shaded on Fig. 24.16. It should have the following properties:

- The core of the impeder should be made of ferrite having the following optimum characteristics:

Initial relative permeability	3000
Saturation flux density	0.35 T
Curie point	180°C.

- The core should be enclosed in a non-metallic sheath and cooled by water or mill coolant.
- Metallic particles must be prevented from passing into the impeder via the coolant, otherwise premature failure due to overheating may occur.

- The ferrite should be 3½ to 4 times the length of the work coil and its downstream end should project approximately 3 mm through the weld-point centre-line.
- The greater the volume of cooled ferrite within the tube, the greater will be its efficacy.
- The smaller the diameter of tube being welded, the more important the efficiency of the impeder.

The sheath of the impeder which retains the ferrite can be made from PTFE (Teflon), resin-bonded paper, resin-bonded fabric, nylon, or any non-metallic material. However, the longest life appears to be obtained when using GRP (glass fibre) tube. The plug in each end may be made from any non-ferrous or non-metallic material, provided it has adequate strength to take the thread of the impeder retaining-tube which carries the coolant supply.

Impeders can be designed to suit tube mills where internal weld bead removal is required. Generally speaking, the smallest practical internal diameter for this operation is 32 mm, but some impeders have been made to fit around internal bead trimmers for tubes as small as 19 mm in diameter.

As with the work coil, a change in the optimum position or size of the impeder results in a change in the profile of the heat-affected zone through the cross-section of the strip being welded. If, for example, the impeder is drawn back from the weld point, the heat-affected zone on the inside of the tube will be wider. Generally speaking, the width of heat-affected zone on the outside surface of the tube is governed by the position and dimensions of the work coil, and the width of the heat-affected zone on the inside surface of the tube is governed by the position and dimensions of the impeder. If both coil and

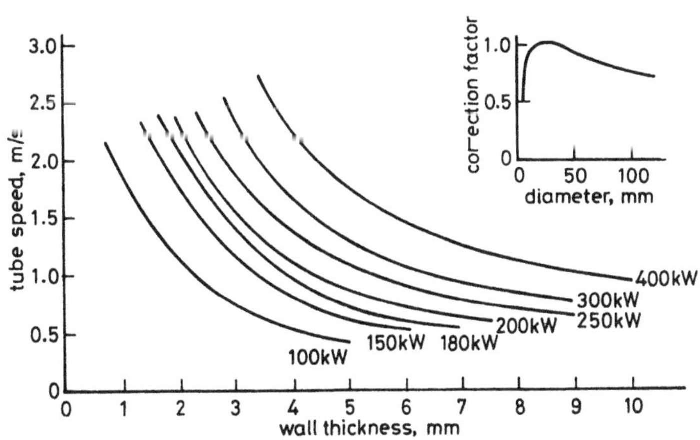

Fig. 24.19 *Welding speeds for various wall thicknesses and input powers, plotted for steel tube of 25 mm diameter. For other diameters, use the correction factor in the insert. For other metals, multiply the speed by: aluminium 1.4; brass 1.1; copper 0.8; stainless steel 0.9 (Thermatool Europe Ltd)*

impeder are pulled back from the weld point, a wider heat-affected zone is obtained and the resulting weld has a larger upset, but more ductile characteristics. A large mass of material will have been raised in temperature and, therefore, more power will have been consumed. It is unlikely that weld speeds shown on the charts will be achieved under these conditions.

The welding speeds that can be obtained from welding generators of various powers are shown in Fig. 24.19. They are plotted for steel tube with diameters in the 20–40 mm range. For other metals, use the correction factor given in the caption; for other diameters, use the factor from the correction curve. These curves are empirical and have been compiled as 'best fits' from experimental data. In particular, the correction curve with diameter is subject to continuous change, as the properties of ferrites improve. The data should, therefore, be treated as approximate and a manufacturer should be consulted for up-to-date figures.

Frequencies in the 300–500 kHz range are used for tube welding. Figure 24.19 was compiled for 400–500 kHz, but is believed to be applicable for the wider range.

24.2.3 Strip

The characteristics of the raw material used in the manufacture of tube are of prime importance if consistent results are to be obtained.

The quality of the incoming material must be of satisfactory standard and the dimensions must be as constant as possible. A variation of plus or minus half a gauge can be tolerated in the thickness of the material, but a variation in width of strip by as little as 0.25% can cause complete loss of the weld. This is obvious considering that, under ideal weld conditions, the strip edges are only pressed together with sufficient load to produce the minimum of upset and to provide a consistent weld throughout the section of the material. These tolerances are quite practical on modern slitting lines. High quality and consistent strip goes a long way towards the production of the best quality tubing.

24.2.4 Seam guide

The seam guide is a simple device for the correction of small variations occurring in the incoming strip. Most tube mills are fitted with a seam-guide unit which is positioned between the last fin pass and the weld point. It retains the weld area on top of the tube so that the weld bead can be removed by the trimming tools.

The seam guide must not be used to compensate for incorrect setting of the forming mill, otherwise considerable reduction in weld efficiency and quality results. When using a seam guide with the high-frequency process, it is necessary to take some basic precautions to prevent, or at least limit, the current flow through the blade. If, for example, a hardened steel blade is used, it should be positioned far enough away from the weld point to eliminate sparking. Any sparks which occur between the strip edges and the seam-guide blade can cause the formation of oxide pockets and inclusions in the subsequent weld. In addition, if current is flowing through the seam guide, then power is being lost and weld efficiency suffers. The most elegant solution to the problem is to use a seam-guide blade made from a ceramic material, such as sintered alumina. This completely eliminates the flow of current between strip and blade but, unfortunately, the ceramic alone is not particularly strong. A good compromise can be obtained by having a hardened steel blade and a ceramic blade clamped together.

The seam guide, apart from locating the current radial position of the weld, can be used to provide control of the approach angle of the weld point. Ideally, this angle should be between 1° and 2° on steel, and between 3° and 7° on non-ferrous materials.

If the approach angle varies, then a change in weld temperature results due to the change in impedance of the open V. It is thus important that the approach angle is kept as constant as possible, if consistent weld temperatures are to be obtained.

24.2.5 Squeeze rolls

Squeeze rolls have two distinct functions:

- to press the edges of the strip together with a controlled force
- to locate and shape the tube to a welded form, dictated by the squeeze-roll profile and the previous forming operations.

The simplest form of squeeze-roll assembly consists of two rolls mounted horizontally with a single control which adjusts both rolls simultaneously to apply the required squeeze load (Fig. 24.20a). The rolls cannot be adjusted

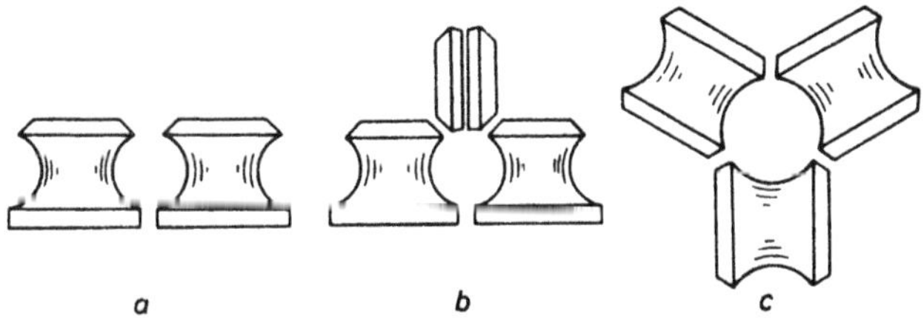

Fig. 24.20 *Roll arrangements for induction tube welding: (a) simple two-roll (b) two-roll, with supplementary roll (c) three-roll (Thermatool Europe Ltd)*

vertically other than by means of shims beneath the bearing assemblies. This system does not allow individual control of the strip edges, which can be a disadvantage where the highest weld quality and strength are required.

An improved and more flexible system is achieved if two additional rolls are added to the basic two-roll assembly (Fig. 24.20b). These rolls are mounted vertically above and to each side of the weld seam and are individually adjustable in a vertical plane. They are of necessity quite narrow in section, but only have to provide a correction force if there is a misalignment of the edges of the strip. In either of the two-roll, squeeze-roll systems, the roll diameter is large in order to allow adequate bearing and spindle diameter to give the required forge load.

A three-roll, squeeze-roll system combines most of the advantages of the other systems with few disadvantages (Fig. 24.20c). The rolls are mounted at 120° to one another, with one roll beneath the tube and the other two either side of the weld seam. Each roll should be individually adjustable and the whole assembly can be moved in a vertical plane. The squeeze rolls are supported in yokes which allow smaller bearing spindles and, therefore, smaller diameter rolls, which is important for tube diameters less than 37 mm, where it is necessary to place the work coil near the weld point for maximum efficiency.

On some installations it has been the practice to insulate the roll from its spindle in an endeavour to reduce roll heating due to induced currents from the work coil. At radio frequencies, however, appreciable current flows through the insulation which acts as the dielectric of a capacitor, and, therefore, some heating must occur. Better thermal transfer takes place if the insulation is not provided.

The squeeze-roll diameter should be as small as possible, consistent with adequate bearing strength, in order that the work coil may be situated in its most efficient position.

24.3 Heating of resin kettles and other vessels

Not all applications of induction heating involve high temperatures. Certain chemical engineering operations need a heated vessel; for example, in the preparation of paints, the resins have to be mixed at a temperature of 200 to 300°C. Oil or gas heating could not be tolerated because a flame would be dangerous in the presence of inflammable vapours, so the choice is between steam or electric heating. If steam heat is available, this may be cheaper, but still involves heat transfer. Up to 7 to 10 kW/m^2,[66] resistive elements are possible and should be considered; above 10 kW/m^2, induction heating is used (Fig. 24.21).

The basic idea is simple; a coil, air- or water-cooled, is wound around the vessel. Air cooling can be used to pressurise the coil region for flame-proofing. Since vessels are naturally cylindrical, the coils can normally be wound on a cylindrical former. (Apart from the ease of manufacture of the cylindrical vessels, they often accommodate rotating scrapers or agitators which go naturally with a cylindrical inner surface.) As seen in previous sections, the gap between coil and vessel should be small, to improve the power factor. Bottom heaters are also used, as shown.

The induction method is very controllable, the heat is produced in the steel vessel without heat transfer problems, and the induction coil does not add to the thermal mass of the vessel. Induction heaters are offered in the range 1 to

500 kW and their use is growing. The vessels are preferably ferromagnetic (usually mild steel) and, for chemical reasons, they are often lined with stainless steel. Occasionally they are completely fabricated from non-magnetic stainless steel, with a resultant lower power-factor and increased heater cost. Although higher frequencies are possible, supply frequency is normal, remembering that the power densities are low compared with billet heating.

For larger installations, say over 30 kW, the coil system would be designed for three-phase operation.

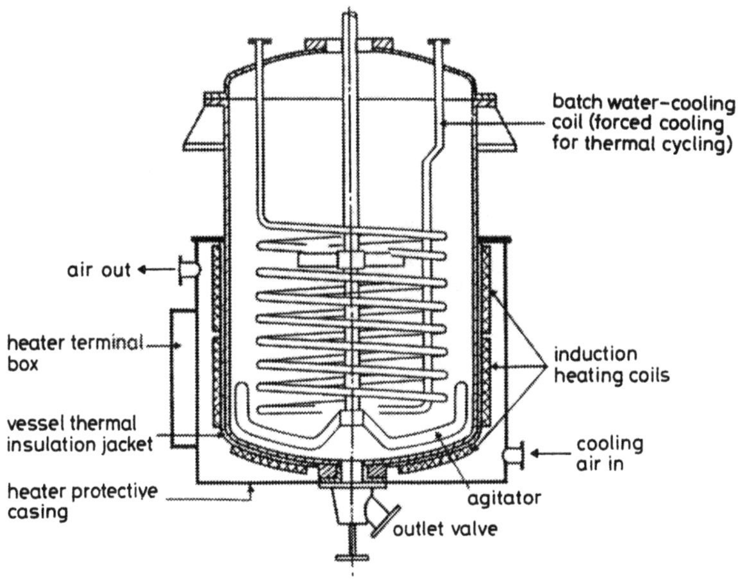

Fig. 24.21 *Basic vessel-heating layout (Cheltenham Induction Heating Ltd)*

If the coils are simple, as shown in Fig. 24.21, the flux has a free path outside the coil. Care must be taken that this flux does not link with surrounding steelwork, otherwise heating and the risk of sparking results. These leakage fluxes can be channelled, using suitable packets of laminations, but this adds to the cost. The electrical efficiency of an air-cooled coil is greater than 90%. Since the vessels are ferromagnetic and heated well below the Curie point, the calculation of losses involves variable permeability. A semi-empirical method, claimed to give accurate results, is found in Thornton[20] and the problem is discussed in detail by Ellett[21] (see Chapter 13).

There are many applications for this technique in chemical engineering. A 500 kW, 50 Hz heater is shown in Fig. 24.22. Rectangular lead-baths are also heated with similar coils.

Fig. 24.22 *A 500 kW, 50 Hz vessel heater (Cheltenham Induction Heating Ltd)*

24.4 Paint drying

In the Soviet Union,[55] induction heating at 50 Hz has been used to heat the metal surface of a railway carriage to dry the paint and varnish more quickly. The scheme is shown diagrammatically in Fig. 24.23.

The coil is wound in broad slots in a magnetic core. When the assembly is placed near the surface to be heated, flux paths are formed, as shown; the pulsating flux heats the metal and dries the paint on its surface. This is a good method of drying paint and varnish as the heat comes from below and drives

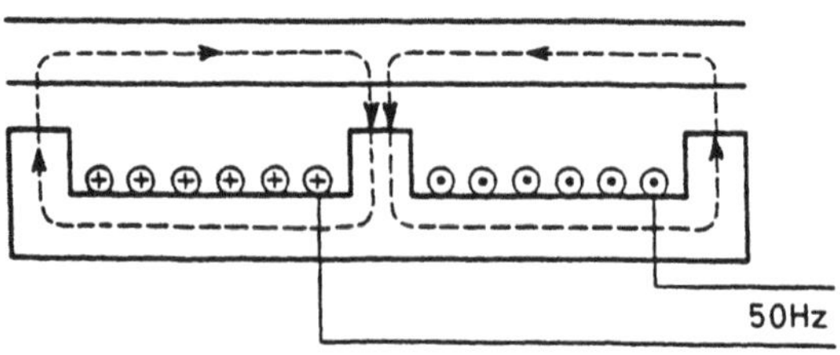

Fig. 24.23 *Heating a metal surface for paint drying*

the volatile constituents towards the surface, in contrast to radiant heating where a skin is formed on the surface, trapping the volatile vapour. Lozinskii gives details in his book;[55] he quotes 15 to 20 min to heat 3 mm of steel up to 100 to 130°C. The power consumption is 0.7 to 1.65 kWh/m^2 of surface, at a power factor of 0.49 to 0.58. Travelling-wave heaters (Sections 22.12 and 24.11) can also be used for this.

24.5 Induction heating in plastic working

Thermosetting plastics are formed by press moulding, injection moulding, or extrusion. Heat and pressure are necessary for polymerisation (curing). The press platens, injection chambers, and extrusion cylinders can be heated by gas, steam or resistive elements, but greater efficiency and more rapid generation of heat come from induction heating, which can be accurately controlled.[67]

Induction coils are built into the body of the platen, as shown in Fig. 24.24. When alternating current at mains frequency is passed through them, heat is generated in the platen. Inductors may also be fitted easily to the appropriate parts of injection moulding machines and extrusion machines to heat the parts in a similar way. Surges of excessive temperature with the consequent waste of power are avoided. The temperature control responds quickly and heat losses are considerably reduced.

Heated platens are available from 150 mm × 150 mm to 1200 mm × 1200 mm, with temperatures of 200°C, 300°C and 350°C. The power densities

Fig. 24.24 *Inductor coils for platen heating (Cheltenham Induction Heating Ltd)*

are approximately 15, 30 and 37.5 kW/m^2, respectively, for the three temperatures and are reasonably constant throughout the size range. The heaters are single-phase up to 400 mm square and three-phase over 750 mm square: between these sizes, both single- and three-phase heaters are offered.

Induction heating can be used for heating the cylinders of extrusion machines, speeding up production by more rapid and uniform polymerisation to give a better product.

Induction heating improves productivity for the following reasons:

- Heat is generated in the metal parts deeply and evenly without unacceptable temperature gradients. There are negligible temperature changes at the moulding surface. This results in better and more accurately-controlled heating of the material and, consequently, more consistent polymerisation.
- The heating units have a high thermal efficiency and heat can be concentrated where it is required. Consequently, there is a wider choice of conditions for processing and demands can be catered for that cannot be met by other methods.
- Conditions for moulding the material are improved, leading to the quick production of good quality products, with less current consumption and negligible maintenance costs.

See also Section 24.11.

24.6 Annealing and stress relieving

A hardened steel may be brittle and therefore unable to withstand mechanical shock. To overcome this, steel is *tempered.* High-carbon steels are tempered at temperatures about 200 to 300°C. The original hardness of the steel is not greatly decreased, but the brittleness is lessened by equalising the internal strains which have accrued during hardening.

Tool steels are tempered at higher temperatures – up to about 650°C. The original hardness is reduced, but the ductility and toughness of the steel are greatly improved. When this treatment is applied to medium-carbon steel, a tough and strong material is produced capable of withstanding sudden shock and having a good fatigue life.

Annealing steel and other metals means heating to a definite temperature for a definite time and then cooling at a suitable rate. The treatment softens the metal and improves its machinability and cold-working properties. It also frees the metal from any internal stresses set up by previous working.

The *normalising* of steel involves heating it to a higher temperature than for annealing and then permitting it to cool naturally in air. The process is sometimes referred to as *grain refining*. It renders the structure of the metal more uniform and improves the mechanical properties. Bars which have been rolled and emerge from the mill at a comparatively low temperature lose much of their ductility, which can then be restored by normalising.

Stress relieving is a heating process which is carried out at a temperature generally lower than that required for annealing or normalising. The sole purpose is to release internal stresses, such as occur due to machining, casting, hot forming, irregular cooling, or welding.

Induction heating can be used to advantage for all the above processes owing to the ease with which it can be applied, its cleanliness, its availability and easy adaptation to special requirements.[67]

24.6.1 Tempering

The induction tempering of steel is quicker than furnace tempering. For example, 25 mm diameter bars can be tempered in 1 min, compared with 30 min or more which would be required by ordinary furnace heating.

Steel bar-stock can be hardened and tempered in a continuous process on a single machine, using induction heating. A uniform and tough material is obtained in a scale-free form suitable for machining into high-quality components. The hardness penetration can be controlled so that a relatively soft core can be left for subsequent removal by drilling or by machining to produce, for example, components such as nuts from hexagon bars, where the hardness is required only in the wall of the finished article.

24.6.2 Annealing

Induction annealing has the advantage that it can be localised in selected parts of the object. For example, annealing is required between successive pressing operations in the manufacture of automobile doors. Induction-heating equipment can restrict heating to the corners of the doors. With conventional methods, a very considerable area would have to be heated, with a consequent waste of power and with the risk of distortion. The work is heated locally in about 20 s by means of four 'pancake' coils connected in series and supplied by a radio-frequency generator rated at 7 kW. The process is semi-automatic and the machine has an output of 120 doors/h.

The ends of shafts made from mild steel which have been carburised can be annealed. The process can be carried out in the production line. Localised annealing is obtained with a minimum scale. The machine can deal with 200 shafts/h, using a 2.75 kW radio-frequency generator. Automobile rear-axle driving pinions are annealed. The component, made from a low-carbon mild steel, is fully machined except for the threading of the shaft end, and is then carburised. The shaft end must then be annealed to obtain the required physical properties. Selective heating by radio-frequency induction allows this to be carried out in the production line. Three sizes of shafts are treated in pairs, one pair at a time. They are inserted by hand into the appropriate inductors and the operator initiates the heating cycle, which is controlled automatically by a process timer. At the end of the cycle, the operator removes the components, which are allowed to cool slowly in a storage receptacle. The process is fed from a 6 kW, 400 kHz radio-frequency generator, and an output of 370 components/h is obtained.

24.6.3 Stress relieving

Stress relieving of large components can be carried out by supply-frequency induction heating, but experience has shown that medium-frequency induction heating is often preferable, in spite of the capital cost being somewhat higher, because of the higher power-density. The lower current in the inductor allows smaller cables to be used, which are easier to handle and can be applied more readily to the component.

Medium-frequency power is used, for example, for stress relieving the heat exchangers for a nuclear-energy power-station. The heat exchanger is built from a number of steel cylinders welded together and finally closed by having domed ends welded to them. The edges of the sections to be welded together can be preheated inductively before welding and, when welding is finished, the whole of the affected area can be inductively heated to the temperature required for stress relieving. The parts to be heated are covered with thermal insulation on the outside and flexible cables of a special construction are laid in place to form the inductor. Heat insulation is then applied outside these, held in place by non-ferrous plates clamped to the cylinders. Similar lagging is applied to the inside of the parts to be heated. The ends of the cables are taken through boxes in the floor and there connected to bus-bars, which are supplied from a 200 kW, 2500 Hz generator.

The use of induction heating for stress relieving and annealing is not confined to large-scale operation but can equally well be applied to small components, e.g. local stress-relief of tractor pins which have previously been furnace-hardened. One machine for this purpose has a production rate of 720 pins an hour.

24.7 Longitudinal flux induction heating

The methods of billet and slab heating described in Chapter 22 can be described as 'longitudinal flux' methods, since the field produced by the solenoid is parallel with the surface in the axial direction of the workpiece. It is possible to use similar methods to heat moving steel strip. In the manufacture of tin-plate, the tin is electrolytically deposited on the steel surface. It is then desirable to 'flow-brighten' the tin by raising its temperature above its melting point (232°C), causing the tin surface to melt and flow into a uniform tightly-adhering layer. This can be done using coils wound around the strip, as shown in Fig. 24.25.

Because tin plate is usually thin (0.18 to 0.36 mm), radio frequencies have to be used, but thicker materials can be heated at 10 kHz. Strip widths up to 1.5 m can be handled. The speed of the strip may be up to 66 m/s. The water-cooled coils are wound to conform closely to the strip, taking electrical and mechanical

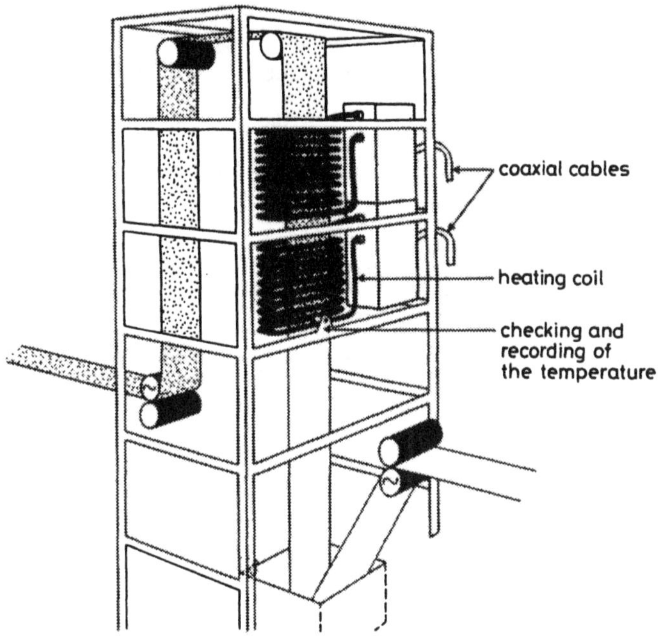

Fig. 24.25 *Longitudinal-flux heater for flow-brightening of tin-plate after electrolytic plating (after Elphiac Ltd)*

clearances into account. Calculation methods are given by Baker.[68] Simpson[69] shows a 2400 kW, 150 kHz line.

It must be emphasised that this method only applies to ferromagnetic materials, where the depth of penetration at radio frequencies is small compared with the thickness.

The power needed to heat a 1.5 m wide (w), 0.25 mm thick (t) steel strip through 250°C (θ) at 5 m/s (v) is

$$P = wtv\theta(c\gamma)$$

$$= (1.5)\frac{0.25}{1000}(5)(250)(c\gamma) = 1.83 \quad \text{MW}$$

where c is the specific heat (496 J/kg K) and γ is the density (7850 kg/m^3). This shows that substantial power is involved.

24.8 Transverse-flux heating

Longitudinal flux is only suitable for heating ferromagnetic strip. For annealing and heat treatment of non-ferrous strip, and for steel above the Curie point, some other method is needed. In 1950, Baker[70–72] proposed a transverse-flux heater, where the flux goes through the strip, instead of passing along it as it did in Section 24.7. The strip passes between an array of magnetic poles. The coils on the poles are arranged to be aided on the two sides of the gap, so that the polarities are as shown in Fig. 24.26.

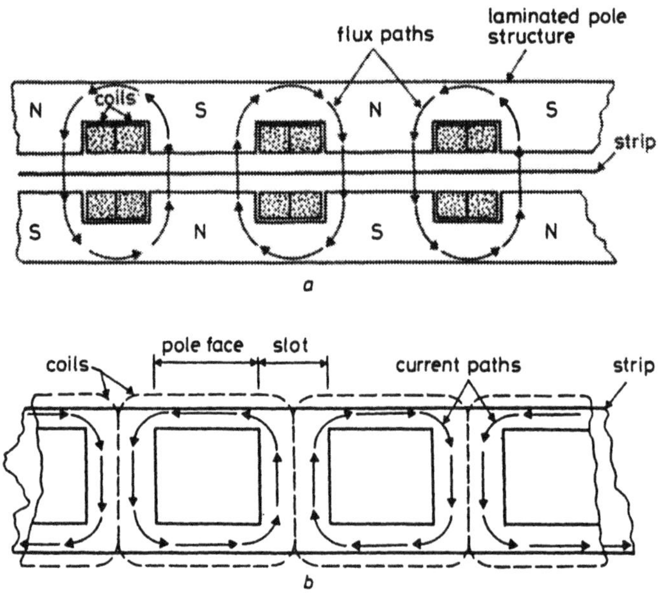

Fig. 24.26 *Transverse-flux heating*

Little seems to have happened to this idea at the time,[73,74] although Baker discusses it in detail. Simpson[69] implies that the method is used, but no ratings are given. The concept has been revived by the Electricity Council Research Centre at Capenhurst, both for moving strip and for slabs. The reader is referred to Jackson and his co-workers[75–77] for details. One problem that had to be overcome was the non-uniform heating, since the currents flow around the poles. Skewing of the poles was tried, but this simply skewed the non-uniformity.

Although, at first sight, the transverse-flux heater looks simple, there are complications. Nevertheless, Davy McKee (Poole) Ltd has developed this, in cooperation with Electricity Council Research Centre, Capenhurst, into a successful commercial device (TFX™). Ireson[78] describes an aluminium annealing line handling 20 tonnes/h, having a power input to the strip of 2125 kW at 75% efficiency (input 2800 kW). The power-density is 1–1.5 MW/m^2. The air gap is about 100 mm. The pair of inductors weighs about 12 tonnes. TFXs have been built for aluminium annealing lines up to 3.1 MW. TFXs use medium frequencies; the actual frequency depends on power level, material and thickness.

The heating currents flow in patterns dictated by the geometry of the gap. Left to themselves, they would not heat the strip uniformly. Ireson says: 'The most difficult single problem . . . is controlling precisely the flux at the strip edges where the field distribution is affected by several factors.' The aim is to get heating uniform to 1–2% and this is done by using 'auxiliary pole-pieces and other trimming devices or flux-modifiers' to adjust the flux-density distribution until the temperature uniformity is achieved, whereupon the settings of these trimmers are stored for future use. Each pair of inductors covers a limited range of strip width etc., so each system is provided with five or six interchangeable pairs stored in a magazine; their deployment is computer controlled and they can be changed automatically in 2–3 minutes.

Despite the megawatts of power input, there is typically only 5 kg of aluminium being heated at any instant, which would melt in 1.5 s if the line stopped, so control is vitally important. In fact, Ireson says that the TFX progress 'is largely due to the development of medium-frequency thyristor converters with rapid-acting electronic control'.

TFX is a very important idea, which is capable of much further development.

24.9 Semiconductor processing

High-frequency induction heating has been extensively used in the refining and subsequent doping of germanium and silicon crystals for semiconductor manufacture. Unlike many high-frequency heating processes, the times are sufficiently long to warrant the use of temperature, rather than time, control. Figure 24.27 shows a typical crystal-growing installation.

The principles of crystal production are complex and are covered in the literature.[79–83] The material (silicon or germanium) is first refined to remove as much impurity as possible. This is usually done by zone-refining, that is by passing a molten zone from one end of a suitably-shaped charge to the other. A very high degree of purity is achieved by successive passes.

Zone-refining concentrates impurities at the interface between the molten and just-frozen material, using the effect of impurities on the melting point of the pure material. If the impurities lower the melting point of the pure material, their concentration in the molten zone is increased. If the melting point of the pure material is raised, the concentration of impurities in the molten zone is lower. Therefore the molten zone, which is caused to travel along the rod of material, builds up impurities, which are then left at one end. The degree of refining and purification is a function of the number of passes made along the rod by the molten zone, but it has a practical limit. Figures of one part per billion of electrically-active impurities in silicon have been quoted.[79]

Germanium is usually refined in a carbon boat (Fig. 24.28a), where the melted zones are caused to move down the material, carrying the impurities with them to the end of the boat. The melting temperature is 950°C.

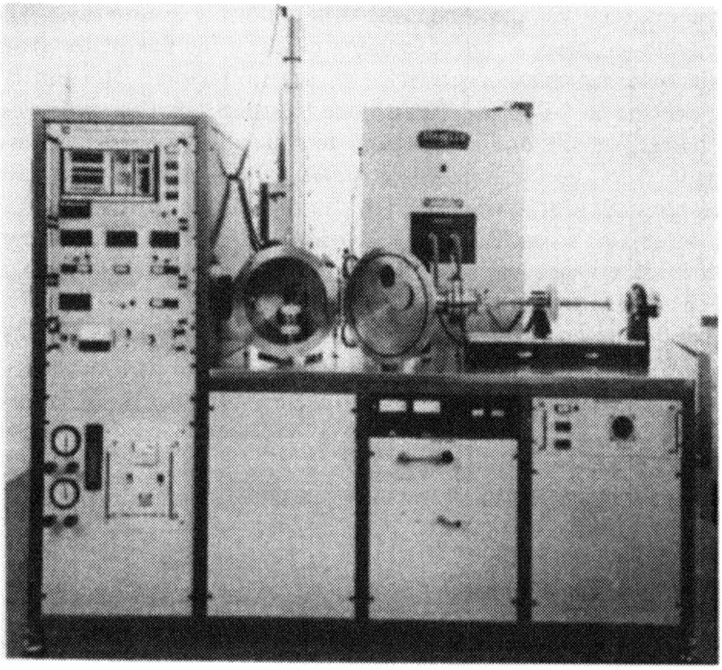

Fig. 24.27 *Crystal growth and zone-refining equipment: a development unit at the Clarendon Laboratory, University of Oxford (Stanelco Ltd)*

Molten silicon has a great affinity for most boat materials, so the floating-zone method shown in Fig. 24.28b is used. Here, a silicon rod is raised and lowered through a coil; it is also rotated. The small molten zone, heated to 1400°C, has enough surface tension to keep the rod in one piece. The impurities are carried to the ends of the rod, as in the germanium method. Rods of 25 mm or more in diameter can be refined using the levitation force which results from the interaction between the coil and the induced currents. The number of passes needed is greater for silicon than germanium, as the distribution of impurities is less favourable.

The basic requirements for floating-zone refining are a radio-frequency heater, an accurate mechanical variable-speed vertical drive for the rod, an automatic recycling mechanism, and possibly a means of temperature control. In general, the radio-frequency power output is time-controlled via a variable anode voltage by a programmed sequence. It is sometimes necessary to preheat the rod by resistance heating in order to lower its resistivity sufficiently so that the radio-frequency power can be induced into the material. If a frequency of 3 to 5 MHz is selected instead of the conventional 450 kHz, the loading problem is easier and the preheating can often be eliminated. The higher frequencies also result in a reduction of mechanical oscillation in the zone, caused by the generator power-supply ripple (usually 300/360 Hz, 6 to 10%) modulating the radio frequency. Most levitation processes use a controlled atmosphere or a vacuum, so that the rod is surrounded by a tube to contain the gas or vacuum. This results in a loosely-coupled high-Q coil, requiring fairly high radio-frequency current and voltage. The higher-frequency range is more suitable for this type of coil.

It is extremely important to maintain the length of the zone, and therefore the power input, constant during zone refining. Variations in the power supply to the oscillator are usually controlled by stabilising circuits. The maximum zone length[81] that can be supported by its own tension is a function of the ratio of the surface tension of the material to its density when it is molten. Silicon and aluminium (also iron and titanium) are highly suitable for the process, because of the combination of their properties.

In general, power efficiencies of the tank circuit used are very low (20% or less). Although a few hundred watts are sufficient for the molten zone, generator power ratings of 5 to 10 kW are usually required.[81]

After refining, the semiconductor material must be a single crystal and must have a strictly controlled amount of impurity added for transistor action.

Three methods of crystal growing are used:[83] crystal-pulling, zone-levelling and levitation. Crystal-pulling is the most widely used (Fig. 24.28c). A small

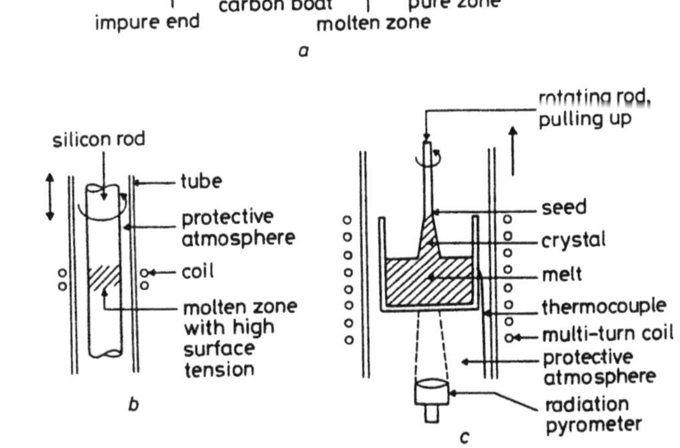

Fig. 24.28 *Semiconductor processing: (a) germanium zone refining (b) silicon refining and crystal growing (c) crystal-pulling*

seed, containing the right amount of impurity, is dipped into molten material and then pulled up by a rotating rod. The crystal grows as the rod is raised. Sometimes the crucible is also rotated to ensure uniformity. This method is used for silicon and germanium.

Germanium crystals can also be grown by the zone-levelling method, which consists essentially of the same equipment as the germanium zone-refining method, with a single, oriented, impure seed added at one end of the boat. The rest of the boat contains purified germanium; as the boat is pulled through the coils, a single crystal containing the required impurity is formed.

The levitation method of silicon zone-refining can also be used for crystal growing if one end of the rod is formed of a single, oriented crystal. Continuous passage of the rod up and down through the coil coarsens the structure until it is a solid crystal.

With all these methods, very accurate temperature and power control is needed to maintain the highly-critical impurity density and crystal growth-rate.

Most processes involve high-impedance, multi-turn coils heating large crucibles. In the levitation method, a narrow zone is heated by a low-impedance coil. One method to accomplish this combines a radio-frequency transformer and a complete tank circuit in a small single unit. This is then capable of scanning up and down the vertical rod, with only low-current leads feeding back to the generator.

24.10 Miscellaneous uses of induction heating

Other uses of induction heating include plastic-wire stripping and the bonding and unbonding of rubber and metal. Fusing plastic parts by the insertion of a thin wire between the two plastic surfaces is achieved by heating the wire with radio-frequency induction, which then melts the plastic locally and fuses the surfaces. Many curing and drying processes, such as paint drying, plastic surfacing, etc., are rapidly accomplished.

Rings of metal can be rapidly heated by induction by making them secondary circuits of a transformer. Figure 24.29 shows a copper ring heater. The magnetic circuit of the transformer is split so that the ring can be inserted, as shown. When the toggle is closed, the magnetic circuit is completed with a tight fit and the pulsating flux induces voltage into the short-circuited secondary formed by the ring. In this example, a 120 mm ID copper ring, 10 mm thick and 32 mm long, was heated to 440°C in 30 s, using a 15 kW, 50 Hz supply. This method should always be considered when the part to be heated is reasonably symmetrical and needs uniform heating. It is useful for expanding rings quickly for shrinking on another piece, e.g. starter gear rings and other engine parts. Ball and roller bearings are often fitted by this method.

Nylon can be directly joined to metal by induction heating the metal. The temperature range is very sensitive, between 250 and 270°C, to melt the nylon

Fig. 24.29 *Ring-type induction heater. Heats 120 mm ID, 10 mm thick copper rings, 32 mm long, to 440°C in 30 s using 15 kW, 50 Hz (Cheltenham Induction Heating Ltd)*

yet not burn it. This can be done elegantly with the precise controllability of induction heating.

In addition to heating, it is possible to use the interaction of the currents induced in the work and the currents in the coil to produce levitation. Using this, metal can be melted without contact with a contaminating crucible, in the presence of inert gas or vacuum, to produce high-purity metal. A simple coil is shown in Fig. 24.1 (item 16) and more complex coils are described in the literature.[63]

Dupont 'Kapton' (Type H) film has found many uses in the insulation of high-temperature electrical devices. It is not coated on conductors[63] but wound on as a high-integrity tape, since it cannot be directly bonded to copper or aluminium. The tape is sealed onto cleaned copper conductor by heating the copper, using a 40 kW, 450 kHz induction heater to produce approximately 300°C at 76 mm/s and then applying pressure. A radiation pyrometer focused on the exit side of the induction heater controls the temperature by varying the heater power.

24.11 Travelling-wave heaters

These devices are described in some detail in Section 22.12 as slab heaters. In smaller sizes, they can be used to heat flat plates for various purposes. For example, they are a valid alternative to the platen heaters described in Section 24.5 and have been supplied commercially for this. The advantages are:

- Balanced three-phase loading and supply are used.
- The core and windings can be water- or air-cooled, keeping the press cool and only heating the platen itself. Conventional platen heaters need substantial thermal insulation to prevent heat being conducted into the press, reducing the usable daylight.
- The windings are not situated in a natural oven, being heated both by their own losses and by heat conducted back from the platen. It is easy to put thermal insulation in the air gap to prevent this.
- Conventional platen heaters work by heat conduction into the work from the hot platen; it follows that the platen temperature has to be higher than the work temperature to provide a temperature gradient. Travelling-wave heaters generate the heat in the workpiece and are themselves cool (say, 50°C).
- Travelling-wave heaters only use power whilst actually heating, whereas the rival methods have themselves to heat before they start to heat the load. For the same reason, travelling-wave heaters are much quicker.
- Because the heaters are never hot and only warm when actually heating, safety and comfort are improved and the windings have a long life.
- Energy is not wasted in heating and cooling.

For the same reasons, these travelling-wave heaters (manufactured by Force Engineering under the trade name of TWIG – travelling wave induction generators) have proved very attractive as mould preheaters, being capable of heating moulds to 250°C at the rate of 1°C/s. A typical 500 mm square mould weighing 350 kg can be heated from room temperature to 180°C in 12 minutes (including soak time) compared with an hour for a conventional platen. These heaters are offered as standard packaged units up to 1 tonne, complete with surface temperature probes, digital indicators, interlocked guard, electric actuator, cycle timer, self-contained cooling system, power-factor correction protection and safety circuits.

Other applications are:

- rapid preheating of steam, oil or electric platens *in situ* in the press after cooling
- heating of extruder dies, which can be fitted cold and heated *in situ*
- preheating metal inserts for rubber-to-metal bonding
- rapid heating of faulty rubber-to-metal bonded parts to recover the metal
- preheating of steel or cast sections before welding.

Obviously, there are many more applications for this device. The author has been closely connected with these developments.

Chapter 25

Induction melting

25.1 Principles of induction melting

There is no basic difference between the use of a pulsating field to heat a solid billet or to melt metal for, say, casting. The principles of induction heating still apply; the pulsating field at the surface of the melt induces voltages in the material, which acts as a short-circuited secondary, causing currents to flow and heat the metal. All the ideas on skin depth can be used without modification for melted material. The relative permeability is obviously unity, except when steel scrap is being heated up to Curie point.

The advantages of induction melting are as follows:[67, 84–91, 103–105]

- Freedom from detrimental gases and combustion products, which cause undesirable metallurgical reactions, such as oxidation, de-oxidation, and sulphur contamination.
- Fast melting. Size for size, induction furnaces have a better daily output than other furnace types.
- Low running costs. Energy costs may be as low as, or only marginally higher than, those of the most efficient designs of fossil-fuel furnaces.
- Improved working conditions. There are no excessive heat losses from the furnace and no smoke, dirt or ashes.
- Reduced metal loss. When melting Cu/Zn alloys, the loss of zinc by oxidation may be less than 0.5% by weight.
- Automatic stirring. (In some circumstances, this may be a disadvantage.)
- Low maintenance costs on refractory linings. No part of the furnace is hotter than the metal itself. Compared with fossil-fuel furnaces, the refractory costs are less in a channel furnace and higher in a coreless furnace.
- Special techniques, such as vacuum melting, produce alloys that can be made in no other way.

Figure 25.1 shows a typical melting installation. Two types of furnace have been used – the 'coreless' furnace and the 'channel' furnace; they are compared in Section 25.4. The modern trend is to de-emphasise the channel furnace, so less space is devoted to this. For a more detailed discussion of these furnaces, see Robiette[91] (Chapters 6 and 7 and the references after those chapters).

25.1.1 Coreless induction furnaces

A section through a coreless induction furnace is shown in Fig. 25.2. The charge is melted inside a rammed refractory lining. A cylindrical water-cooled copper coil surrounds the outside of the refractory and produces a magnetic field pulsating in the vertical direction when supplied with power. Outside the coil, there are packets of magnetic laminations which provide return paths for the

Fig. 25.1 *A typical modular packaged melting installation. This is pre-assembled in the factory and delivered ready to couple to water and electricity. Here, two 4 tonne steel shell hydraulic tilting furnaces with a melting capability of 5 t/h to 1500°C are fed from a 2500 kW, 500 Hz power and control system for melting, and a 350 kW power and control system for holding (Inductotherm Europe Ltd)*

flux, improving the power factor and reducing the risk of flux linking with surrounding metalwork to cause stray heating. The whole furnace is contained in a mechanically-rigid structure and mounted so that it can be tilted for pouring.

Coreless furnaces have been made for various frequencies:

- 50/60 Hz mains frequency
- 150/180 Hz 'triple' frequency (medium frequency)
- 250 to 5000 Hz (medium frequency)
- higher frequencies (25 kHz and above) for small units and special applications.

The economics favour the medium-frequency furnaces, because of improvements in conversion efficiency since the introduction of solid-state generators in place of motor-alternators.

The main use for the coreless furnace is in the foundry, used as a remelting unit. It is capable of taking a wide variety of charge material, and heating it to the correct temperature with the correct composition with little or no contamination. Since the melt is fluid, there will be hydrodynamic forces set up by the interaction of the fluxes and currents which produce vigorous stirring; while this stirring takes place in all induction melting, it is especially useful in the coreless type, as the whole bath is stirred. This allows a wide range of charge materials to be melted economically with minimal loss of expensive alloying additions. The final mix is homogeneous and the composition can be measured and changed quickly and easily.

Fig. 25.2 *A section through a coreless induction furnace (Inductotherm Europe Ltd)*

The coreless induction furnace allows a wide range of different alloys to be treated in the same foundry. At the end of each melt, the furnace can be emptied completely and a different alloy started (provided that it is of the same refractory lining group, see Section 25.2.1). From the theory, it will be appreciated that it is difficult to heat small pieces of metal at *mains frequency*. If one starts

a

b

c

Fig. 25.3 *(a) Front view of tilted 1.5 t coreless furnace for steel melting. In use, the cover will be lifted hydraulically and the spout will be lined with refractory cement. (b) Coil assembly showing ceramic-coated coil, which is clamped in the vertical direction. The magnetic lamination packs also serve as radial constraints. Cooling hoses omitted for clarity. (c) Complete coil assembly in tilting frame, including water cooling (Radyne Ltd)*

with an empty furnace at 50/60 Hz, either a 'molten heel' has to be provided or relatively large pieces of stock have to be heated until the melt is under way, after which smaller scrap can be used. At low frequencies, the high currents cause vigorous stirring, so these furnaces are very effective for melting light scrap, e.g. borings and swarf, ferrous or non-ferrous, quickly and with little oxidation, as the metal is instantly submerged by the stirring action.

Medium-frequency furnaces need no starting plug. The stirring is more modest, but present practice finds it acceptable for a wide range of non-ferrous and steel applications. They are more suitable for high-grade steel production, where a molten slag covering is used to prevent contamination. This would be submerged by the powerful stirring of the mains-frequency furnace. Also, steels are not produced in mains-frequency furnaces because their limited power-densities would need long times at high temperature, with consequent wear on the refractories.

Higher frequencies are sometimes used for small induction furnaces to melt laboratory quantities of precious metals, such as gold, silver, and platinum.

Small coreless furnaces use separate crucibles, i.e. the refractory is not rammed, and is not part of the furnace. This is expensive, but has the advantage of being able to use separate crucibles for specific alloys, avoiding risk of contamination. These are also called 'crucible induction furnaces' (see Section 25.5). Small furnaces are also built without lamination packets; the structure must then be made of non-magnetic materials. While unscreened furnaces have been built up to 10 t capacity at 500 Hz, modern practice usually restricts their use to small furnaces, up to 2 tonnes, because of constructional difficulties.

All coreless furnaces can be started quickly from both empty and cold conditions, once the fritting melts are complete (see Section 25.2.1). The coreless furnace can have higher output levels than other furnace types.[92] Figure 25.3 shows aspects of coreless furnaces.

25.1.2 Channel induction furnaces

The channel induction furnace is especially useful in the production of a single class of alloy and has been widely used to melt and hold brass, nickel, copper, silver, zinc, and aluminium. It is also used to melt and hold ferrous materials. Figure 25.4 shows an installed furnace.

The 'channel' from which this type of furnace gets its name is shown in Fig. 25.5. A loop of molten metal is formed by a channel which is connected to the main bath at both ends, making a complete electrical conducting circuit. This circuit acts as the secondary of a transformer. The core of the transformer is made up of steel laminations which link the secondary, as shown. The primary is wound on the same core. The loop of metal in the channel is heated by acting as the short-circuited secondary of the transformer, so the ratio of resistance to reactance is important.

Whenever currents flow in a fluid, 'pinch-effect' exists. The interaction of the currents with the flux produced by them results in forces which constrict the circuit and eventually interrupt it at a certain critical value. If the channel were shallow and horizontal, there would be a constant making and breaking of the circuit. This is avoided in the modern channel furnace by using the principle, first developed by J. R. Wyatt, of putting the channel deep below the bath surface. By this means, the hydrostatic pressure of the melt counteracts the pinch effect and the current can flow continuously.

Fig. 25.4 *One of a pair of 70 tonne useful capacity vertical channel furnaces duplexing from a range of coreless induction melting furnaces (Inductotherm Europe Ltd)*

The supply frequency for channel furnaces is always 50/60 Hz. Because of the laminated core and closer coupling to the load, the inherent power factor is 0.5 to 0.7 for iron melting, compared with 0.15 to 0.25 for a coreless furnace. For non-ferrous melting, the power factor may be 0.25–0.5. Their electrical efficiency is high, 90 to 95%, from which the thermal and water-cooling losses must be subtracted, giving 85 to 90% overall.

Channel furnaces are only suitable for continuous use and, for best possible lining life, should not be shut down between lining replacements. Because of the constraint of the channel itself, the channel furnace is restricted in power/furnace capacity ratio.

After this brief introduction to the principles of operation of the two main types of induction furnace, each is described in more detail. In Section 25.4, their advantages and disadvantages are compared.

25.2 The coreless induction furnace

25.2.1 Refractory linings

The normal way to make the crucible of non-conducting refractory material in which the metal is melted is to form it *in situ* from granulated refractory

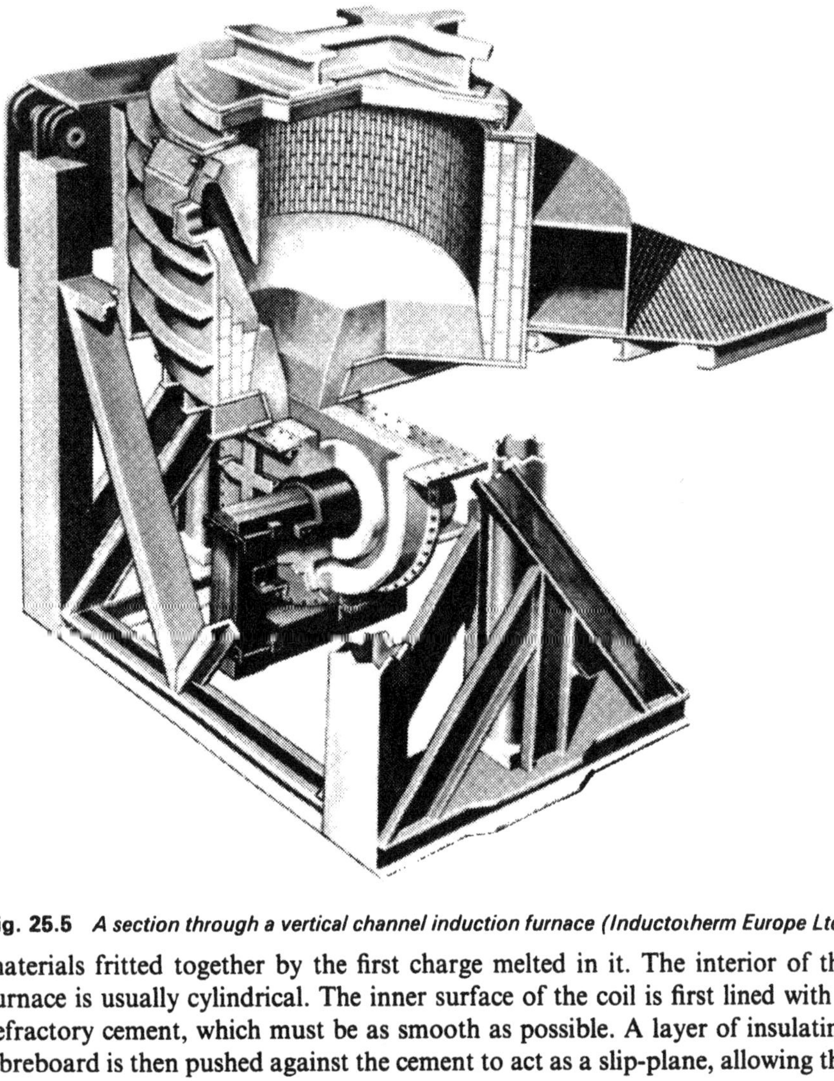

Fig. 25.5 *A section through a vertical channel induction furnace (Inductotherm Europe Ltd)*

materials fritted together by the first charge melted in it. The interior of the furnace is usually cylindrical. The inner surface of the coil is first lined with a refractory cement, which must be as smooth as possible. A layer of insulating fibreboard is then pushed against the cement to act as a slip-plane, allowing the crucible, when formed, to expand longitudinally and assisting its removal when it has to be replaced. A metal former, having the inside dimensions of the crucible, is located centrally in the inductor and the space behind it is dry-rammed with the refractory mixture, using hand or pneumatic tools. When the first melt takes place, the container melts away and the heat glazes or 'frits' the inner face of the refractory. Ultimately, at the end of its life, the refractory frits right through.

The material of the refractory lining must be chosen with the metal to be melted in mind; the linings can be acid, basic, or neutral. *Acid linings* are used for high-speed tool steels, plain carbon steels, and low-alloy steels, for special cast irons, and for non-ferrous metals. *Basic linings* are suitable for stainless steel, high-manganese steel, and any steels where special slag-reaction processes are to be carried out. Basic material is preferred for nickel-iron alloys. Lining materials commonly used are:

- acid – high silica quartzite
- basic – a magnesite compound
- neutral – alumina or zircon.

The designer of the refractory lining has to compromise between two contradictory requirements:

- Thicker linings increase the leakage flux which passes between the winding and the charge, reducing the heating of the charge and resulting in a poorer power factor.
- Thinner linings increase both the risk of breakthrough of the molten material to the coil and the furnace heat loss.

Every effort should be made to improve the furnace performance with a minimal lining consistent with reasonable lining life and furnace efficiency. A typical steel furnace has, for example, a lining thickness which is just over 10% of the internal diameter.

The aim of the designer must be to keep the coil side of the lining in powder form as long as possible. The temperature at the interface between the lining and coil is important; in one example, reducing this temperature from 200/250°C to 100/140°C by design extended the lining life from a few heats to 50–60 heats.

It is also important to taper the refractory lining. The weakest part is at the bottom, so a taper should be incorporated from the bottom up to 25/35% of the height. From there, it is parallel. Care should also be taken to avoid a thermal-stress region by putting a radius where the conical side joins the bottom, which may be flat or sometimes dome-shaped. The rammed lining must be particularly strong where the liquid metal is at its highest temperature and where the turbulence and scrubbing is greatest.

The lining is supported by a structural layer of fireclay bottom bricks and the top of the lining is also formed by fireclay bricks.

Modern techniques, including lining push-out and vibration of the former, have reduced relining times. Typically, relining can be completed in a single shift. This, together with liquid-metal priming, means that furnaces can be back in production within 12 h.

Linings have to be maintained.[93] Lining wear on modern medium-frequency furnaces is small, since time at temperature and stirring are both reduced. Slag-erosion demands weekly patching. The lining is also eroded by chemical action with the metal; it can be patched every few weeks or left untouched for 12 to 14 weeks, depending on the alloy and power rating of the furnace. Electrical readings give reliable indications of lining condition. Most maintenance work can be planned in advance for a weekend. The linings of small-capacity furnaces (0.5 to 50 kg) are often preformed and prefired.

In Britain, it is a requirement of HM Inspectorate that the charge should be solidly connected to earth, which is done by embedding a 'spider' wire in the base of the crucible. Also, a protective device must be fitted to switch out if the insulation breaks down between the coil supply and the earthed charge. An earth-leakage device is fitted to monitor the leakage current at all times.[94] This is displayed on an indicating meter with a 0–100 scale, fitted with contacts which de-energise the system if an 'unhealthy' state is detected, so that the furnace can be emptied before a major break-out occurs.

Another consequence of the trend to medium-frequency furnaces, with their high power-densities and shorter heating times, is that thermal losses through the lining have become comparatively unimportant.

25.2.2 Windings

The windings are made from high-conductivity hollow water-cooled copper. For 50/60 Hz, this was normally of D-shaped cross-section, where the flat side of the D was wound on the inside of the coil and made thick enough to carry the load currents (typically 10 mm thick for 50/60 Hz; see Chapter 7). The outside carries the water for cooling.[95] This D-section has been superseded by a rectangular equivalent, where the thick flat section to carry the current is backed by a rectangular water channel, which gives better cooling.

The electrical insulation of the coil to the frame must be adequate for the maximum value of the RMS voltage applied continuously, and for transients of five to seven times this value, which occur when switching the furnace load and the capacitor units. The inter-turn voltages depend on power density and frequency; they are 300–400 V in present-day furnaces. The main criterion is that the insulation maintains its quality throughout the life of the coil under hot conditions. The coil insulation must be designed to avoid the entrapment of water. During the first few heats of a newly-rammed lining, water is driven off, and this can be trapped in the insulation if care is not taken in the design. In the past, half-lap taping with glass tape, impregnated with high-temperature varnish or encapsulated in epoxy resin, has been tried, but there is still a tendency to crack and allow the ingress of water, together with the formation of metallic oxides. Spacing by bonded asbestos and fibrous materials has been used, and glazed ceramic spacing-washers have been successful. In essence, the water must be able to escape from the lining without being trapped in the cold coil.

The coils are rigidly clamped, both axially and radially, to prevent motion that might result in chafing and damage to the insulation. Many firms braze studs to the individual turns, which are bolted back to hardwood coil supports. Whilst expensive, this ensures reliability. In addition to the $I^2 R$ and supplementary losses in the coil copper (due to skin effect), heat reaches the coil from the melt and has to be cooled away. About 30% of the power leaves in the water cooling. However, this cold barrier of copper maintains a steep temperature gradient through the refractory wall, which is good for refractory life.

Some manufacturers put additional cooling at the top and bottom of the active coil to provide uniform cooling, in an attempt to reduce the thermal stress on the linings. It is important that this is done properly, as examples have occurred where the whole refractory has split at the junction between the active and the cooling coils, resulting in metal breakthrough.

It must always be remembered that the lining will expand and contract, axially and radially, with temperature. The lining must be free to move with respect to the coil and damage can occur if a badly-wound coil interferes with this motion.

25.2.3 Flux guides

Packets of low-loss electrical steel are placed behind the coil with the laminations in the radial direction (see Fig. 25.2). They form a well-laminated path to channel the main flux in the region outside the coil. If they were not there, the flux would spread all around the back of the coil and link with the surrounding metalwork. The flux guides (or 'screening') also improve the efficiency of the coil by 1–2%, partly by reducing the stray losses, but also by reducing the magnetising current.

The packets can be used to support the coil while being kept insulated from it. If they are not firmly clamped, the punchings will tend to vibrate and be noisy, so care must be taken in the design. The packets must also be securely connected to the main frame of the furnace. Arrangements are provided to screw the packets in towards the coil: the position has to be adjusted after commissioning, and the packets have to be moved outwards when the coil is replaced.

In principle, the coil should be completely surrounded by flux guides, but most practical designs rely on a limited number of packets of punchings uniformly distributed around the coil. Ideally, they should also be curved to the radius of the coil, but are usually flat.

Since the flux is dominantly parallel with the axis of the coil, it is sensible to use grain-oriented sheet steel for these punchings. The sheets are clamped together with electrically-insulated studs passing through holes: non-magnetic materials are used for the studs, washers, and nuts.

The length of the packets, compared with the coil length, is important to prevent stray flux from linking with adjacent steelwork. To contain the stray flux, non-magnetic bottom plates, copper rings, and water-cooled rings at top and bottom have been used, but these absorb energy and lower the efficiency. Furnaces have been built with packets of iron laminations at 90° to the main pack, at the bottom end. This is expensive, but is worth doing in high-powered furnaces, as it improves efficiency and avoids heating the steelwork at the bottom.

25.2.4 Frame

The coil and flux guides are often mounted in a subframe, similar to the stator frame of an electric motor, which is readily detachable. This subassembly is known as the 'gut' and allows a spare to be held and put in place quickly. It fits into an open supporting frame. It is important that a breakthrough of molten metal is not entrapped by the frame, but can run directly down to the pit below, which, for good practice, should be lined with refractory bricks to protect the expensive foundations. Sometimes the coil and flux guides are built into a cylindrical shell that is not readily detachable, but this could result in a longer period of shut-down and great inconvenience when breakthrough occurs. Where a cylindrical shell is used, it is normal to have large cutouts to allow water to escape into the pit below, if a leak occurs.

All furnaces have to tilt to pour the melt, so the frame has to be designed to take the forces at various angles during pouring. The furnace usually tilts about an axis near the pouring spout to ensure the minimum change of spout position during pouring ('lip-axis pouring'). The whole moving frame is connected to a strong girder structure which is fastened to the ground and which supports the trunnions and the hydraulic tilting-mechanism. Arrangements can also be made for the furnace to be tilted backwards to help with slag removal. The hydraulic rams are usually duplicated to avoid twisting of the frame. A hand pump is also fitted so that, if power fails, the furnace can be emptied before the charge freezes. The presence of an intense magnetic field must never be forgotten when designing the frame. Leakage flux can induce voltages in closed electric circuits, producing losses and heating.

25.2.5 Electrical connections

Electrical connections to the coil are made through flexible cables, which allow the furnace to tilt. It is usual to supply both the electricity and the coil-cooling water through the same water-cooled cable.

The supply arrangements and power-factor correction are similar to those for other induction-heating systems at the appropriate frequency and vary with furnace power. The voltage range from the power source is different, as the power range is very much greater. The power must be able to vary from full-load to 10% full-load (needed when fritting a new lining). 'Turndown' power levels fall to around 20 to 30% for holding. The power-factor correction requirements are also substantially different from the billet-heater supply. For melting, using a fixed frequency supply, a fixed bank of 40% of the capacitors can be switched on- or off-load; modern practice uses fully-automatic power-factor and power control, which is readily provided on solid-state systems. Inverter power-supplies can usually tune over a frequency range to match all load and lining conditions without capacitor switching.

25.2.6 *Choice of frequency*

Inverter medium-frequency power supplies are now cheaper than their mains-frequency equivalent.

Cost The advent of solid-state equipment, with the possibility of fully-automatic matching of the load impedance to the supply, has changed the economics considerably in the last decade. Until then, 50/60 Hz was desirable, if all other factors were right, and the disadvantage of a large molten heel was accepted. The heel meant that a scrap-drying plant had to be installed, so that the scrap was completely dry before entering the furnace, otherwise there was a danger of explosions, with liquid metal being thrown into the shop. Today, 10 to 15 tonne furnaces can be supplied at the lower end of the medium-frequency range (150 Hz, or thereabouts) with high electrical inputs and good melting rates, without the need for a molten heel, using an empty furnace and wet scrap.

Number of turns From $E = 4.44 f \Phi N$, the turns are inversely proportional to frequency, other things being equal. Too low a frequency means a large number of turns with small conductors and, hence, small water-passages, risking clogging and high coil maintenance. Too high a frequency means a high number of volts per turn, with problems of insulation.

Stirring The turbulence in the charge increases as frequency drops. This stirring is valuable when recarburising steel scrap or absorbing finely-divided materials, such as cast-iron swarf, turnings, etc. with the minimum of metal loss. Too high a frequency may produce too little stirring for proper alloying. On the other hand, turbulence means shortened lining life and the risk of refractory inclusions in the metal. It may also mean gassy metal.

Starting It is often difficult (or impossible) to start a melt at low frequencies. The melting rate is low at 50/60 Hz (because of poor ratio of particle size to skin depth) until 50–60% of the charge has been melted. To improve this, it is normal to start with a 'starting plug' from a previous operation of 15–20% of the total capacity. With medium frequencies, starting is not a problem.

Efficiency The poor starting at lower frequencies reduces the efficiency of melt. Against this must be set the converter efficiency, the higher capacitor losses, and supplementary losses in the coil, together with higher voltage drops in the bus-bars, at higher frequencies. Modern solid-state systems, with their automatic optimisation of frequency and their high efficiency, make medium frequency increasingly attractive, especially if one takes into account melting time lost in the adjustment of tappings and the losses in the control gear.

Size of charge Lower frequencies are not suitable for small sizes of scrap because of the poor ratio of particle size to skin depth, until the melt has progressed sufficiently to absorb these materials into the molten metal. If finely-divided materials are used, starting with an empty furnace, a medium-frequency furnace is preferable.

Power/weight ratio The high stirring produced by 50/60 Hz means that the power/weight ratio must be kept low. If a 6 t furnace powered at 1500 kW gives 2.5 t/h output, an increase to 5 t/h, needing 3000 kW, would need a 15 t furnace. As the frequency rises, the power/weight ratio improves; for example, 2.5 t/h at 1500 kW needs only a 3 t furnace at 150 Hz – half the size of a 50 Hz furnace.[96]

The decision about frequency is the usual engineering balance of advantages and disadvantages. The problems must be carefully discussed with the furnace manufacturer, taking into account the melting rate, the output rate, the amount of liquid metal to be poured at any one time, the holding capacity, the range of scrap, and type of metal to be melted.[88] This will generally lead to a medium-frequency furnace nowadays.

The standard frequencies available are 50/60 Hz, and 1, 3, 4 and 10 kHz. Triplers (150/180 Hz) were used but they have become less attractive relative to medium-frequency solid-state sources. Triplers have to be very well-engineered to avoid harmonic problems with the supply and the original cost attraction is no longer valid.

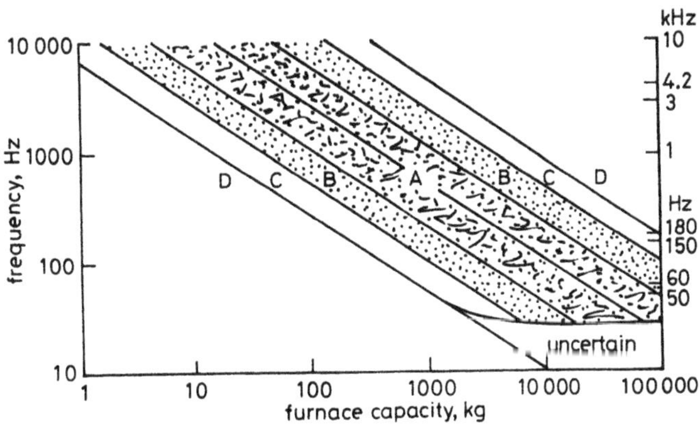

Fig. 25.6 *Choice of frequency with furnace size: A, recommended zone; B, zones to be used with caution; C, zones in which furnaces have been worked but should not be used; D, zones must not be used (after Inductotherm Europe Ltd)*

25.2.7 Sizes of furnace

Figure 25.6 is a summary of 1971 practice[96] and indicates the ranges of furnace that were available at various frequencies. It is substantially valid today. Naturally, a point on that chart does not mean that a certain rating exists, or has been made, but it does indicate feasibility.

25.3 The channel furnace

In the coreless furnace, the whole charge is surrounded by the coil. The refractory is at the heated surface and is scoured by the moving molten metal. In contrast, the channel and throat of the channel furnace are the only parts with intense heat input and the metal movement (apart from convection). The device can, therefore, be considered as a large storage vessel, receiving additions of metal to be heated, or held at temperature, which is tapped from time to time, as needed, and is often known as a 'holding' furnace. The furnace is normally fed with molten metal from an arc furnace, a cupola, or an induction coreless furnace, although it can be used for primary melting. It is sometimes used in conjunction with cheaper 'off-peak' supplies of electricity to melt or store metal ready for pouring in a single day-shift. Channel furnaces are used for the correction of temperature and metallurgy, receiving additions for alloying after being fed from a melting furnace.

The power can be adjusted easily to give an accurate pouring temperature. The gentle stirring action ensures thorough mixing and uniform composition. Whether melting or holding, the product is of consistently high quality, causing few rejects. The channel furnace is normally well sealed to prevent atmospheric contamination and so the metal can be held for long periods without change of composition. For the same reason, it is also good for the shop environment.

The temperature time-constant of the channel furnace is long, both because of the isolation of the heating channel from the main bath and because of the lower power/volume ratio. With vertical channel-furnaces, the stirring at the metal surface is almost nil.

Although the main use of the channel furnace is as a holding furnace, they are sometimes used to 'superheat', i.e. raise the metal temperature by 60 to 100°C from, say 1300°C, but the temperature rise is slow. If used for this purpose, the furnace capacity should be smaller and the inductor larger, e.g. a 20 t furnace with a 2 MW inductor.

25.3.1 The main vessel

The main furnace chamber may be of the drum type, with a cylindrical shell and rectangular shallow bath, or with a rectangular deep bath. The description that follows is of the horizontal-axis type. The body of the furnace is of structural steel. The rotation of the vessel is accomplished either by substantial stub-shafts at each end moving in large bearings, or by machined rings at each end which rest on rollers in two-roller frames mounted on the concrete floor. These furnaces are substantial structures and the foundation loads are high: 29 t for a 5 t capacity, and 300 t for 100 t capacity.

Figure 25.7 shows one manufacturer's lining construction. The lining has three layers:

- an outer layer of insulating blocks and bricks
- a back-up lining
- the main hot-face lining.

These linings provide thermal insulation, as well as metallurgical protection for the melt. To assist relining, substantial access openings are provided. Relining the hot face takes several weeks; a large furnace uses 160 t of lining bricks. This can last up to 5 years, depending on working conditions.

The main lining is either *basic* or *neutral*. The latter is used for acid or neutral slags from the melting furnace, whilst the basic lining is more suitable for basic slags. These slags are used to reduce oxidation, rather than being active metallurgical ingredients.

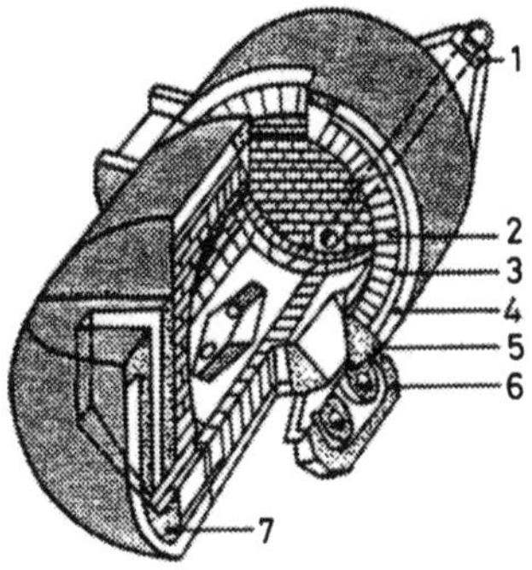

Fig. 25.7 *Construction of a channel furnace: (1) teapot-style tapping spout of castable compound (2) hot-face lining (3) back-up lining (4) layers of insulating blocks and bricks (5) inductor throat of castable compound (6) inductor lining of castable or rammed compound (7) teapot-type hot-metal receiver of castable compound (after ASEA)*

25.3.2 The inductor

The inductor unit is the heater element for the whole furnace. Individual inductors are rated up to 2 MW. For additional power, two or more inductors can be fitted to the same vessel. When melting or holding conditions require their use, double-channel inductors are available, rated 1.1, 1.5, and 2 MW, depending on the maker. These are two single-channel inductors joined together to form one unit, with a shared central channel.

Figure 25.5 shows the elements of the inductor. The coil is wound around a laminated core of oriented transformer sheet. The coil is either air- or water-cooled,[95] depending on the manufacturer and on the metal being melted. It is important to prevent the heat from the channel reaching the coil and it is usual to interpose a split metal shield between the coil and the lining, air-cooled for low-temperature metals and water-cooled for ferrous metals. This allows the lining to be made thinner and improves the electrical and heating efficiency. It also acts as a metal 'freezing' barrier for minor leaks in the lining.

The inductor is factory-built and bolts directly on to the frame, with special attention being paid to a leak-proof joint. Spare inductors are normally kept on site. Designers and users have successfully changed inductors while the furnace is hot, and even while still full, but this presents many practical problems. More often, a second furnace is available and can be used while the faulty inductor is replaced.

Inductor units are lined with neutral rammed- or cast-material or with cast basic linings. Dry-lining materials now predominate. About 700 kg of lining material is used for a 200 kW inductor and 3 t for a 1100 kW inductor. The design of high-power inductors is based on achieving good energy-transfer between the channel and the bath, so channel temperatures are very near the bath temperature. Modern lining materials of the Al_2O_3 and MgO (basic) types have been very successful at the power levels used.

The reactance of the inductor must be kept low by making the axial length of the channel as long as possible and minimising the radial distance between the coil and the channel. High reactance means a larger core, larger inductor, and poor power-factor; the power-factor depends on the metal and power-level, ranging from 0.2 for high-conductivity metals to about 0.6 lag for ferrous metals. The cross-sectional area and shape of the channel control the active power. The radial distance from coil to channel is determined by heat flow and lining life. The cooling jacket allows the use of a thinner wall.

In the channel furnace, the split metal shield is monitored to check that it is not shorted by seepage of molten metal. It is also possible to monitor the reactance[97] of the lining to act as a more precise indicator of inductor condition.

The frame of the inductor is water-cooled when used for ferrous materials. It must be electrically split around the loops and must be strong enough to support the relatively weak refractories.

For ferrous metals, the life of an inductor averages nine months, and a year is not an unreasonable life to expect. Open-channel furnaces have been built and extensively used in Italy for iron.

Channel furnaces are extensively used for non-ferrous metals, and inductors with ratings as low as 20 kW are used for zinc and aluminium holding furnaces for die-casting. When dealing with aluminium, it is important to provide facilities to clear the dross that accumulates in the channel. This may be necessary every five or six hours (see Section 25.4.1). The Ajax-Wyatt furnace, with its V-shaped inductor, has been very successful in the brass industries. At lower temperatures, 'dry-type' inductors are possible, where air-cooling is used. When used for casting, channel furnaces supply consistent quality and specification metal which is gas-free, resulting in extremely low rejection rates and the ability to produce thin-walled castings.

25.3.3 Loading and pouring

Details of exact arrangements vary with manufacturer. It is desirable to be able to charge the furnace at any time (even during pouring) without oxidising the main charge. This can be done by the so-called 'teapot' arrangement, where the spout communicates with the bottom of the melt. This is also used for pouring, so preventing the slag being poured with the melt. Side- and end-pouring arrangements are used.

The input and output spouts are lined with a castable compound.

25.3.4 Choice of size

Continuous charging from elsewhere (duplexing): If the furnace is continuously charged from a cupola furnace, an arc furnace, or a coreless furnace, and is continuously tapped, it is acting as a buffer to ensure homogeneity of melt and is levelling-out the temperature and composition of the charge. Experience has shown that, under these conditions, a capacity of about one hour's supply is adequate.

Used as a store: When the channel furnace is charged with metal produced during off-peak or other periods for use during the day, then the capacity must be the total quantity needed during the day.

When charged from another furnace: The channel furnace is also used as a store when charged from, say, an electric crucible furnace. Here the capacity must obviously be at least sufficient to take a whole melt from the supplying furnace, together with any left over from previous loadings.

When iron at 1500°C is supplied to the furnace and has to be superheated by θ°C, the net power needs to be

$$P = 0.23(\text{throughput})(\theta)$$

where the throughput is measured in tonnes per hour and the net power is in kilowatts. This power level allows for the heat supplied to the iron, and to it must be added the losses from the furnace and from the inductor. The heat losses through the lining of the shell and lid are between 2.5 and 3 kW/m^2.

If the furnace is simply being used to hold the iron at 1500°C, without superheat, then only the losses will have to be supplied.

The above figures will vary with each manufacturer, but can be taken as typical of good practice.

The gross inductor power is, therefore, related to the superheat power plus the losses in the furnace. When the inductors are big compared with the furnaces they are heating, the net power into the melt is 70–77% of the inductor rating. (Small units in large furnaces will obviously be less effective.)

When used as a melting furnace: The size here depends on working needs. Typical melting figures are 470 kWh/t for a 300 kW inductor falling to 440 kWh/t for 400 to 800 kW inductors.

25.4 Comparison of coreless and channel furnaces

In addition to coreless and channel furnaces, which work by induction, arc furnaces must be considered for the melting and refining of steel scrap, but it is not appropriate to discuss them in this book. They are robust melters capable of reliable performance with all kinds of scrap over long periods. They can cope with various raw materials, from heavy scrap to swarf, even in a heavily-oxidised, wet, oily condition. This allows the cheapest available charge material to be used. Melting losses are higher and the uniformity of batch temperature is not so controllable.

Both channel and coreless furnaces are very suitable for holding and superheating, with precise temperature control, uniform temperature distribution, small melting losses, and small heat losses. From the previous sections, it will have been seen that both types are capable of doing the same melting job.

The market in Europe is now concentrated on inverter-fed coreless furnaces. Techniques have evolved which permit inverter-fed furnaces to operate as melters over a wide range of scrap types. If mixing is needed, inverters at 100 Hz and below are usually cheaper than 50 Hz. The more exacting refractory requirements and the high first cost has meant that very few new channel-furnaces have been installed.

The channel furnace was extensively used for the holding and superheating of metals, but it is no longer used, as high power-density furnaces can meet production-line needs more closely, without storing large volumes of molten metal.

The disadvantages of the channel furnace are:

- It is necessary to prime and preheat the channel furnace. This is not a problem if another melting source is used, but is difficult in small foundries with only one furnace.
- The power density is comparatively low.
- Mixing is inferior to the coreless furnace, which is why coreless furnaces are often used to feed channel holding furnaces.
- Accurate power input does not necessarily mean accurate metal temperature control throughout the bath, because of the poorer mixing.
- It cannot be emptied and restarted easily.

25.4.1 Combination methods

Apart from capital costs, energy costs are increasingly important. Combinations of furnaces are often economic. If a cupola is still used, it is sensible to use either a channel furnace or a special design of coreless furnace in its holding role to act as a buffer to accept the continuous output of the cupola and to hold it until the metal is needed.

A channel furnace can be operated overnight on preferential rates and the metal stored for use during the day, using only the low holding-power at daytime rates; however, the operating costs are difficult to justify.

Another operating mode is to melt and refine in a small coreless-furnace continuously, storing the output in a bigger furnace for daytime casting. This minimises the size of the melting furnace and the overall power rating of the plant.

In all methods, the most efficient performance is obtained by pouring and charging quickly and always running with the furnace cover in position. Hot metal can be directly returned to either type of furnace.

25.5 Special features in melting furnaces

25.5.1 Furnaces using separate crucibles

There is no necessity for the coil and flux guides of a coreless furnace to be integral with the crucible or lining. If changes of composition may not be compatible, it may be advantageous to reserve a crucible for a given mix. It also saves time to melt metal in a crucible and then to transfer it directly to the point of pouring, rather than transfer it to an intermediate vessel. Naturally, these advantages are limited to small melts, but these exist in many small foundries, especially those dealing with more expensive non-ferrous castings. There are several forms of induction melting furnaces suitable for separate crucibles. All use two crucibles; when one charge is molten and ready to be lifted away, the other crucible is presented to the second coil and power is reapplied. By alternately charging two crucibles, molten metal is constantly available. Power is transferred by a two-position transfer switch in the furnace enclosure. This is interlocked for safety.

The 'lift-swing' or 'lift-coil' employed one coil which was lowered over two working positions (Fig. 25.8a). Whilst one crucible was being heated, the other

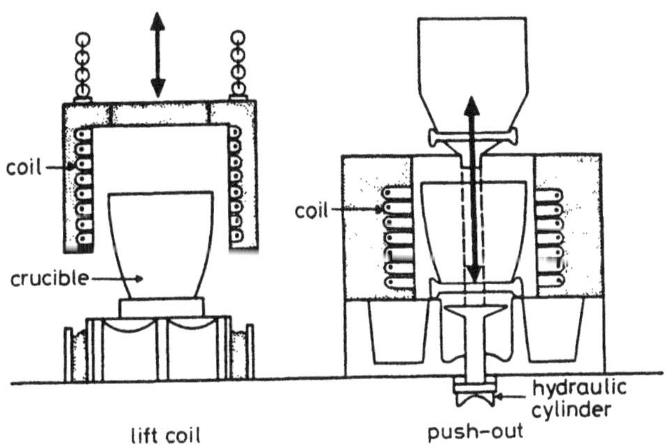

Fig. 25.8 *Sections through lift-coil and push-out furnaces*

was being prepared. When the first was melted, the coil was raised, swung across, and lowered over the cold charge. The 'double-push-out' furnace (Fig. 25.8b) had two separate coils fixed in the body of the furnace. A separate crucible could be lowered into each coil alternately. The crucible stood on a hydraulically-operated platform. This scheme is more expensive, involving two coils, two hydraulic systems and two holes in the ground. Table 25.1 shows melting times for these two furnace types.

The preferred modern variant of this is the 'drop-down' furnace, shown in Fig. 25.9, where the crucible stands on a fixed central pillar and the coil-box is raised to the heating position by a pneumatic cylinder, which is mounted at the back of the frame for protection from spilled molten metal. This design avoids

Fig. 25.9 *Copper-melting drop-down furnace: 200 kW, 1 kHz, 200 kg (Radyne Ltd)*

holes in the ground and is more self-contained than the 'lift-swing'. Figure 25.9 shows a 200 kW, 1 kHz, 200 kg copper-melting installation of this type.

All three systems deal with crucibles of average capacity from 50 kg in steps up to 200 kg. Two types of crucible – conducting and non-conducting – are used. With conducting crucibles, it is important to heat them adequately to avoid chilling in the spout when pouring; the frequency is chosen so that a modest proportion of the power heats the crucible itself, with the rest of the power going into the charge, both for efficiency and for long crucible life by preventing overheating. Non-conductive crucibles are needed for some materials (such as gold, and particularly silver, powders) to prevent contamination. Here, power is provided by solid-state generators at 25 or 50 kHz. At the start of the melt, it is necessary to initiate heating by the operator impacting the powder with a

Table 25.1 *Melting times in minutes (double-push-out and lift-swing furnaces) (after Inductotherm (Europe) Ltd)*

Average capacity (kg)	Power					
	2500/3000 Hz			900/1200 Hz		
	60 kW	100 kW	150 kW	125 kW	175 kW	250 kW
55	26/21					
70	33/27					
80	39/31	18				
95	45/36	21				
110		24	17	18		
126		28	19	22		
135		32	22	25		
170		40	27	32	23	
200			35	43	30	21

Times are given for red brass with second heat in crucible. For other metals, multiply times by: gold, 0.4; copper, 1.1; yellow brass, 0.85; aluminium, 2.0; silver, 0.7.

ramming-rod to produce local arcing between the particles, causing a local hot-spot, which quickly spreads until power is being induced in the whole charge.

These furnaces are designed for safety in case of spillage; even if the crucible breaks, the entire melt is contained safely. The coils are lined with refractory cement to protect against metal spillage and reduce heat losses.

Clay-graphite or silicon-carbide crucibles are used with these furnaces, especially for low-resistivity materials, such as copper and silver, which are difficult to heat by direct induction. These crucibles are used to melt the more esoteric metals, e.g. nickel-based alloys for investment castings, platinum, and silver-based alloys. The coil power-factor is lower and the heat loss is greater than with other coreless furnaces.

25.5.2 Vacuum furnaces

Since induction melting is independent of the surrounding atmosphere, it is equally valid under vacuum. The melting furnace is entirely contained in a vacuum vessel (typically 10^{-4} torr). The electromagnetic stirring, combined with the vacuum, degasses the metal and there is no possibility of oxidation during either the melting or the casting. The inclusions are few in number and are uniformly distributed in the product. Additives can be put in during the melt, without fear of reactions with the atmosphere, so that precise repeatability is guaranteed. These features ensure metals and alloys of very high purity, with high resistance to corrosion, fatigue, and temperature-stress cracking.

Vacuum melting is not cheap and can only be used where the special qualities of the end-product command an enhanced price. The overall melt takes longer because of the time taken to produce the vacuum and to do all the working, such as charging and pouring, under vacuum by remote control.

In recent years, however, there have been considerable developments in the investment casting industry in low-volume vacuum vessels with heating through a non-metallic vessel wall. These not only permit fast pump-down times, but also provide a cheap vacuum-melting technique when additives are not needed in the charge. Typically, a 2 kg charge is melted in less than a minute, using a 100 kW, 3 kHz generator.

Typical uses of vacuum furnaces are:

- heat-resistant nickel-based and cobalt-based superalloys (Nimonic, Udimet, Incoloy, Waspalloy)
- steels with special properties (managing steels, stainless steels, high-speed steels)
- high-purity metals (nickel, cobalt, uranium, beryllium)
- alloys of basic metals (cupro-chrome, cupro-titanium, ferro-titanium, ferro-chrome, ferro-nickel)
- materials with controlled coefficients of expansion (Invar, Unispan)
- high-purity magnetic materials (Mumetal, Supermendur)
- light and reactive metals (titanium, aluminium, zirconium).

The additions must be made from outside, unless they can be put into the chamber before the start of the melt. A charging lock is provided to allow

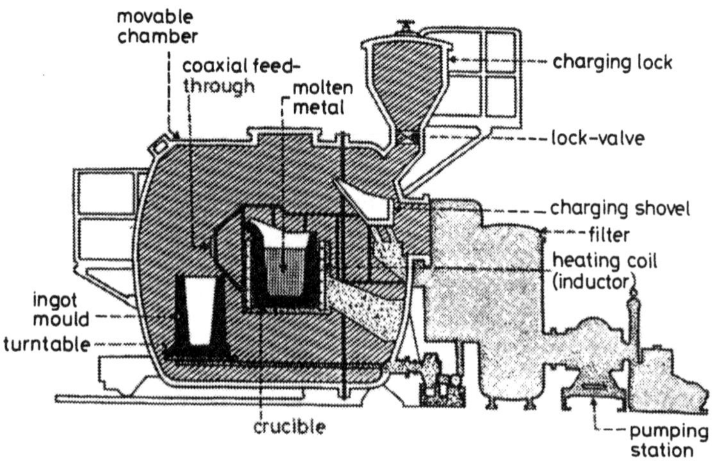

Fig. 25.10 *Section through a vacuum furnace (after Elphiac-Balzers Ltd)*

additions to be made after the start of the melt, without exposing the melt to the atmosphere. The materials are dropped into a shovel, to await the appropriate time to add them to the melt. Large vacuum furnaces are available (300, 600, 1200, 2500, 4000 and 5000 kg of steel). The vacuum vessels themselves are made of stainless steel: because of this, flux-screening packets are not normally used, except on larger furnaces.

Vacuum furnaces are often provided with mould locks, which permit the moulds to be prepared outside the vacuum chamber and then introduced to the main chamber after being evacuated in the lock. This, together with the charging lock, allows the furnace to be used in a continuous manner, without constantly remaking the vacuum.

Figure 25.10 shows a section through a vacuum-melting furnace.

Fig. 25.11 *A rollover furnace (Radyne Ltd)*

25.5.3 Rollover furnaces

Rollover furnaces are extensively used for casting by the lost-wax and investment casting processes. These are small, medium-frequency furnaces (up to 30 kg). The furnace body is mounted so that it can roll over through 180° after the melt is done. A mould is fixed to the top of the furnace before the melt starts, being clamped securely in place, either manually or pneumatically, to two vertical non-magnetic stainless-steel columns by an adjustable clamp bar. An inert gas can be pumped into the space between the mould and the charge to prevent oxidation. When the melt is complete, the whole furnace and mould is rotated, in a controlled manner, pneumatically or manually, and the melt passes into the mould. The system is very effective, since the exact weight of material can be melted and there is no risk of contamination, as it goes directly into the mould (Fig. 25.11).

Care has to be taken in the construction of these furnaces that the stray fields of the coil cannot link with the support frame, or with any other closed electrical circuits, to avoid losses and heating.

Chapter 26

Direct resistance heating

26.1 Introduction

This chapter is necessarily vaguer than Chapters 22–25 because there are fewer established uses of direct resistance heating (DRH). This is sad because, as shown in Chapters 1–6, there are great advantages, especially if DC is used. Before looking at DC, let us discuss some reasons for the neglect.

- The method is not well understood. As stated earlier, the theory was developed for conductors, not for heating. Part I of this book attempts to correct this.
- Although a heater may process a large throughput of metal once it has been developed and installed, there are very few standard machines available 'off-the-shelf', so every application involves both manufacturer and customer in risk, which usually is not accepted for understandable fear of the unknown. The manufacturer fears that he may be involved in development costs which he cannot recoup by spreading them over future orders; the customer fears that his process line may be held up for teething troubles. It is hoped that the examples and advantages described in Section 26.5 will show that these fears are unfounded in well-conceived systems.
- For many years, cheap fuel oil allowed the use of inefficient methods and discouraged development. Also, there was much prejudice about electricity as an 'expensive' energy source, without calculating total process costs. It is to be hoped that attitudes have now changed.
- Except for strip heating, DRH tends to be discontinuous, which is sometimes a disadvantage.
- Some contacts were unreliable; this led to the rumour that all contacts were unreliable, which was untrue.
- Modern semiconductor rectifiers were not available when the original developments took place and it is doubtful whether they have all been rethought since.
- At the practical level, some previous contacts required that the material be shotblasted before heating, which was unacceptable to many users. This must be eliminated.

- Some radial contacts produced 'cold ends', which prevented forging.

Most of the progress in DRH has been abroad, not in Britain. European firms have developed many contact arrangements. In the last decade, however, there has been a modest revival of interest in Britain, mainly inspired by the Midlands Electricity Board Development Centre.

Direct resistance heating is ideally suited to heating steel pieces having a uniform cross-section which are long compared to their other dimensions. Very rapid heating is possible, avoiding scale formation and surface decarburisation, and giving high production rates. Efficiencies higher than 90% have been obtained. Heating times of less than a minute can be expected.

26.2 AC versus DC for DRH

We saw, in Chapter 1, that DRH needs high currents (e.g. Example 1.1, 1220 A; Example 1.2, 85 000 A). These are often at low voltages, so the use of step-down transformers is natural and suggests that AC is desirable.

In Sections 3.4 and 4.3, it is shown that the quantity p_r, which controls power, falls off rapidly with increasing $2b/\delta$ or R/δ, so that AC can be disadvantageous. Obviously, at $2b/\delta < 1$ or $R/\delta < 1$, there is little fall-off in power, but above these values the loss of power is considerable and DC is preferable.

Also, when AC is being conducted to the two ends of a workpiece, it is likely that inductance effects will become important, and great care must be taken in the placing of the 'go-return' paths to avoid worsening the power factor even further than the values given in Chapters 3 and 4. DC also gives uniform heating across the section.

For all these reasons, it is technically desirable to use DC rather than AC when $2b/\delta$ (or $R/\delta > 1$). The decision then becomes an economic one: is it cheaper to pay for the initial installation of the rectification or to accept longer heating times and poorer temperature distribution (with perhaps the need to soak for a time to get more uniform temperature)?

In previous chapters, we compared the loss on the basis of the same values of E and J at the surface. Clearly the heating time can always be speeded up by increasing E, but with a corresponding increase of reactive volt-amperes. Here, we consider the same effective E, i.e. DC volts = RMS AC volts, and simple rectification. If the power is great enough to justify polyphase transformation and rectification, the arguments are strengthened in favour of DC.

26.2.1 Unrectified AC

Let the RMS applied voltage be V, giving an instantaneous value

$$v = (\sqrt{2})V \sin \omega t$$

V produces J_s, so that, for a round wire of radius R, from eqn 4.15,

$$P_w = (\pi R^2 l) J_s^2 \{p_r\}$$

$$= (\pi R^2 l) \frac{V^2}{\pi l^2} \{p_r\} = \left(\frac{\pi R^2}{\pi l}\right) V^2 \{p_r\}$$

where p_r is obtained from Fig. 4.5.

26.2.2 Rectified AC

The Fourier series for the simple rectified wave is

$$y = \frac{4\hat{V}}{\pi}\left[\frac{1}{2} - \frac{\cos\theta}{3} - \frac{\cos 2\theta}{15} - \frac{\cos 3\theta}{35} - \ldots - \frac{\cos n\theta}{4n^2 - 1}\right]$$

So we have, for the applied voltage V,

$$v = \frac{4\sqrt{2}V}{\pi}\left[\frac{1}{2} - \frac{\cos 2\omega t}{3} - \frac{\cos 4\omega t}{15} - \frac{\cos 6\omega t}{35} - \ldots\right]$$

DC component:

$$DC \text{ voltage } V_{dc} = \frac{2\sqrt{2}}{\pi} V$$

$$E_{dc} = V_{dc}/l$$

$$P_W = (\pi R^2 l) \frac{V_{dc}^2}{\rho l^2} = \left(\frac{\pi R^2}{\rho l}\right) \frac{8}{\pi^2} V^2$$

This is less than the unrectified loss when $p_r > 8/\pi^2$, i.e. 0.81, corresponding to $d/\delta \lesssim 2.5$, from Fig. 4.5. The heating effect of DC is uniform through the cylinder, so no soaking is needed for this, the dominant component.

AC components:

Second harmonic

$$\text{RMS 2nd harmonic voltage} = (4/3\pi)V$$

Again, from eqn 4.15,

$$P_{w2} = (\pi R^2 l/\rho) E_2^2 \{p_{r2}\}$$

p_{r2} is found using $(\sqrt{2})(d/\delta)$ from Fig. 4.5.

Fourth harmonic

Similarly,

$$\text{RMS 4th harmonic voltage} = \text{2nd harmonic voltage}/5 = (4/15\pi)V$$

and

$$P_{w4} = (\pi R^2 l/\rho)(E_2^2/25)\{p_{r4}\}$$

p_{r4} is found using $2(d/\delta)$ from Fig. 4.5.

Sixth harmonic
Similarly,

$$\text{RMS 6th harmonic voltage} = \text{2nd harmonic voltage} \times (3/35)$$
$$= (4/35\pi)V$$

and

$$P_{w6} = (\pi R^2 l/\rho)(E_2^2)(3/35)^2\{p_{r6}\}$$

p_{r6} is found using $(\sqrt{6})(d/\delta)$ from Fig. 4.5.

Table 26.1 shows that sixth harmonics and above can be neglected.

26.2.3 Comparison: single phase

$$\text{unrectified loss} = \kappa\{p_{r1}\}$$

$$\text{total rectified loss} = \kappa\left[\frac{8}{\pi^2} + \left(\frac{4}{3\pi}\right)^2\{p_{r2}\} + \left(\frac{4}{15\pi}\right)^2\{p_{r4}\}\ldots\right]$$

These figures are shown for various R/δ in Table 26.1. For the higher values of R/S in this table, the result is a linear function of R/δ.

The results in this section suggest that rectification is desirable, unless p_{r1} is close to unity. For ferromagnetic materials, where μ_r is important, δ is small at 50 Hz, and it will be well worth paying for rectification. Table 26.1 tells us the extra size of the transformer needed if raw AC is used.

Note the relatively-small ripple heating: this occurs at the surface, where it tends to offset radiation. There is no advantage in using filters and chokes to reduce this ripple, as it contributes usefully to the heating.

Table 26.1

R/δ	0	0.5	1	1.5	2	2.5	3	3.5	
p_{r1}	1	0.995	0.926	0.736	0.537	0.410	0.336	0.287	
p_{r2}	1	0.98	0.775	0.5	0.36	0.28	0.24	0.20	
p_{r4}	1	0.926	0.537	0.336	0.251	0.201	0.167	0.143	
$8/\pi^2$	0.811	0.811	0.811	0.811	0.811	0.811	0.811	0.811	
2nd term	0.180	0.177	0.140	0.09	0.06	0.05	0.04	0.04	
3rd term	0.007	0.007	0.004	0.002	0.002	0.001	0.001	0.001	(neglect)
Σ	0.998	0.995	0.955	0.90	0.87	0.86	0.85	0.85	
Ratio Σ/p_{r1}	1	1	1.031	1.22	1.62	2.10	2.53	2.96	

Example 26.1
A straight black bar, 5 m long and 50 mm diameter, is to be heated from 20°C to 900°C for winding into springs. Compare single-phase 50 Hz power, rectified and unrectified.

Solution
Assume a value of $\mu_r = 10$; this can be checked and corrected later, if necessary (see Section 26.3).

θ (°C)	20	200	400	600	800	900
ρ ($\mu\Omega$ m)	0.16	0.29	0.48	0.755	1.095	1.135
μ_r	10	10	10	10	1	1
δ (mm)	9.0	12.1	15.6	19.6	74.5	75.8
R/δ	2.8	2.06	1.60	1.28	0.34	0.33
Σ/p_r	2.4	1.7	1.3	1.1	1	1

(Table 26.1: approximate values)

In this example, no advantage is gained by rectification at the higher temperatures; it is then a matter of economics whether it is desirable overall. For larger diameters, it becomes increasingly attractive.

26.2.4 Polyphase AC

In the last section, we discussed simple full-wave rectification of single-phase AC. The use of polyphase rectification is obviously even more desirable, if economic:

- because it gives better phase-balance, compared with either raw single-phase or rectified single-phase
- because it is so near to DC.

It is not worthwhile to do a Fourier series for even the simple three-phase rectifier, as clearly the effect of ripple is negligible. Simple three-phase rectification is greatly superior to both single-phase rectification and to unrectified AC. If any of the more sophisticated rectifier arrangements are used, e.g. six-phase, the advantage is even greater.

These discussions of rectification have not taken power factor into account: a study of Fig. 4.5 shows how bad this can be and provides an additional argument for rectification.

In these sections, we have discussed rectification on the basis of equal *E*. This can be changed by transformer tappings. Also, the configuration is important – how the workpiece is connected to the supply, which brings in the inductance of the bus-bars, which does not matter with DC. If two workpieces can be heated at the same time, arranged in a 'go-and-return' path, the mutual inductance is reduced and the bus-bars are simplified.

26.3 Value of μ in DRH

If it is possible to supply DC, depth of penetration is unimportant but, if magnetic material is being fed with AC, then μ comes into play. In Example 1.1, 12 mm diameter rod was fed with 1220 A; this gives a typical J of about 10^7 A/m^2 for material being heated to forging temperature. Using $\oint H\,dl = I$, the surface H is $(1220/\pi 12)1000 = 32.4$ kA/m; from Fig. 7.3, this corresponds to $\mu = 60$. The larger diameter material of Example 1.2, with a similar J, has $H = 270$ kA/m, giving $\mu = 7.5$. These orders of magnitude are similar to those used in induction heating, as one might expect, since both are doing similar jobs. They suggest that one should heat slowly at low current-densities to keep the depth of penetration low, if one wants to heat small-diameter work.

26.4 Contacts

Where the material to be heated is reasonably plane or cylindrical and free from scale, it is possible to use make contact over an area fairly simply. However, if the metal being heated has cropped ends and it would be expensive to clean up the rough face, it is better to accept 'line' contact and to arrange that the current is applied in two discrete steps. The first current level is 50% (or less, according to experience) of the final value: this, combined with substantial axial thrust, ensures that good contact is made because the metal softens and 'gives' at the hot points without risk of thermal runaway: after one second, it is safe to go to full current.

When the bar reaches 600/700°C, the axial pressure can be reduced considerably. Arrangements must also be made to offset the effects of bar expansion during heating. In the Midlands Electricity Board (MEB) design, the compensating contact system consisted of sliding-fit, taper-faced, copper-alloy contact bars, equispaced within a water-cooled brass block. The non-tapered end of each contact bar reacted against hydraulic plungers connected to a common hydraulic circuit, so that the first contact to touch loaded the rest of the contacts via the hydraulics. Suitable electrical terminations and secondary water cooling were provided in the brass block.

26.4.1 Contact materials

Ideally, the properties of the contact material would be

- low resistivity
- high surface hardness
- high melting point
- high thermal conductivity
- high tensile strength
- low cost

- good machinability
- good corrosion resistance
- readily available (contacts are consumable items)
- no health or environmental problems.

These properties must be maintained over a temperature range up to 1300°C. Most contact materials are copper alloys; the common alloying elements are beryllium, cobalt, chrome and tungsten, each of which improves some properties at the expense of others. As usual, the result is a compromise. For example, tungsten improves the mechanical properties but increases the resistivity considerably, causing extra losses in, and heating of, the contacts. Beryllium copper seems to give the best compromise. This presents problems, as it is getting very scarce, because it is carcinogenic and difficult to work with. It has to be machined wet and disposed of under controlled conditions.

26.4.2 Liquid metal contacts

Some years ago, British Steel reported work using low melting-point metals to make good contact with steel strip, combined with an induction heating system, but this was abandoned because of carry-over of the liquid metal, amongst other reasons. The concept has been revived at the Power Technology Centre of MEB, in combination with DRH, using their expertise with heated air-knives to eliminate carry-over. The report on prototype experiments is optimistic and suggests that high system energy-conversion efficiencies are possible by combining it with a liquid-metal quench, the waste heat from this being used to preheat the strip and also to heat the molten-metal contacts.

26.5 Practical applications: case studies

26.5.1 Spring winding

User: West Bromwich Spring Company
Manufacturer: Newelco (Ukside) Ltd

In the manufacture of large springs for heavy vehicles, agricultural machinery, railways, etc., steel bars from 12 to 76 mm in diameter and up to 9 m long are hot coiled. This shape is excellent for DRH. The previous method was a batch process, where bars were heated in a furnace to 920°C in about 30 minutes and then transferred to the winding lathe by the furnace operator, where two further operators wound the spring and adjusted the length before quenching. When all of the bars were wound, the furnace was reloaded, and coiling restarted about 30 minutes later. The process had high production costs, high energy costs, variable spring quality, poor hearth utilisation and low output.

In the new process, commercially-straight black bars are loaded into a magazine and fed automatically to the 1000 kW DRH. At a signal from the operator, a bar is moved to the heating position and the two sets of contacts close hydraulically – one set being mounted on a motorised trolley which is positioned automatically at the correct bar length. A radiation pyrometer monitors temperature and controls the heating time, typically 30–60 seconds, depending on bar size. When up to temperature, the bar is transferred to a new fully-automated winding lathe which coils it and forms the spring ends; the lathe then reverses, ejecting the spring onto a conveyor which transfers it to the quench. Thus DRH has changed this process to a flowline, needing only one operator to supervise the whole process, giving a twelvefold increase in operator productivity.

The gas furnace had to be able to heat the longest bars used (9 m) even though 70% of the bars are less than 5 m long. This was very wasteful and resulted in an average energy consumption of 115 therms/tonne. By contrast, the average consumption of the DRH is 300 kWh/tonne (10.2 therms/tonne), which represented at least a 60% reduction in energy costs alone (1981 prices).

Routine maintenance is minimal – a daily wire-brushing of the contacts. The contacts have to be replaced from time to time; this takes 30 minutes. The worn contacts can be remachined for next time.

Accurate temperature control is a bonus since the springs are much more uniform metallurgically. Also, the short heating times virtually eliminate the decarburisation and heavy scaling which occurred in the old furnace.

Working conditions are greatly improved by DRH. Not only has the hard physical labour been eliminated but also a comfortable working environment has replaced the heat, fumes and noise of the gas furnace.

26.5.2 Warm cropping

User: John Stokes Ltd, Walsall
Manufacturer: Inductotherm Europe Ltd

Some high-quality steel forgings are made from bought-in rolled sections, which are chopped up ('cropped') to the billet size needed. The sections are up to 75 mm square and are supplied in 6 m lengths. Many of these forgings are used in safety-critical components for the vehicle industry. Much of the material is medium carbon and alloy steel, which is prone to form hairline cracks ('post-crop cracking') on the billet end face, if cold sheared. These cracks may appear up to three days after shearing and can result in unacceptable forgings. Where the problem exists, and also to give a cleaner shear, one-third to one-half of the material is warm-cropped, usually at temperatures between 80°C and 450°C. As in Section 26.5.1, a fuel-fired furnace has been replaced with a 600 kVA direct resistance heater, giving the following advantages:

- uniform temperature across the bar section
- immediate availability
- clean heating, with an improved working environment (no combustion products)
- economy, because energy is only used whilst the bar is heating
- compactness: the heater is sited in a limited space.

26.5.3 Heating for forging
User: Neill Tools Ltd, Britool ADF Division, Cannock, West Midlands
Manufacturer: Inductotherm Europe Ltd

Here, the billets are precut to size. They vary from 12 to 27 mm diameter and 300 to 450 mm long; they are heated in their 'as purchased' condition, without removal of mill scale. In this application – forgings for small tools – DRH is being substituted for oil-fired slot furnaces.

The pieces are delivered at regular intervals from a large-capacity hopper to the 200 kVA heater, where current is passed through the steel via pneumatically-operated contacts. Current levels and heating times are accurately controlled to preset values to maintain constant workpiece temperature.

The electrical consumption is 270 kWh/tonne, one-third of the energy cost of the oil-fired furnaces.

After heating, the workpiece falls onto a short conveyor for delivery to the press. Because the operator receives billets at regular intervals, without leaving the press, productivity is increased. The operators prefer the new system because the work is less strenuous, cooler and quieter. When the material dimensions change, the hammer operator simply repositions the heating contacts, adjusts the sides of the hopper and resets the timing and current controls to predetermined values.

All the advantages listed for Sections 26.5.1 and 26.5.2 apply here. In addition, the reduced scaling, coupled with accurate temperature control, has increased die life by at least 50%.

Appendix A.1

Resistivity, specific heat and thermal conductivity as functions of temperature

Stainless steel (19.11% chromium, 8.14% nickel, 0.60% tungsten)

Temperature (°C)	ρ (Ωm)	$(c\gamma)$ (W s/m^3 K)	k (W/m K)
20	0.695×10^{-6}	4.04×10^{6}	15.9
100	0.776×10^{-6}	4.15×10^{6}	16.3
200	0.850×10^{-6}	4.24×10^{6}	17.2
300	0.915×10^{-6}	4.36×10^{6}	18.0
400	0.976×10^{-6}	4.51×10^{6}	19.7
500	1.030×10^{-6}	4.83×10^{6}	21.4
600	1.072×10^{-6}	4.84×10^{6}	23.0
700	1.111×10^{-6}	4.87×10^{6}	24.3
800	1.141×10^{-6}	4.90×10^{6}	26.0
900	1.171×10^{-6}	4.86×10^{6}	27.0
1000	1.196×10^{-6}	4.87×10^{6}	28.0
1100	1.220×10^{-6}	4.93×10^{6}	29.0
1200	1.241×10^{-6}		29.8
1300	1.257×10^{-6}		30.5

Mild steel (0.23% carbon)

Temperature (°C)	ρ (Ωm)	$(c\gamma)$ (W s/m³ K)	k (W/m K)	$\rho(c\gamma)$ (ΩJ/m² K)
20	0.160×10^{-6}	3.65×10^{6}	52.0	0.584
100	0.220×10^{-6}	3.85×10^{6}	51.0	0.847
200	0.290×10^{-6}	4.10×10^{6}	49.0	1.189
300	0.380×10^{-6}	4.40×10^{6}	46.0	1.672
400	0.483×10^{-6}	4.77×10^{6}	43.0	2.304
500	0.610×10^{-6}	5.19×10^{6}	39.3	3.166
600	0.755×10^{-6}	5.66×10^{6}	35.5	4.273
700	0.922×10^{-6}	6.66×10^{6}	31.5	6.141
800	1.095×10^{-6}	6.73×10^{6}	26.0	7.369
900	1.135×10^{-6}	5.94×10^{6}	26.5	6.742
1000	1.168×10^{-6}	5.09×10^{6}	27.3	5.945
1100	1.195×10^{-6}	5.10×10^{6}	28.5	6.095
1200	1.22×10^{-6}		29.7	
1300	1.24×10^{-6}		31.2	
1400	1.26×10^{-6}	5.08×10^{6}	32.8	

Copper

Temperature (°C)	ρ (Ωm)	$(c\gamma)$ (W s/m³ K)	k (W/m K)
20	0.017×10^{-6}	3.39×10^{6}	395
100	0.022×10^{-6}	3.48×10^{6}	387
200	0.033×10^{-6}	3.57×10^{6}	380
300	0.037×10^{-6}	3.65×10^{6}	373
400	0.044×10^{-6}	3.72×10^{6}	366
500	0.052×10^{-6}	3.77×10^{6}	360
600	0.060×10^{-6}	3.82×10^{6}	353
700	0.068×10^{-6}	3.87×10^{6}	347
800	0.077×10^{-6}	3.92×10^{6}	341
900	0.086×10^{-6}	3.96×10^{6}	335
1000	0.097×10^{-6}	3.99×10^{6}	330
1100	0.215×10^{-6}	3.66×10^{6}	324
1200	0.222×10^{-6}		319
1300	0.228×10^{-6}		313

Aluminium

Temperature (°C)	ρ (Ωm)	$(c\gamma)$ (W s/m³ K)	k (W/m K)
20	0.027×10^{-6}	2.52×10^{6}	211
100	0.0364×10^{-6}	2.59×10^{6}	219
200	0.0478×10^{-6}	2.65×10^{6}	224
300	0.0599×10^{-6}	2.71×10^{6}	223
400	0.073×10^{-6}	2.78×10^{6}	216
500	0.087×10^{-6}	2.84×10^{6}	209
600	0.104×10^{-6}	2.89×10^{6}	200
700	0.21×10^{-6}	2.51×10^{6}	92
800	0.225×10^{-6}	2.52×10^{6}	88
900	0.235×10^{-6}	2.54×10^{6}	

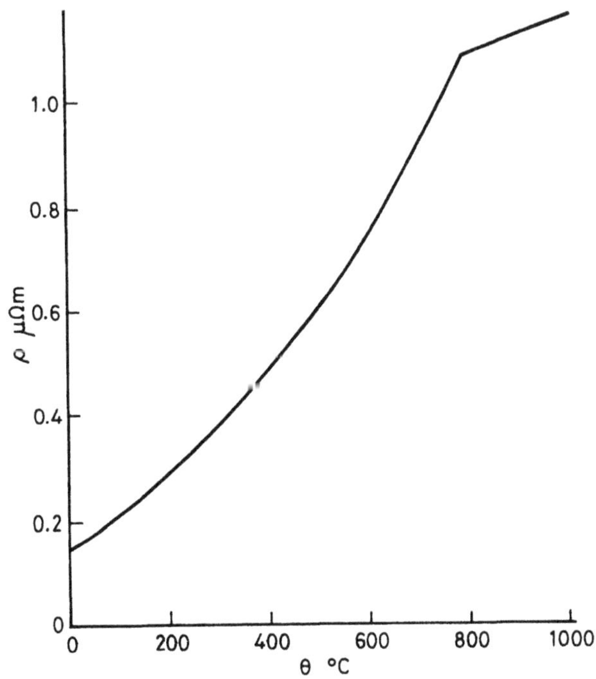

Fig. A.1.1 *Resistivity of mild steel (0.23%C) with temperature (Davies and Simpson[7])*

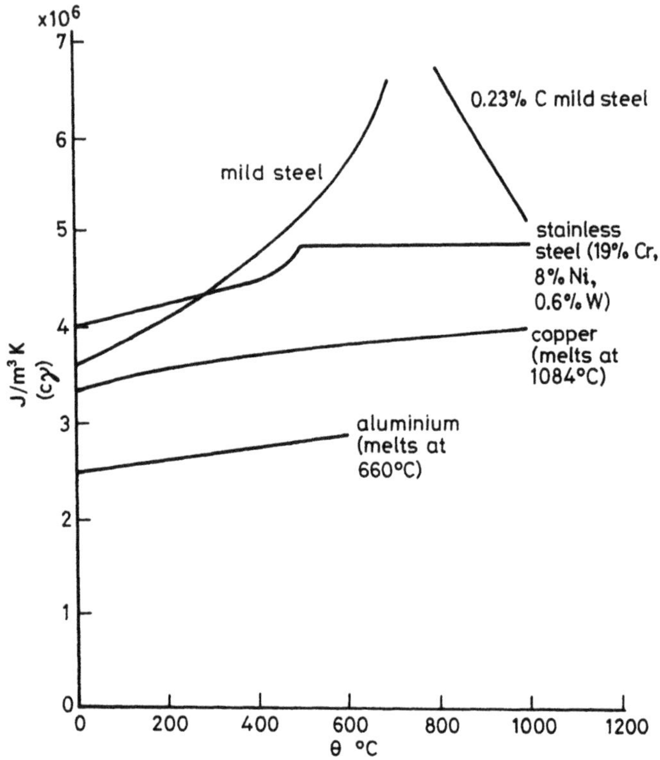

Fig. A.1.2 *Variation of volume specific heat ($c\gamma$) with temperature (Davies and Simpson[7])*

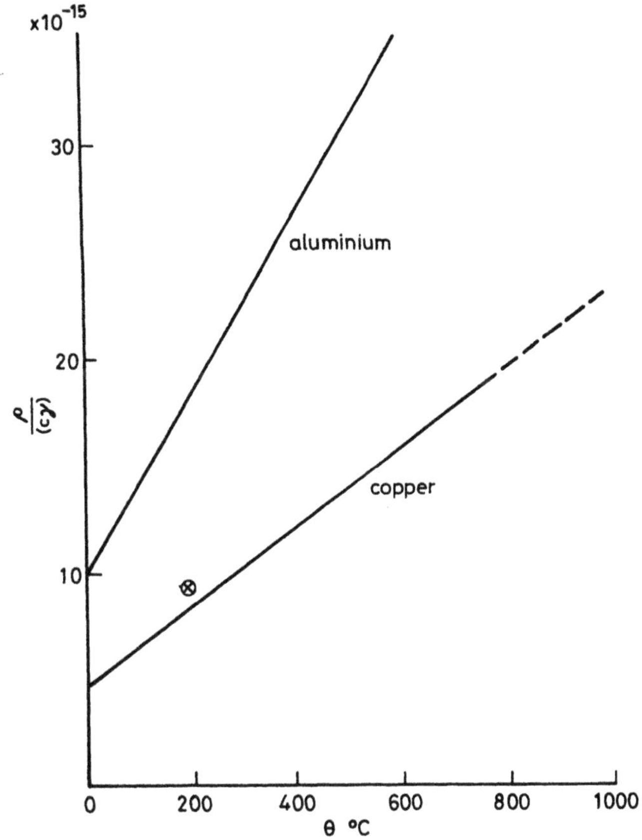

Fig. A.1.3 *Plots of $\rho/(c\gamma)$: copper, 0–800°C, regression line $y = (4.5 + 0.0187\theta)10^{-15}$ (200°C point excluded); aluminium, 0–500°C, regression line $y = (9.84 + 0413\theta)10^{-15}$ (Davies and Simpson[7])*

Appendix A.2

Relations for tanh $(1 \pm j)\theta$

$$\tanh(1 \pm j)\theta = \frac{\sinh(1 \pm j)\theta}{\cosh(1 \pm j)\theta}$$

$$= \frac{\sinh\theta\cos\theta \pm j\cosh\theta\sin\theta}{\cosh\theta\cos\theta \pm j\sinh\theta\sin\theta} = \frac{A \pm jB}{C \pm jD}$$

$$= \frac{(AC + BD) \pm j\,(BC - AD)}{C^2 + D^2}$$

$$= \frac{\sinh\theta\cosh\theta\,(\cos^2\theta + \sin^2\theta) \pm j\sin\theta\cos\theta\,(\cosh^2\theta - \sinh^2\theta)}{C^2 + D^2}$$

$$= \frac{(\sinh 2\theta)/2 \pm j(\sin 2\theta)/2}{C^2 + D^2} \qquad \begin{cases} \cos^2 + \sin^2 = 1 \\ \cosh^2 - \sinh^2 = 1 \end{cases}$$

$$= \frac{\sinh 2\theta \pm j\sin 2\theta}{2\cosh^2\theta\cos^2\theta + 2\sinh^2\theta\sin^2\theta}$$

$$= \frac{\sinh 2\theta \pm j\sin 2\theta}{\cosh^2\theta\,(1 + \cos 2\theta) + \sinh^2\theta\,(1 - \cos 2\theta)}$$

$$= \frac{\sinh 2\theta \pm j\sin 2\theta}{(\cosh^2\theta + \sinh^2\theta) + \cos 2\theta\,(\cosh^2\theta - \sinh^2\theta)}$$

$$= \frac{\sinh 2\theta \pm j\sin 2\theta}{\cosh 2\theta + \cos 2\theta}$$

Appendix A.3

Bessel functions and values of M_0, M_1, N_0 and N_1 for $x\sqrt{2}$

x	ber $(x\sqrt{2})$	bei $(x\sqrt{2})$	ber′ $(x\sqrt{2})$	bei′ $(x\sqrt{2})$	$M_0\,(x\sqrt{2})$	$M_1\,(x\sqrt{2})$
0.01	+1.00000000	+0.00005000	−0.00000018	+0.00707107	+1.00000000	+0.00707107
0.02	+1.00000000	+0.00020000	−0.00000141	+0.01414214	+1.00000002	+0.01414214
0.03	+0.99999995	+0.00045000	−0.00000477	+0.02121320	+1.00000005	+0.02121320
0.04	+0.99999984	+0.00080000	−0.00001131	+0.02828427	+1.00000016	+0.02828427
0.05	+0.99999961	+0.00125000	−0.00002210	+0.03535534	+1.00000039	+0.03535534
0.06	+0.99999919	+0.00180000	−0.00003818	+0.04242639	+1.00000081	+0.04242641
0.07	+0.99999850	+0.00245000	−0.00006063	+0.04949745	+1.00000150	+0.04949748
0.08	+0.99999744	+0.00320000	−0.00009051	+0.05656849	+1.00000256	+0.05656856
0.09	+0.99999590	+0.00405000	−0.00012887	+0.06363952	+1.00000410	+0.06363965
0.10	+0.99999375	+0.00500000	−0.00017678	+0.07071052	+1.00000625	+0.07071074
0.11	+0.99999085	+0.00605000	−0.00023529	+0.07778151	+1.00000915	+0.07778186
0.12	+0.99998704	+0.00719999	−0.00030547	+0.08485245	+1.00001296	+0.08485300
0.13	+0.99998215	+0.00844998	−0.00038838	+0.09192334	+1.00001785	+0.09192416
0.14	+0.99997599	+0.00979997	−0.00048508	+0.09899416	+1.00002401	+0.09899534
0.15	+0.99996836	+0.01124996	−0.00059662	+0.10606490	+1.00003164	+0.10606658
0.16	+0.99995904	+0.01279994	−0.00072408	+0.11313554	+1.00004096	+0.11313785
0.17	+0.99993780	+0.01444992	−0.00086850	+0.12020606	+1.00005220	+0.12020920
0.18	+0.99993439	+0.01619988	−0.00103096	+0.12727644	+1.00006561	+0.12728061
0.19	+0.99991855	+0.01804984	−0.00121251	+0.13434664	+1.00008145	+0.13435211
0.20	+0.99990000	+0.01999978	−0.00141421	+0.14141663	+1.00010000	+0.14142370
0.21	+0.99987845	+0.02204970	−0.00163712	+0.14848641	+1.00012154	+0.14849543
0.22	+0.99985359	+0.02419961	−0.00188230	+0.15555590	+1.00014640	+0.15556728
0.23	+0.99982510	+0.02644949	−0.00215082	+0.16262508	+1.00017489	+0.16263931
0.24	+0.99979264	+0.02879934	−0.00244373	+0.16969390	+1.00020734	+0.16971149
0.25	+0.99975586	+0.03124915	−0.00276210	+0.17676231	+1.00024411	+0.17678389
0.26	+0.99971439	+0.03379892	−0.00310698	+0.18383025	+1.00028557	+0.18385651
0.27	+0.99966785	+0.03644866	−0.00347943	+0.19089770	+1.00033210	+0.19092940
0.28	+0.99961584	+0.03919833	−0.00388052	+0.19796454	+1.00038409	+0.19800257
0.29	+0.99955795	+0.04204793	−0.00431130	+0.20503074	+1.00044196	+0.20507607
0.30	+0.99949375	+0.04499746	−0.00477284	+0.21209622	+1.00050614	+0.21214992
0.31	+0.99942280	+0.04804692	−0.00526619	+0.21916094	+1.00057705	+0.21922420
0.32	+0.99934465	+0.05119628	−0.00579241	+0.22622475	+1.00065518	+0.22629889
0.33	+0.99925881	+0.05444552	−0.00635256	+0.23328759	+1.00074097	+0.23337406
0.34	+0.99916481	+0.05779464	−0.00694771	+0.24034939	+1.00083492	+0.24044978

0.35	+0.99906213	+0.06124362	−0.00757891	+0.24741001	+1.00093752	+0.24752606
0.36	+0.99895027	+0.06479244	−0.00824721	+0.25446937	+1.00104930	+0.25460298
0.37	+0.99882869	+0.06844110	−0.00895369	+0.26152737	+1.00117078	+0.26168059
0.38	+0.99869684	+0.07218954	−0.00969939	+0.26858385	+1.00130250	+0.26875893
0.39	+0.99855416	+0.07603778	−0.01048537	+0.27563873	+1.00144503	+0.27583810
0.40	+0.99840007	+0.07998579	−0.01131270	+0.28269188	+1.00159894	+0.28291814
0.41	+0.99823399	+0.08403350	−0.01218243	+0.28974310	+1.00176480	+0.28999910
0.42	+0.99805529	+0.08818094	−0.01309562	+0.29679232	+1.00194324	+0.29708110
0.43	+0.99786338	+0.09242806	−0.01405332	+0.30383937	+1.00213485	+0.30416419
0.44	+0.99765759	+0.09677480	−0.01505659	+0.31088403	+1.00234028	+0.31124843
0.45	+0.99743729	+0.10122117	−0.01610648	+0.31792623	+1.00256016	+0.31833396
0.46	+0.99720181	+0.10576711	−0.01720406	+0.32496573	+1.00279516	+0.32542082
0.47	+0.99695046	+0.11041257	−0.01835038	+0.33200234	+1.00304594	+0.33250908
0.48	+0.99668255	+0.11515754	−0.01954649	+0.33903592	+1.00331319	+0.33959891
0.49	+0.99639736	+0.12000193	−0.02079344	+0.34606621	+1.00359761	+0.34669033
0.50	+0.99609417	+0.12494574	−0.02209229	+0.35309306	+1.00389992	+0.35378352
0.51	+0.99577225	+0.12998890	−0.02344410	+0.36011623	+1.00422083	+0.36087854
0.52	+0.99543082	+0.13513136	−0.02484991	+0.36713549	+1.00456110	+0.36797552
0.53	+0.99506913	+0.14037304	−0.02631077	+0.37415058	+1.00492147	+0.37507454
0.54	+0.99468637	+0.14571391	−0.02782775	+0.38116131	+1.00530270	+0.38217578
0.55	+0.99428177	+0.15115387	−0.02940187	+0.38816738	+1.00570559	+0.38927931
0.56	+0.99385449	+0.15669291	−0.03103421	+0.39516857	+1·00613091	+0.39638532
0.57	+0.99340371	+0.16233092	−0.03272581	+0.40216459	+1.00657948	+0.40349391
0.58	+0.99292858	+0.16806780	−0.03447770	+0.40915513	+1.00705211	+0.41060520
0.59	+0.99242824	+0.17390355	−0.03629094	+0.41613995	+1.00754963	+0.41771939
0.60	+0.99190182	+0.17983802	−0.03816658	+0.42311871	+1.00807288	+0.42483660
0.61	+0.99134843	+0.18587114	−0.04010566	+0.43009112	+1.00862272	+0.43195699
0.62	+0.99076716	+0.19200282	−0.04210923	+0.43705685	+1.00920000	+0.43908071
0.63	+0.99015709	+0.19823293	−0.04417831	+0.44401552	+1.00980561	+0.44620792
0.64	+0.98951729	+0.20456143	−0.04631396	+0.45096686	+1.01044042	+0.45333883
0.65	+0.98884681	+0.21098818	−0.04851722	+0.45791048	+1.01110535	+0.46047359
0.66	+0.98814470	+0.21751305	−0.05078911	+0.46484599	+1.01180130	+0.46761236
0.67	+0.98740995	+0.22413596	−0.05313069	+0.47177305	+1.01252918	+0.47475539
0.68	+0.98664160	+0.23085678	−0.05554299	+0.47869127	+1.01328994	+0.48190285
0.69	+0.98583862	+0.23767534	−0.05802703	+0.48560019	+1.01408449	+0.48905489
0.70	+0.98500000	+0.24459157	−0.06058386	+0.49249948	+1.01491381	+0.49621179
0.71	+0.98412470	+0.25160533	−0.06321452	+0.49938869	+1.01577885	+0.50337376
0.72	+0.98321168	+0.25871636	−0.06591998	+0.50626728	+1.01668056	+0.51054090
0.73	+0.98225985	+0.26592464	−0.06870134	+0.51313495	+1.01761993	+0.51771358
0.74	+0.98126814	+0.27322997	−0.07155958	+0.51999116	+1.01859795	+0.52489197
0.75	+0.98023547	+0.28063218	−0.07449576	+0.52683544	+1.01961561	+0.53207631
0.76	+0.97916071	+0.28813109	−0.07751087	+0.53366731	+1.02067391	+0.53926685
0.77	+0.97804275	+0.29572653	−0.08060593	+0.54048627	+1.02177385	+0.54646383
0.78	+0.97688046	+0.30341824	−0.08378193	+0.54729173	+1.02291645	+0.55366745
0.79	+0.97567266	+0.31120617	−0.08703996	+0.55408328	+1.02410274	+0.56087809
0.80	+0.97441820	+0.31909004	−0.09038098	+0.56086033	+1.02533374	+0.56809597
0.81	+0.97311590	+0.32706965	−0.09380601	+0.56762231	+1.02661049	+0.57532135
0.82	+0.97176456	+0.33514477	−0.09731605	+0.57436866	+1.02793404	+0.58255453

0.83	+0.97036298	+0.34331519	−0.10091211	+0.58109880	+1.02930541	+0.58979578
0.84	+0.96890992	+0.35158067	−0.10459520	+0.58781211	+1.03072567	+0.59704542
0.85	+0.96740416	+0.35994097	−0.10836630	+0.59450799	+1.03219587	+0.60430374
0.86	+0.96584445	+0.36839576	−0.11222638	+0.60118576	+1.03371705	+0.61157099
0.87	+0.96422950	+0.37694494	−0.11617650	+0.60784488	+1.03529031	+0.61884762
0.88	+0.96255805	+0.38558815	−0.12021761	+0.61448464	+1.03691669	+0.62613389
0.89	+0.96082880	+0.39432512	−0.12435070	+0.62110437	+1.03859727	+0.63343013
0.90	+0.95904044	−0.40315556	+0.12857675	+0.62770337	+1.04033311	+0.64073669
0.91	+0.95719165	+0.41207917	−0.13289674	+0.63428095	+1.04212528	+0.64805390
0.92	+0.95528109	+0.42109565	−0.13731164	+0.64083638	+1.04397486	+0.65538214
0.93	+0.95330743	+0.43020468	−0.14182242	+0.64736892	+1.04588294	+0.66272175
0.94	+0.95126930	+0.43940585	−0.14643001	+0.65387778	+1.04785056	+0.67007306
0.95	+0.94916531	+0.44869897	−0.15113546	+0.66036230	+1.04987883	+0.67743656
0.96	+0.94699408	+0.45808362	−0.15593967	+0.66682162	+1.05196881	+0.68481257
0.97	+0.94475420	+0.46755944	−0.16084361	+0.67325497	+1.05412159	+0.69220150
0.98	+0.94244425	+0.47712605	−0.16584822	+0.67966151	+1.05633822	+0.69960374
0.99	+0.94006283	+0.48678299	−0.17095440	+0.68604036	+1.05861977	+0.70701965
1.00	+0.93760846	+0.49653001	−0.17616318	+0.69239078	+1.06096733	+0.71444976

x	ker $(x\sqrt{2})$	kei $(x\sqrt{2})$	ker′ $(x\sqrt{2})$	kei′ $(x\sqrt{2})$	N_0 $(x\sqrt{2})$	N_1 $(x\sqrt{2})$
0.01	+4.3746	−0.7851	−70.7051	+0.0345	+4.4445	+70.7051
0.02	+3.6815	−0.7845	−35.3442	+0.0591	+3.7642	+35.3443
0.03	+3.2763	−0.7835	−23.5536	+0.0801	+3.3686	+23.5537
0.04	+2.9889	−0.7822	−17.6555	+0.0987	+3.0895	+17.6558
0.05	+2.7661	−0.7807	−14.1145	+0.1155	+2.8741	+14.1149
0.06	+2.5842	−0.7789	−11.7519	+0.1308	+2.6990	+11.7527
0.07	+2.4305	−0.7770	−10.0629	+0.1450	+2.5517	+10.0639
0.08	+2.2976	−0.7749	−8.7947	+0.1582	+2.4247	+8.7962
0.09	+2.1805	−0.7725	−7.8072	+0.1705	+2.3133	+7.8091
0.10	+2.0758	−0.7700	−7.0161	+0.1820	+2.2141	+7.0185
0.11	+1.9814	−0.7674	−6.3679	+0.1928	+2.1248	+6.3708
0.12	+1.8952	−0.7646	−5.8269	+0.2030	+2.0436	+5.8304
0.13	+1.8162	−0.7616	−5.3683	+0.2126	+1.9694	+5.3725
0.14	+1.7431	−0.7586	−4.9745	+0.2217	+1.9010	+4.9794
0.15	+1.6752	−0.7554	−4.6325	+0.2303	+1.8376	+4.6382
0.16	+1.6119	−0.7521	−4.3326	+0.2384	+1.7787	+4.3392
0.17	+1.5525	−0.7486	−4.0675	+0.2461	+1.7236	+4.0749
0.18	+1.4967	−0.7451	−3.8312	+0.2533	+1.6719	+3.8396
0.19	+1.4440	−0.7415	−3.6193	+0.2602	+1.6233	+3.6287
0.20	+1.3942	−0.7377	−3.4282	+0.2668	+1.5774	+3.4385
0.21	+1.3470	−0.7339	−3.2548	+0.2730	+1.5339	+3.2662
0.22	+1.3021	−0.7300	−3.0967	+0.2789	+1.4928	+3.1093
0.23	+1.2593	−0.7260	−2.9520	+0.2845	+1.4536	+2.9657
0.24	+1.2185	−0.7220	−2.8190	+0.2898	+1.4164	+2.8338
0.25	+1.1795	−0.7178	−2.6962	+0.2948	+1.3808	+2.7123
0.26	+1.1422	−0.7136	−2.5826	+0.2996	+1.3468	+2.5999
0.27	+1.1065	−0.7094	−2.4771	+0.3041	+1.3143	+2.4957
0.28	+1.0721	−0.7050	−2.3788	+0.3083	+1.2832	+2.3987
0.29	+1.0391	−0.7007	−2.2870	+0.3124	+1.2533	+2.3082
0.30	+1.0074	−0.6962	−2.2011	+0.3162	+1.2246	+2.2236

0.31	+0.9769	−0.6917	−2.1204	+0.3198	+1.1970	+2.1444
0.32	+0.9474	−0.6872	−2.0445	+0.3232	+1.1704	+2.0699
0.33	+0.9190	−0.6826	−1.9730	+0.3264	+1.1448	+1.9999
0.34	+0.8916	−0.6779	−1.9055	+0.3294	+1.1201	+1.9338
0.35	+0.8651	−0.6733	−1.8417	+0.3322	+1.0962	+1.8714
0.36	+0.8395	−0.6685	−1.7812	+0.3349	+1.0732	+1.8124
0.37	+0.8147	−0.6638	−1.7237	+0.3374	+1.0509	+1.7564
0.38	+0.7907	−0.6590	−1.6691	+0.3397	+1.0293	+1.7034
0.39	+0.7675	−0.6542	−1.6172	+0.3419	+1.0085	+1.6529
0.40	+0.7450	−0.6493	−1.5676	+0.3439	+0.9882	+1.6249
0.41	+0.7231	−0.6444	−1.5204	+0.3457	+0.9686	+1.5592
0.42	+0.7020	−0.6395	−1.4752	+0.3475	+0.9496	+1.5156
0.43	+0.6814	−0.6346	−1.4320	+0.3491	+0.9312	+1.4739
0.44	+0.6614	−0.6297	−1.3906	+0.3505	+0.9132	+1.4341
0.45	+0.6421	−0.6247	−1.3509	+0.3518	+0.8958	+1.3960
0.46	+0.6232	−0.6197	−1.3139	+0.3530	+0.8789	+1.3595
0.47	+0.6049	−0.6147	−1.2763	+0.3541	+0.8624	+1.3245
0.48	+0.5871	−0.6097	−1.2411	+0.3551	+0.8464	+1.2909
0.49	+0.5698	−0.6047	−1.2073	+0.3559	+0.8309	+1.2587
0.50	+0.5530	−0.5996	−1.1747	+0.3567	+0.8157	+1.2277
0.51	+0.5366	−0.5946	−1.1433	+0.3573	+0.8009	+1.1979
0.52	+0.5206	−0.5895	−1.1130	+0.3578	+0.7865	+1.1692
0.53	+0.5051	−0.5845	−1.0838	+0.3583	+0.7725	+1.1415
0.54	+0.4900	−0.5794	−1.0556	+0.3586	+0.7588	+1.1148
0.55	+0.4752	−0.5743	−1.0283	+0.3588	+0.7455	+1.0891
0.56	+0.4609	−0.5693	−1.0019	+0.3590	+0.7324	+1.0643
0.57	+0.4469	−0.5642	−0.9764	+0.3591	+0.7197	+1.0404
0.58	+0.4333	−0.5591	−0.9517	+0.3590	+0.7073	+1.0172
0.59	+0.4200	−0.5540	−0.9278	+0.3589	+0.6952	+0.9948
0.60	+0.4070	−0.5489	−0.9046	+0.3588	+0.6834	+0.9731
0.61	+0.3944	−0.5439	−0.8821	+0.3585	+0.6718	+0.9522
0.62	+0.3821	−0.5388	−0.8603	+0.3582	+0.6605	+0.9318
0.63	+0.3700	−0.5337	−0.8391	+0.3577	+0.6495	+0.9122
0.64	+0.3583	−0.5287	−0.8185	+0.3573	+0.6387	+0.8931
0.65	+0.3469	−0.5236	−0.7985	+0.3567	+0.6281	+0.8746
0.66	+0.3357	−0.5186	−0.7791	+0.3561	+0.6178	+0.8567
0.67	+0.3249	−0.5136	−0.7603	+0.3554	+0.6077	+0.8392
0.68	+0.3142	−0.5085	−0.7419	+0.3547	+0.5978	+0.8223
0.69	+0.3039	−0.5035	−0.7241	+0.3539	+0.5881	+0.8059
0.70	+0.2937	−0.4985	−0.7067	+0.3530	+0.5786	+0.7899
0.71	+0.2839	−0.4936	−0.6898	+0.3521	+0.5694	+0.7744
0.72	+0.2742	−0.4886	−0.6733	+0.3511	+0.5603	+0.7593
0.73	+0.2648	−0.4836	−0.6572	+0.3501	+0.5514	+0.7447
0.74	+0.2556	−0.4787	−0.6416	+0.3490	+0.5427	+0.7304
0.75	+0.2467	−0.4738	−0.6263	+0.3479	+0.5341	+0.7165
0.76	+0.2379	−0.4688	−0.6115	+0.3467	+0.5258	+0.7029
0.77	+0.2294	−0.4639	−0.5970	+0.3455	+0.5176	+0.6898
0.78	+0.2210	−0.4591	−0.5828	+0.3442	+0.5095	+0.6769

0.79	+0.2129	−0.4542	−0.5691	+0.3429	+0.5016	+0.6644
0.80	+0.2049	−0.4494	−0.5556	+0.3416	+0.4939	+0.6522
0.81	+0.1972	−0.4445	−0.5425	+0.3402	+0.4863	+0.6403
0.82	+0.1896	−0.4397	−0.5296	+0.3388	+0.4789	+0.6287
0.83	+0.1822	−0.4350	−0.5171	+0.3373	+0.4716	+0.6174
0.84	+0.1750	−0.4302	−0.5049	+0.3358	+0.4644	+0.6063
0.85	+0.1679	−0.4255	−0.4929	+0.3342	+0.4574	+0.5955
0.86	+0.1610	−0.4208	−0.4812	+0.3327	+0.4505	+0.5850
0.87	+0.1543	−0.4161	−0.4698	+0.3311	+0.4438	+0.5747
0.88	+0.1477	−0.4114	−0.4587	+0.3294	+0.4371	+0.5647
0.89	+0.1413	−0.4067	−0.4477	+0.3278	+0.4306	+0.5549
0.90	+0.1351	−0.4021	−0.4371	+0.3261	+0.4242	+0.5453
0.91	+0.1290	−0.3975	−0.4267	+0.3243	+0.4179	+0.5359
0.92	+0.1230	−0.3929	−0.4165	+0.3226	+0.4117	+0.5268
0.93	+0.1172	−0.3884	−0.4065	+0.3208	+0.4057	+0.5178
0.94	+0.1115	−0.3839	−0.3967	+0.3190	+0.3997	+0.5091
0.95	+0.1060	−0.3794	−0.3872	+0.3172	+0.3939	+0.5005
0.96	+0.1006	−0.3749	−0.3778	+0.3153	+0.3882	+0.4921
0.97	+0.0953	−0.3705	−0.3687	+0.3134	+0.3825	+0.4839
0.98	+0.0901	−0.3660	−0.3598	+0.3115	+0.3770	+0.4759
0.99	+0.0851	−0.3616	−0.3510	+0.3096	+0.3715	+0.4681
1.00	+0.0802	−0.3573	−0.3424	+0.3077	+0.3662	+0.4604

x	ber $(x\sqrt{2})$	bei $(x\sqrt{2})$	ber′ $(x\sqrt{2})$	bei′ $(x\sqrt{2})$	M_0 $(x\sqrt{2})$	M_1 $(x\sqrt{2})$
0.1	+1.0000	+0.0050	−0.0002	+0.0707	+1.0000	+0.0707
0.2	+0.9999	+0.0200	−0.0014	+0.1414	+1.0001	+0.1414
0.3	+0.9995	+0.0450	−0.0048	+0.2121	+1.0005	+0.2121
0.4	+0.9984	+0.0800	−0.0113	+0.2827	+1.0016	+0.2829
0.5	+0.9961	+0.1249	−0.0221	+0.3531	+1.0039	+0.3538
0.6	+0.9919	+0.1798	−0.0382	+0.4231	+1.0081	+0.4248
0.7	+0.9850	+0.2446	−0.0606	+0.4925	+1.0149	+0.4962
0.8	+0.9744	+0.3191	−0.0904	+0.5609	+1.0253	+0.5681
0.9	+0.9590	+0.4032	−0.1286	+0.6277	+1.0403	+0.6407
1.0	+0.9376	+0.4965	−0.1762	+0.6924	+1.0610	+0.7144
1.1	+0.9087	+0.5989	−0.2341	+0.7541	+1.0883	+0.7896
1.2	+0.8709	+0.7096	−0.3033	+0.8120	+1.1234	+0.8667
1.3	+0.8224	+0.8283	−0.3845	+0.8647	+1.1672	+0.9464
1.4	+0.7615	+0.9539	−0.4786	+0.9110	+1.2206	+1.0291
1.5	+0.6864	+1.0856	−0.5862	+0.9494	+1.2844	+1.1158
1.6	+0.5951	+1.2220	−0.7076	+0.9780	+1.3592	+1.2071
1.7	+0.4855	+1.3616	−0.8434	+0.9947	+1.4456	+1.3042
1.8	+0.3558	+1.5027	−0.9935	+0.9975	+1.5442	+1.4079
1.9	+0.2039	+1.6430	−1.1579	+0.9837	+1.6556	+1.5194
2.0	+0.0277	+1.7800	−1.3362	+0.9506	+1.7802	+1.6398
2.1	−0.1747	+1.9108	−1.5275	+0.8954	+1.9188	+1.7706
2.2	−0.4049	+2.0321	−1.7307	+0.8149	+2.0720	+1.9129
2.3	−0.6647	+2.1399	−1.9443	+0.7057	+2.2408	+2.0684
2.4	−0.9553	+2.2301	−2.1661	+0.5643	+2.4261	+2.2384

2.5	−1.2776	+2.2979	−2.3936	+0.3872	+2.6292	+2.4247
2.6	−1.6324	+2.3378	−2.6234	+0.1707	+2.8513	+2.6289
2.7	−2.0196	+2.3441	−2.8516	−0.0888	+3.0941	+2.8530
2.8	−2.4386	+2.3105	−3.0735	−0.3951	+3.3594	+3.0988
2.9	−2.8883	+2.2300	−3.2836	−0.7515	+3.6490	+3.3685
3.0	−3.3665	+2.0954	−3.4755	−1.1612	+3.9654	+3.6643
3.1	−3.8701	+1.8989	−3.6420	−1.6271	+4.3109	+3.9889
3.2	−4.3950	+1.6325	−3.7747	−2.1515	+4.6884	+4.3448
3.3	−4.9358	+1.2876	−3.8645	−2.7362	+5.1009	+4.7351
3.4	−5.4856	+0.8557	−3.9011	−3.3819	+5.5519	+5.1629
3.5	−6.0361	+0.3281	−3.8731	−4.0886	+6.0450	+5.6318
3.6	−6.5774	−0.3036	−3.7683	−4.8549	+6.5844	+6.1457
3.7	−7.0977	−1.0477	−3.5733	−5.6780	+7.1746	+6.7089
3.8	−7.5832	−1.9120	−3.2739	−6.5537	+7.8205	+7.3259
3.9	−8.0181	−2.9036	−2.8550	−7.4755	+8.5276	+8.0022
4.0	−8.3843	−4.0282	−2.3009	−8.4350	+9.3018	+8.7432
4.1	−8.6617	−5.2906	−1.5951	−9.4211	+10.1497	+9.5552
4.2	−8.8276	−6.6935	−0.7211	−10.4202	+11.0783	+10.4452
4.3	−8.8569	−8.2377	+0.3380	−11.4156	+12.0957	+11.4206
4.4	−8.7225	−9.9212	+1.5986	−12.3871	+13.2103	+12.4899
4.5	−8.3945	−11.7391	+3.0766	−13.3111	+14.4317	+13.6620
4.6	−7.8413	−13.6826	+4.7869	−14.1600	+15.7703	+14.9472
4.7	−7.0290	−15.7391	+6.7426	−14.9020	+17.2374	+16.3565
4.8	−5.9221	−17.8908	+8.9549	−15.5012	+18.8455	+17.9019
4.9	−4.4837	−20.1149	+11.4320	−15.9169	+20.6085	+19.5969
5.0	−2.6759	−22.3820	+14.1785	−16.1039	+22.5414	+21.4561
5.1	−0.4607	−24.6566	+17.1947	−16.0122	+24.6609	+23.4957
5.2	+2.1999	−26.8953	+20.4755	−15.5873	+26.9851	+25.7334
5.3	+5.3426	−29.0468	+24.0094	−14.7700	+29.5341	+28.1887
5.4	+9.0019	−31.0514	+27.7775	−13.4970	+32.3299	+30.8830
5.5	+13.2090	−32.8397	+31.7525	−11.7010	+35.3967	+33.8399
5.6	+17.9909	−34.3329	+35.8972	−9.3114	+38.7611	+37.0852
5.7	+23.3680	−35.4420	+40.1632	−6.2551	+42.4523	+40.6474
5.8	+29.3536	−36.0672	+44.4900	−2.4574	+46.5024	+44.5579
5.9	+35.9511	−36.0985	+48.8033	+2.1571	+50.9469	+48.8509
6.0	+43.1525	−35.4151	+53.0135	+7.6625	+55.8244	+53.5644
6.1	+50.9357	−33.8859	+57.0149	+14.1302	+61.1776	+58.7398
6.2	+59.2630	−31.3699	+60.6843	+21.6272	+67.0536	+64.4230
6.3	+68.0776	−27.7174	+63.8795	+30.2132	+73.5038	+70.6642
6.4	+77.3010	−22.7706	+66.4387	+39.9381	+80.5850	+77.5188
6.5	+86.8308	−16.3657	+68.1795	+50.8391	+88.3596	+85.0474
6.6	+96.5369	−8.3347	+68.8985	+62.9372	+96.8961	+93.3171
6.7	+106.2594	+1.4921	+68.3706	+76.2332	+106.2699	+102.4014
6.8	+115.8047	+13.2827	+66.3500	+90.7040	+116.5640	+112.3812
6.9	+124.9431	+27.2000	+62.5700	+106.2973	+127.8696	+123.3456
7.0	+133.4058	+43.3971	+56.7447	+122.9274	+140.2869	+135.3924
7.1	+140.8822	+62.0122	+48.5703	+140.4693	+153.9263	+148.6294
7.2	+147.0178	+83.1628	+37.7284	+158.7535	+168.9092	+163.1751
7.3	+151.4121	+106.9391	+23.8887	+177.5600	+185.3689	+179.1598

7.4	+153.6174	+133.3961	+6.7138	+196.6125	+203.4522	+196.7271
7.5	+153.1379	+162.5452	−14.1356	+215.5717	+223.3208	+216.0346
7.6	+149.4302	+194.3454	−38.9939	+234.0298	+245.1521	+237.2561
7.7	+141.9045	+228.6926	−68.1829	+251.5043	+269.1416	+260.5827
7.8	+129.9269	+265.4088	−102.0022	+267.4323	+295.5044	+286.2245
7.9	+112.8230	+304.2305	−140.7187	+281.1650	+324.4769	+314.4129
8.0	+89.8842	+344.7961	−184.5540	+291.9636	+356.3194	+345.4026
8.1	+60.3742	+386.6329	−233.6706	+298.9955	+391.3183	+379.4736
8.2	+23.5390	+429.1435	−288.1555	+301.3315	+429.7886	+416.9344
8.3	−21.3814	+471.5927	−348.0033	+297.9454	+472.0771	+458.1242
8.4	−75.1386	+513.0927	−413.0965	+287.7144	+518.5653	+503.4166
8.5	−138.4584	+552.5907	−483.1842	+269.4222	+569.6729	+553.2226
8.6	−212.0203	+588.8554	−557.8596	+241.7645	+625.8620	+607.9946
8.7	−296.4344	+620.4656	−636.5363	+203.3578	+687.6416	+668.2312
8.8	−392.2120	+645.7989	−718.4198	+152.7518	+755.5703	+734.4795
8.9	−499.7374	+663.0245	−802.4844	+88.4446	+830.2644	+807.3435
9.0	−619.2314	+670.0967	−887.4428	+8.9037	+912.4018	+887.4874
9.1	−750.7124	+664.7522	−971.7190	−87.4088	+1002.7286	+975.6424
9.2	−893.9541	+644.5117	−1053.4199	−202.0050	+1102.0659	+1072.6134
9.3	−1048.4382	+606.6862	−1130.3081	−336.3344	+1211.3177	+1179.2867
9.4	−1213.3042	+548.3882	−1199.7754	−491.7404	+1331.4791	+1296.6378
9.5	−1387.2951	+466.5512	−1258.8199	−669.4100	+1463.6454	+1425.7410
9.6	−1568.7004	+357.9547	−1304.0247	−870.3159	+1609.0222	+1567.7787
9.7	−1755.2958	+219.2601	−1331.5419	−1095.1509	+1768.9371	+1724.0532
9.8	−1944.2819	+47.0547	−1337.0823	−1344.2542	+1944.8512	+1895.9981
9.9	−2132.2212	−162.0926	−1315.9114	−1617.5297	+2138.3735	+2085.1919
10.0	−2314.9753	−411.5629	−1262.8544	−1914.3556	+2351.2751	+2293.3728

x	ker $(x\sqrt{2})$	kei $(x\sqrt{2})$	ker′ $(x\sqrt{2})$	kei′ $(x\sqrt{2})$	N_0 $(x\sqrt{2})$	N_1 $(x\sqrt{2})$
0.1	+2.07584764	−0.77003354	−7.01611880	+0.18200210	+2.21406754	+7.01847903
0.2	+1.39421570	−0.73774442	−3.42818303	+0.26679823	+1.57737258	+3.43854914
0.3	+1.00741954	−0.69620777	−2.20105406	+0.31619472	+1.22458131	+2.22364972
0.4	+0.74497246	−0.64932550	−1.56763790	+0.34388222	+0.98823457	+1.60491232
0.5	+0.55297232	−0.59964196	−1.17472471	+0.35666809	+0.81568920	+1.22767678
0.6	+0.40701281	−0.54894718	−0.90458434	+0.35873003	+0.68337676	+0.97312611
0.7	+0.29374773	−0.49853937	−0.70667385	+0.35301578	+0.57864430	+0.78994181
0.8	+0.20494391	−0.44936716	−0.55559211	+0.34158132	+0.49389559	+0.65219659
0.9	+0.13507534	−0.40211780	−0.43708181	−0.32605738	+0.42419816	+0.54530168
1.0	+0.08019765	−0.35727739	−0.34244304	+0.30769832	+0.36616772	+0.46037538
1.1	+0.03736270	−0.31517429	−0.26600105	+0.28749332	+0.31738117	+0.39167457
1.2	+0.00428945	−0.27601187	−0.20384538	+0.26622643	+0.27604519	+0.33530501
1.3	−0.02083288	−0.23989468	−0.15315551	+0.24451874	+0.24079756	+0.28852387
1.4	−0.03946957	−0.20684907	−0.11181708	+0.22285930	+0.21058106	+0.24933778
1.5	−0.05282271	−0.17684020	−0.07819387	+0.20162891	+0.18456081	+0.21626026
1.6	−0.06188815	−0.14978573	−0.05098557	+0.18111870	+0.16206760	+0.18815821
1.7	−0.06749599	−0.12556779	−0.02913711	+0.16154570	+0.14255868	+0.16415232
1.8	−0.07034076	−0.10404214	−0.01177755	+0.14306520	+0.12558897	+0.14354916
1.9	−0.07100429	−0.08504634	+0.00182123	+0.12578138	+0.11079030	+0.12579456
2.0	−0.06997381	−0.06840636	+0.01227380	+0.10975651	+0.09785584	+0.11044065

2.1	−0.06765632	−0.05394183	+0.02010257	+0.09501841	+0.08652802	+0.09712163
2.2	−0.06439050	−0.04147042	+0.02575357	+0.08156702	+0.07658938	+0.08553610
2.3	−0.06045668	−0.03081155	+0.02960850	+0.06938025	+0.06785545	+0.07543396
2.4	−0.05608512	−0.02178896	+0.03199443	+0.05841861	+0.06016893	+0.06660614
2.5	−0.05146329	−0.01423299	+0.03319130	+0.04862938	+0.05339521	+0.05887681
2.6	−0.04674193	−0.00798216	+0.03343834	+0.03995015	+0.04741860	+0.05209738
2.7	−0.04204061	−0.00288430	+0.03293931	+0.03231175	+0.04213943	+0.04614160
2.8	−0.03745226	+0.00120260	+0.03186706	+0.02564083	+0.03747157	+0.04090185
2.9	−0.03304735	+0.00441006	+0.03036739	+0.01986178	+0.03334030	+0.03628593
3.0	−0.02887741	+0.00685881	+0.02856250	+0.01489856	+0.02968077	+0.03221465
3.1	−0.02497824	+0.00865892	+0.02655397	+0.01067601	+0.02643652	+0.02861975
3.2	−0.02137247	+0.00990991	+0.02442536	+0.00712092	+0.02355820	+0.02544221
3.3	−0.01807213	+0.01070117	+0.02224472	+0.00416299	+0.02100279	+0.02263091
3.4	−0.01508057	+0.01111238	+0.02006657	+0.00173539	+0.01873255	+0.02014147
3.5	−0.01239429	+0.01121404	+0.01793382	−0.00022480	+0.01671446	+0.01793523
3.6	−0.01000442	+0.01106806	+0.01587946	−0.00177621	+0.01491946	+0.01597849
3.7	−0.00789804	+0.01072835	+0.01392801	−0.00297311	+0.01332203	+0.01424180
3.8	−0.00605930	+0.01024150	+0.01209689	−0.00386525	+0.01189972	+0.01269941
3.9	−0.00447031	+0.00964733	+0.01039754	−0.00449787	+0.01063272	+0.01132871
4.0	−0.00311190	+0.00897960	+0.00883646	−0.00491190	+0.00950353	+0.01010988
4.1	−0.00196434	+0.00826660	+0.00741613	−0.00514400	+0.00849678	+0.00902550
4.2	−0.00100770	+0.00753169	+0.00613583	−0.00522679	+0.00759880	+0.00806026
4.3	−0.00022243	+0.00679391	+0.00499220	−0.00518911	+0.00679755	+0.00720062
4.4	−0.00041049	+0.00606847	+0.00397998	−0.00505619	+0.00608233	+0.00643470
4.5	+0.00090917	+0.00536726	+0.00309242	−0.00484997	+0.00544371	+0.00575198
4.6	+0.00129068	+0.00469926	+0.00232173	−0.00458931	+0.00487328	+0.00514317
4.7	+0.00157096	+0.00407101	+0.00165942	−0.00429030	+0.00436361	+0.00460003
4.8	+0.00176473	+0.00348693	+0.00109666	−0.00396653	+0.00390807	+0.00411534
4.9	+0.00188542	+0.00294972	+0.00062447	−0.00362930	+0.00350081	+0.00368263
5.0	+0.00194520	+0.00246062	+0.00023393	−0.00328790	+0.00313663	+0.00329621
5.1	+0.00195501	+0.00201962	−0.00008369	−0.00294982	+0.00281086	+0.00295100
5.2	+0.00192457	+0.00162584	−0.00033675	−0.00262099	+0.00251939	+0.00264254
5.3	+0.00186244	+0.00127764	−0.00053317	−0.00230598	+0.00225855	+0.00236681
5.4	+0.00177610	+0.00097281	−0.00068040	−0.00200815	+0.00202507	+0.00212029
5.5	+0.00167198	+0.00070872	−0.00078533	−0.00172988	+0.00181599	+0.00189980
5.6	+0.00155565	+0.00048252	−0.00085432	−0.00147270	+0.00162876	+0.00170256
5.7	+0.00143176	+0.00029115	−0.00089313	−0.00123745	+0.00146107	+0.00152609
5.8	+0.00130421	+0.00013148	−0.00090697	−0.00102432	+0.00131082	+0.00136815
5.9	+0.00117619	+0.00000041	−0.00090050	−0.00083309	+0.00117619	+0.00122676
6.0	+0.00105028	−0.00010517	−0.00087785	−0.00066315	+0.00105553	+0.00110018
6.1	+0.00092850	−0.00018814	−0.00084263	−0.00051359	+0.00094737	+0.00098682
6.2	+0.00081239	−0.00025134	−0.00079798	−0.00038329	+0.00085038	+0.00088526
6.3	+0.00070310	−0.00029741	−0.00074662	−0.00027100	+0.00076341	+0.00079428
6.4	+0.00060140	−0.00032878	−0.00069086	−0.00017535	+0.00068541	+0.00071277
6.5	+0.00050782	−0.00034772	−0.00063262	−0.00009495	+0.00061546	+0.00063971
6.6	+0.00042253	−0.00035628	−0.00057352	−0.00002834	+0.00055269	+0.00057422
6.7	+0.00034558	−0.00035632	−0.00051487	+0.00002589	+0.00049638	+0.00051552
6.8	+0.00027684	−0.00034949	−0.00045769	+0.00006909	+0.00044585	+0.00046287

6.9	+0.00021602	−0.00033725	−0.00040280	+0.00010259	+0.00040050	+0.00041566
7.0	+0.00016277	−0.00032086	−0.00035079	+0.00012762	+0.00035978	+0.00037328
7.1	+0.00011664	−0.00030147	−0.00030212	+0.00014536	+0.00032325	+0.00033527
7.2	+0.00007715	−0.00028004	−0.00025709	+0.00015686	+0.00029047	+0.00030117
7.3	+0.00004376	−0.00025735	−0.00021582	+0.00016313	+0.00026104	+0.00027054
7.4	+0.00001594	−0.00023410	−0.00017839	+0.00016504	+0.00023464	+0.00024303
7.5	−0.00000689	−0.00021084	−0.00014477	+0.00016343	+0.00021095	+0.00021833
7.6	−0.00002521	−0.00018802	−0.00011486	+0.00015898	+0.00018970	+0.00019613
7.7	−0.00003955	−0.00016597	−0.00008851	+0.00015230	+0.00017062	+0.00017615
7.8	−0.00005040	−0.00014500	−0.00006554	+0.00014399	+0.00015351	+0.00015821
7.9	−0.00005824	−0.00012529	−0.00004573	+0.00013450	+0.00013816	+0.00014207
8.0	−0.00006347	−0.00010701	−0.00002887	+0.00012424	+0.00012441	+0.00012755
8.1	−0.00006652	−0.00009019	−0.00001468	+0.00011354	+0.00011207	+0.00011449
8.2	−0.00006774	−0.00007490	−0.00000291	+0.00010271	+0.00010099	+0.00010275
8.3	−0.00006746	−0.00006114	+0.00000667	+0.00009196	+0.00009104	+0.00009220
8.4	−0.00006596	−0.00004888	+0.00001430	+0.00008148	+0.00008210	+0.00008272
8.5	−0.00006348	−0.00003807	+0.00002023	+0.00007140	+0.00007402	+0.00007421
8.6	−0.00006029	−0.00002865	+0.00002467	+0.00006185	+0.00006675	+0.00006659
8.7	−0.00005656	−0.00002055	+0.00002783	+0.00005294	+0.00006018	+0.00005980
8.8	−0.00005247	−0.00001366	+0.00002990	+0.00004467	+0.00005422	+0.00005375
8.9	−0.00004814	−0.00000790	+0.00003106	+0.00003709	+0.00004879	+0.00004838
9.0	−0.00004372	−0.00000312	+0.00003149	+0.00003022	+0.00004383	+0.00004364
9.1	−0.00003928	+0.00000070	+0.00003131	+0.00002405	+0.00003928	+0.00003948
9.2	−0.00003488	+0.00000371	+0.00003063	+0.00001857	+0.00003508	+0.00003582
9.3	−0.00003061	+0.00000599	+0.00002961	+0.00001377	+0.00003120	+0.00003266
9.4	−0.00002651	+0.00000764	+0.00002833	+0.00000960	+0.00002759	+0.00002992
9.5	−0.00002262	+0.00000873	+0.00002688	+0.00000606	+0.00002425	+0.00002755
9.6	−0.00001893	+0.00000937	+0.00002532	+0.00000308	−0.00002112	+0.00002550
9.7	−0.00001546	+0.00000963	+0.00002371	+0.0000064	+0.00001822	+0.00002372
9.8	−0.00001222	+0.00000958	+0.00002212	−0.00000133	+0.00001553	+0.00002216
9.9	−0.00000921	+0.00000929	+0.00002057	−0.00000282	+0.00001308	+0.00002077
10.0	−0.00000640	+0.00000879	+0.00001908	−0.00000389	+0.00001087	+0.00001948

Appendix A.4

Calculations for curves of p_{tube} and q_{tube} (Figs 10.4 and 10.5)

Relative permeability $\mu_r = 1$

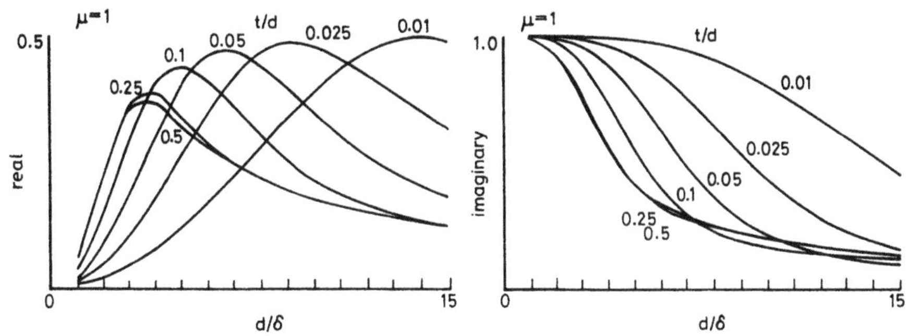

Real part of function (p)

d/δ	t/d 0.01	0.025	0.05	0.1	0.25	0.5
1	0.005	0.012	0.021	0.037	0.058	0.062
2	0.019	0.046	0.005	0.144	0.215	0.224
3	0.044	0.103	0.186	0.291	0.361	0.359
4	0.077	0.179	0.303	0.408	0.388	0.369
5	0.120	0.266	0.404	0.442	0.340	0.323
6	0.169	0.352	0.460	0.410	0.285	0.278
7	0.225	0.422	0.466	0.353	0.244	0.244
8	0.283	0.467	0.347	0.297	0.215	0.218
9	0.339	0.484	0.392	0.250	0.194	0.197
10	0.391	0.476	0.345	0.212	0.178	0.180
11	0.434	0.452	0.301	0.183	0.164	0.165
12	0.466	0.420	0.262	0.161	0.152	0.153
13	0.487	0.385	0.230	0.144	0.142	0.142
14	0.496	0.351	0.203	0.131	0.133	0.132
15	0.487	0.315	0.179	0.120	0.124	0.124

Imaginary part of function (q)

d/δ	t/d					
	0.01	0.025	0.05	0.1	0.25	0.5
1	0.999	0.999	0.999	0.998	0.996	0.995
2	0.999	0.998	0.992	0.976	0.936	0.926
3	0.998	0.989	0.962	0.891	0.758	0.736
4	0.994	0.966	0.889	0.728	0.549	0.537
5	0.985	0.920	0.768	0.541	0.406	0.410
6	0.970	0.848	0.620	0.390	0.326	0.336
7	0.946	0.752	0.477	0.290	0.279	0.287
8	0.911	0.642	0.361	0.228	0.247	0.251
9	0.865	0.531	0.274	0.191	0.222	0.223
10	0.808	0.430	0.213	0.167	0.201	0.201
11	0.742	0.345	0.171	0.151	0.183	0.182
12	0.671	0.276	0.142	0.141	0.168	0.167
13	0.597	0.221	0.122	0.133	0.154	0.154
14	0.528	0.182	0.108	0.126	0.143	0.143
15	0.456	0.151	0.098	0.121	0.134	0.134

Variable modifier real part

d/δ	t/d					
	0.01	0.025	0.05	0.1	0.25	0.5
1	1.002	1.002	1.002	1.002	1.001	1.000
2	1.025	1.030	1.034	1.032	1.008	1.000
3	1.119	1.140	1.155	1.133	1.026	1.000
4	1.324	1.377	1.388	1.276	1.031	1.000
5	1.625	1.701	1.637	1.339	1.015	1.000
6	1.963	2.015	1.771	1.289	0.995	1.000
7	2.302	2.249	1.768	1.194	0.985	1.000
8	2.628	2.380	1.681	1.104	0.986	1.000
9	2.935	2.413	1.563	1.036	0.991	1.000
10	3.204	2.369	1.445	0.988	0.997	1.000
11	3.421	2.274	1.338	0.958	1.000	1.000
12	3.580	2.154	1.247	0.940	1.002	1.000
13	3.673	2.021	1.171	0.932	1.002	1.000
14	3.717	1.907	1.112	0.931	1.001	1.000
15	3.650	1.780	1.061	0.935	1.001	1.000

Constant modifier

d/δ	Real	Imaginary
1	0.062	0.995
2	0.224	0.926
3	0.359	0.736
4	0.369	0.537
5	0.323	0.410
6	0.278	0.336
7	0.244	0.287
8	0.218	0.251
9	0.197	0.223
10	0.180	0.201
11	0.165	0.182
12	0.153	0.167
13	0.142	0.154
14	0.132	0.143
15	0.124	0.134

Variable modifier imaginary part

d/δ	t/d					
	0.01	0.025	0.05	0.1	0.25	0.5
1	0.058	0.051	0.041	0.025	0.004	0.000
2	0.228	0.200	0.159	0.095	0.012	0.000
3	0.486	0.416	0.311	0.158	0.010	0.000
4	0.768	0.614	0.390	0.117	−0.012	0.000
5	0.990	0.692	0.303	−0.023	−0.029	0.000
6	1.121	0.620	0.094	−0.133	−0.026	0.000
7	1.171	0.438	−0.122	−0.216	−0.013	0.000
8	1.153	0.202	−0.283	−0.225	−0.001	0.000
9	1.068	−0.039	−0.379	−0.205	0.005	0.000
10	0.920	−0.254	−0.427	−0.174	0.006	0.000
11	0.718	−0.424	−0.440	−0.140	0.004	0.000
12	0.480	−0.547	−0.433	−0.107	0.002	0.000
13	0.217	−0.635	−0.414	−0.076	0.001	0.000
14	−0.027	−0.687	−0.387	−0.051	−0.000	0.000
15	−0.252	−0.700	−0.354	−0.029	−0.000	0.000

Relative permeability $\mu_r = 2.5$

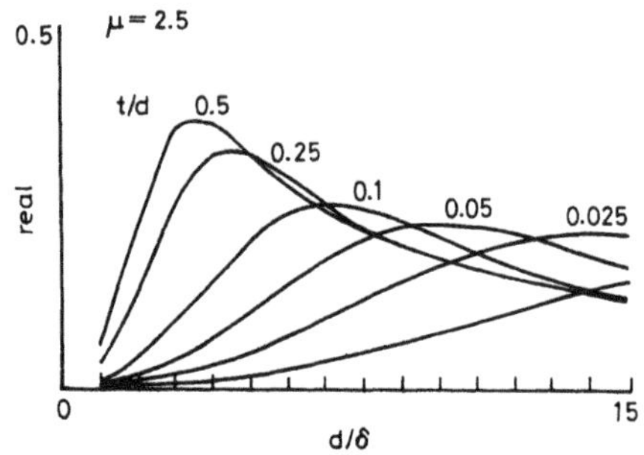

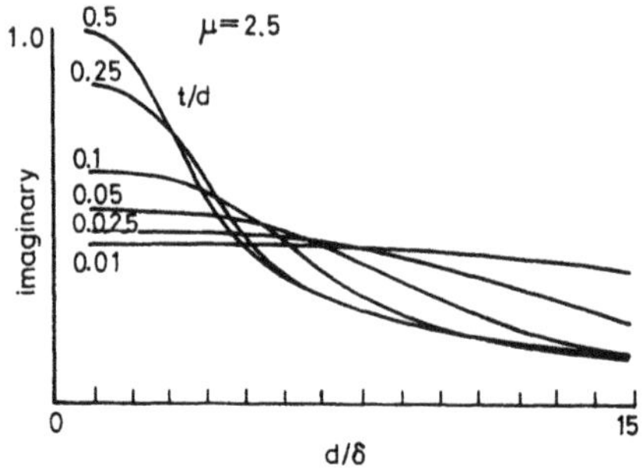

Real part of function (p)

d/δ	t/d 0.010	0.025	0.05	0.1	0.25	0.5
1	0.001	0.002	0.005	0.011	0.034	0.062
2	0.003	0.009	0.019	0.042	0.132	0.224
3	0.007	0.019	0.041	0.092	0.254	0.359
4	0.013	0.034	0.072	0.153	0.326	0.369
5	0.021	0.053	0.109	0.209	0.326	0.323
6	0.030	0.075	0.147	0.246	0.291	0.278
7	0.040	0.100	0.182	0.258	0.253	0.244
8	0.052	0.125	0.208	0.250	0.221	0.218
9	0.065	0.150	0.224	0.232	0.197	0.197
10	0.079	0.172	0.229	0.209	0.179	0.180
11	0.094	0.190	0.224	0.187	0.164	0.165
12	0.110	0.203	0.214	0.168	0.152	0.153
13	0.125	0.212	0.200	0.151	0.142	0.142
14	0.139	0.214	0.185	0.138	0.132	0.132
15	0.152	0.211	0.170	0.126	0.124	0.124

Imaginary part of function (q)

d/δ	t/d					
	0.01	0.025	0.05	0.1	0.25	0.5
1	0.424	0.458	0.514	0.616	0.848	0.995
2	0.424	0.458	0.513	0.613	0.823	0.925
3	0.424	0.458	0.510	0.599	0.733	0.736
4	0.423	0.456	0.502	0.566	0.585	0.537
5	0.423	0.452	0.486	0.509	0.444	0.410
6	0.422	0.445	0.460	0.436	0.347	0.336
7	0.420	0.434	0.424	0.359	0.287	0.287
8	0.417	0.418	0.379	0.293	0.248	0.251
9	0.413	0.397	0.331	0.242	0.221	0.223
10	0.408	0.372	0.283	0.204	0.199	0.201
11	0.401	0.343	0.241	0.178	0.182	0.182
12	0.392	0.311	0.205	0.158	0.167	0.167
13	0.382	0.279	0.175	0.144	0.154	0.154
14	0.370	0.249	0.152	0.133	0.143	0.143
15	0.353	0.218	0.133	0.125	0.134	0.134

Constant modifier

d/δ	Real	Imaginary
1	0.062	0.995
2	0.224	0.926
3	0.359	0.736
4	0.369	0.537
5	0.323	0.410
6	0.278	0.336
7	0.244	0.287
8	0.218	0.251
9	0.197	0.223
10	0.180	0.201
11	0.165	0.182
12	0.153	0.167
13	0.142	0.154
14	0.132	0.143
15	0.124	0.134

Variable modifier real part

d/δ	t/d 0.01	0.025	0.05	0.1	0.25	0.5
1	0.424	0.459	0.515	0.617	0.851	1.000
2	0.433	0.470	0.528	0.636	0.872	1.000
3	0.469	0.513	0.582	0.707	0.940	1.000
4	0.547	0.606	0.698	0.849	1.023	1.000
5	0.661	0.743	0.861	1.015	1.056	1.000
6	0.789	0.897	1.029	1.131	1.040	1.000
7	0.919	1.050	1.171	1.172	1.015	1.000
8	1.051	1.197	1.273	1.161	1.000	1.000
9	1.186	1.335	1.331	1.125	0.995	1.000
10	1.325	1.455	1.350	1.084	0.995	1.000
11	1.466	1.553	1.338	1.047	0.997	1.000
12	1.607	1.621	1.305	1.017	0.999	1.000
13	1.747	1.663	1.262	0.995	1.000	1.000
14	1.878	1.682	1.216	0.981	1.000	1.000
15	1.986	1.664	1.169	0.973	1.000	1.000

Variable modifier imaginary part

d/δ	t/d 0.01	0.025	0.05	0.1	0.25	0.5
1	0.026	0.026	0.027	0.028	0.019	0.000
2	0.101	0.105	0.108	0.109	0.069	0.000
3	0.219	0.224	0.228	0.220	0.114	0.000
4	0.352	0.353	0.346	0.300	0.096	0.000
5	0.471	0.457	0.414	0.290	0.036	0.000
6	0.565	0.519	0.414	0.202	−0.007	0.000
7	0.641	0.545	0.360	0.095	−0.019	0.000
8	0.704	0.540	0.273	0.009	−0.014	0.000
9	0.755	0.507	0.170	−0.045	−0.006	0.000
10	0.791	0.446	0.069	−0.073	−0.001	0.000
11	0.810	0.362	−0.019	−0.081	0.001	0.000
12	0.812	0.265	−0.088	−0.077	0.002	0.000
13	0.794	0.158	−0.139	−0.066	0.001	0.000
14	0.765	0.061	−0.169	−0.053	0.001	0.000
15	0.708	−0.033	−0.186	−0.039	0.000	0.000

Relative permeability $\mu_r = 5$

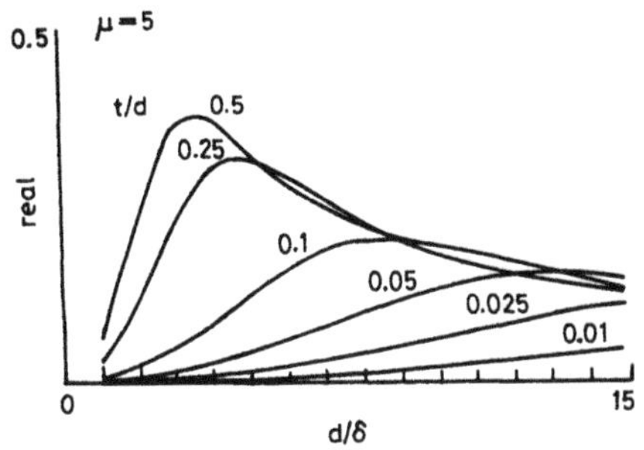

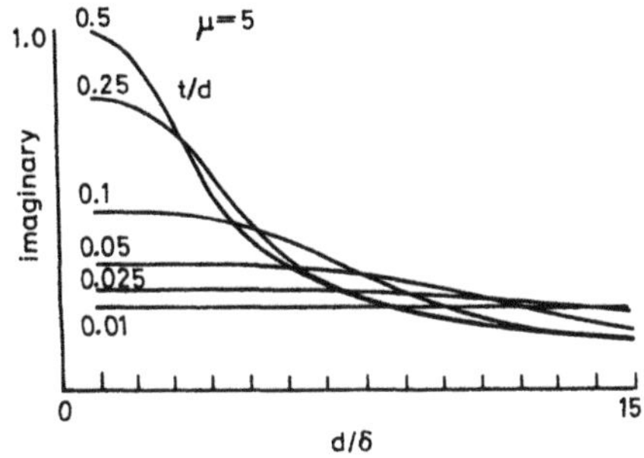

Real part of function (p)

d/δ	t/d 0.01	0.025	0.05	0.1	0.25	0.5
1	0.000	0.001	0.002	0.005	0.028	0.062
2	0.001	0.003	0.007	0.022	0.109	0.224
3	0.002	0.006	0.016	0.048	0.217	0.359
4	0.004	0.011	0.028	0.083	0.297	0.369
5	0.006	0.017	0.043	0.122	0.315	0.323
6	0.008	0.024	0.061	0.158	0.291	0.278
7	0.011	0.033	0.080	0.185	0.256	0.244
8	0.014	0.042	0.099	0.198	0.225	0.218
9	0.018	0.052	0.117	0.200	0.199	0.197
10	0.022	0.063	0.131	0.193	0.180	0.180
11	0.027	0.074	0.142	0.102	0.165	0.165
12	0.032	0.085	0.148	0.168	0.152	0.153
13	0.037	0.095	0.151	0.155	0.141	0.142
14	0.042	0.103	0.149	0.142	0.132	0.132
15	0.048	0.111	0.145	0.131	0.124	0.124

Imaginary part of function (q)

d/δ	t/d					
	0.01	0.02	0.05	0.10	0.25	0.49
1	0.232	0.278	0.352	0.488	0.799	0.995
2	0.232	0.278	0.352	0.487	0.781	0.925
3	0.232	0.278	0.351	0.482	0.713	0.736
4	0.232	0.278	0.349	0.470	0.591	0.537
5	0.232	0.277	0.346	0.447	0.459	0.418
6	0.231	0.276	0.339	0.411	0.358	0.336
7	0.231	0.274	0.329	0.366	0.292	0.287
8	0.231	0.271	0.315	0.317	0.249	0.251
9	0.230	0.266	0.296	0.271	0.228	0.223
10	0.229	0.261	0.275	0.232	0.199	0.281
11	0.228	0.254	0.251	0.288	0.181	0.182
12	0.227	0.245	0.227	0.176	0.167	0.167
13	0.225	0.235	0.283	0.157	0.154	0.154
14	0.223	0.223	0.182	0.142	0.143	0.143
15	0.219	0.289	0.162	0.131	0.134	0.134

Constant modifier

d/δ	Real	Imaginary
1	0.062	0.995
2	0.224	0.926
3	0.359	0.736
4	0.369	0.537
5	0.323	0.410
6	0.278	0.336
7	0.244	0.287
8	0.218	0.251
9	0.197	0.223
10	0.180	0.201
11	0.165	0.182
12	.0153	0.167
13	0.142	0.154
14	0.132	0.143
15	0.124	0.134

Variable modifier real part

d/δ	t/d 0.01	0.025	0.05	0.1	0.25	0.5
1	0.232	0.278	0.353	0.489	0.802	1.000
2	0.237	0.284	0.361	0.502	0.824	1.000
3	0.255	0.308	0.394	0.555	0.899	1.000
4	0.296	0.360	0.466	0.667	1.006	1.000
5	0.355	0.437	0.572	0.817	1.065	1.000
6	0.421	0.523	0.689	0.958	1.060	1.000
7	0.487	0.610	0.804	1.058	1.032	1.000
8	0.553	0.699	0.911	1.113	1.010	1.000
9	0.621	0.788	1.007	1.129	0.999	1.000
10	0.690	0.878	1.086	1.121	0.995	1.000
11	0.763	0.967	1.145	1.100	0.996	1.000
12	0.836	1.052	1.182	1.074	0.998	1.000
13	0.912	1.131	1.201	1.050	0.999	1.000
14	0.986	1.198	1.202	1.029	1.000	1.000
15	1.059	1.254	1.191	1.013	1.000	1.000

Variable modifier imaginary part

d/δ	t/d 0.01	0.025	0.05	0.1	0.25	0.5
1	0.014	0.017	0.020	0.025	0.022	0.000
2	0.056	0.066	0.080	0.098	0.082	0.000
3	0.122	0.142	0.171	0.205	0.144	0.000
4	0.197	0.228	0.269	0.304	0.139	0.000
5	0.266	0.303	0.345	0.347	0.071	0.000
6	0.324	0.361	0.388	0.323	0.010	0.000
7	0.375	0.404	0.403	0.256	−0.017	0.000
8	0.422	0.438	0.394	0.175	−0.019	0.000
9	0.466	0.460	0.364	0.099	−0.012	0.000
10	0.506	0.471	0.316	0.040	−0.006	0.000
11	0.541	0.469	0.257	−0.001	−0.001	0.000
12	0.573	0.453	0.192	−0.025	0.001	0.000
13	0.598	0.425	0.128	−0.037	0.001	0.000
14	0.618	0.390	0.072	−0.040	0.001	0.000
15	0.625	0.338	0.023	−0.037	0.000	0.000

Relative permeability $\mu_r = 10$

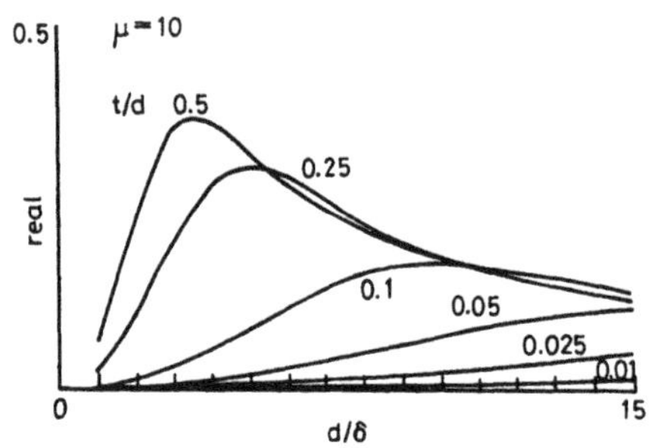

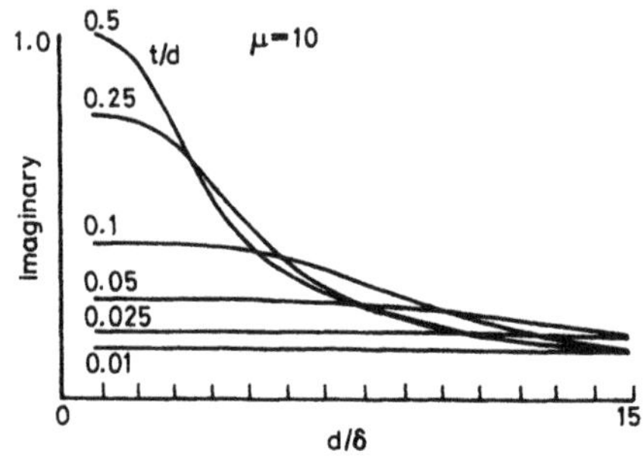

Real part of function (p)

d/δ	t/d 0.01	0.025	0.05	0.1	0.25	0.5
1	0.000	0.000	0.001	0.004	0.025	0.062
2	0.000	0.001	0.003	0.014	0.098	0.224
3	0.001	0.002	0.008	0.032	0.199	0.359
4	0.001	0.004	0.014	0.056	0.280	0.369
5	0.002	0.006	0.021	0.084	0.307	0.323
6	0.002	0.009	0.031	0.113	0.290	0.278
7	0.003	0.012	0.041	0.139	0.258	0.244
8	0.004	0.016	0.052	0.159	0.226	0.218
9	0.006	0.020	0.064	0.170	0.201	0.197
10	0.007	0.025	0.075	0.174	0.181	0.180
11	0.008	0.030	0.086	0.170	0.165	0.165
12	0.010	0.035	0.095	0.163	0.152	0.153
13	0.011	0.040	0.103	0.154	0.141	0.142
14	0.013	0.045	0.108	0.144	0.132	0.132
15	0.016	0.051	0.112	0.134	0.124	0.124

Imaginary part of function (q)

d/δ	t/d					
	0.01	0.025	0.05	0.1	0.25	0.5
1	0.136	0.188	0.271	0.424	0.774	0.994
2	0.136	0.188	0.271	0.423	0.759	0.925
3	0.136	0.188	0.271	0.421	0.702	0.736
4	0.136	0.188	0.270	0.415	0.592	0.537
5	0.136	0.188	0.269	0.403	0.467	0.410
6	0.136	0.187	0.267	0.382	0.365	0.336
7	0.136	0.187	0.263	0.354	0.296	0.287
8	0.136	0.186	0.258	0.320	0.251	0.251
9	0.135	0.185	0.251	0.283	0.221	0.223
10	0.135	0.184	0.242	0.248	0.198	0.201
11	0.135	0.182	0.231	0.217	0.181	0.182
12	0.135	0.180	0.219	0.190	0.166	0.167
13	0.135	0.177	0.205	0.168	0.154	0.154
14	0.134	0.173	0.190	0.151	0.143	0.143
15	0.133	0.169	0.176	0.137	0.134	0.134

Constant modifier

d/δ	Real	Imaginary
1	0.062	0.995
2	0.224	0.926
3	0.359	0.736
4	0.369	0.537
5	0.323	0.410
6	0.278	0.336
7	0.244	0.287
8	0.218	0.251
9	0.197	0.223
10	0.180	0.201
11	0.165	0.182
12	0.153	0.167
13	0.142	0.154
14	0.132	0.143
15	0.124	0.134

Variable modifier real part

d/δ	t/d					
	0.01	0.025	0.05	0.1	0.25	0.5
1	0.136	0.188	0.271	0.425	0.777	1.000
2	0.138	0.192	0.277	0.436	0.799	1.000
3	0.149	0.207	0.301	0.479	0.876	1.000
4	0.172	0.241	0.354	0.573	0.993	1.000
5	0.206	0.290	0.430	0.705	1.066	1.000
6	0.243	0.344	0.516	0.841	1.069	1.000
7	0.280	0.400	0.604	0.956	1.042	1.000
8	0.317	0.455	0.690	1.041	1.017	1.000
9	0.353	0.512	0.775	1.093	1.002	1.000
10	0.391	0.571	0.857	1.117	0.996	1.000
11	0.430	0.630	0.932	1.119	0.995	1.000
12	0.469	0.691	0.998	1.107	0.997	1.000
13	0.510	0.753	1.053	1.088	0.998	1.000
14	0.549	0.810	1.093	1.068	0.999	1.000
15	0.592	0.871	1.121	1.049	1.000	1.000

Variable modifier imaginary part

d/δ	t/d					
	0.01	0.025	0.05	0.1	0.25	0.5
1	0.008	0.011	0.016	0.023	0.023	0.000
2	0.033	0.045	0.064	0.090	0.088	0.000
3	0.072	0.098	0.136	0.190	0.157	0.000
4	0.117	0.158	0.218	0.291	0.161	0.000
5	0.158	0.213	0.287	0.352	0.093	0.000
6	0.194	0.258	0.337	0.360	0.022	0.000
7	0.226	0.296	0.370	0.327	−0.013	0.000
8	0.257	0.330	0.390	0.269	−0.020	0.000
9	0.287	0.361	0.398	0.201	−0.016	0.000
10	0.316	0.386	0.391	0.134	−0.008	0.000
11	0.344	0.407	0.372	0.078	−0.003	0.000
12	0.371	0.422	0.341	0.034	−0.000	0.000
13	0.396	0.432	0.301	0.004	0.001	0.000
14	0.418	0.435	0.257	−0.015	0.001	0.000
15	0.432	0.426	0.208	−0.024	0.001	0.000

Relative permeability $\mu_r = 100$

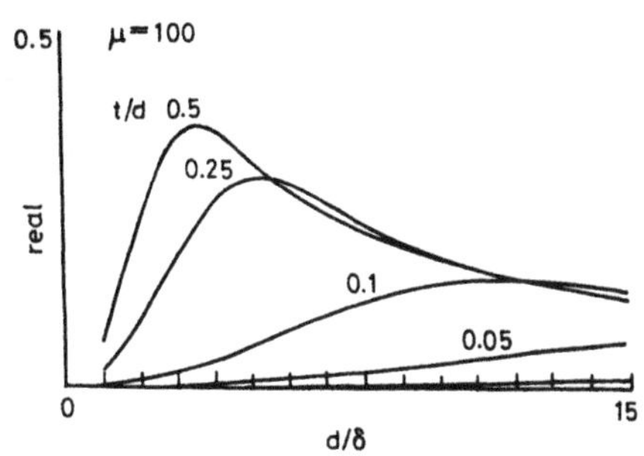

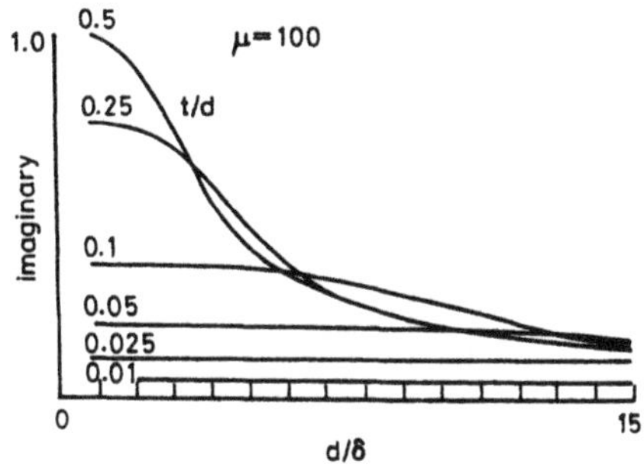

Real part of function (p)

d/δ	t/d					
	0.01	0.025	0.05	0.1	0.25	0.5
1	0.000	0.000	0.000	0.002	0.023	0.062
2	0.000	0.000	0.001	0.009	0.089	0.224
3	0.000	0.000	0.003	0.020	0.183	0.359
4	0.000	0.001	0.005	0.036	0.264	0.369
5	0.000	0.001	0.009	0.055	0.298	0.323
6	0.000	0.002	0.012	0.076	0.288	0.278
7	0.000	0.003	0.017	0.098	0.259	0.244
8	0.000	0.003	0.021	0.117	0.228	0.218
9	0.000	0.004	0.027	0.133	0.202	0.197
10	0.001	0.005	0.033	0.144	0.181	0.180
11	0.001	0.006	0.039	0.149	0.165	0.165
12	0.001	0.007	0.045	0.150	0.152	0.153
13	0.001	0.009	0.051	0.147	0.141	0.142
14	0.001	0.010	0.057	0.141	0.132	0.132
15	0.003	0.012	0.063	0.134	0.124	0.124

Imaginary part of function (q)

d/δ	t/d 0.01	0.025	0.05	0.1	0.25	0.5
1	0.049	0.107	0.198	0.366	0.752	0.994
2	0.049	0.107	0.198	0.366	0.739	0.925
3	0.049	0.107	0.198	0.365	0.690	0.736
4	0.049	0.107	0.198	0.362	0.592	0.537
5	0.049	0.107	0.198	0.356	0.473	0.410
6	0.049	0.106	0.197	0.346	0.371	0.336
7	0.049	0.106	0.196	0.330	0.300	0.287
8	0.049	0.106	0.195	0.310	0.253	0.251
9	0.049	0.106	0.194	0.285	0.221	0.223
10	0.049	0.106	0.191	0.258	0.198	0.201
11	0.049	0.106	0.188	0.231	0.181	0.182
12	0.049	0.106	0.185	0.206	0.166	0.167
13	0.049	0.106	0.180	0.183	0.154	0.154
14	0.049	0.105	0.174	0.164	0.143	0.143
15	0.049	0.105	0.168	0.147	0.134	0.134

Constant modifier

d/δ	Real	Imaginary
1	0.062	0.995
2	0.224	0.926
3	0.359	0.736
4	0.369	0.537
5	0.323	0.410
6	0.278	0.336
7	0.244	0.287
8	0.218	0.251
9	0.197	0.223
10	0.180	0.201
11	0.165	0.182
12	0.153	0.167
13	0.142	0.154
14	0.132	0.143
15	0.124	0.134

Variable modifier real part

d/δ	t/d 0.01	0.025	0.05	0.1	0.25	0.5
1	0.049	0.107	0.198	0.367	0.754	1.000
2	0.050	0.109	0.202	0.376	0.776	1.000
3	0.054	0.117	0.219	0.411	0.855	1.000
4	0.062	0.135	0.255	0.489	0.979	1.000
5	0.074	0.162	0.308	0.601	1.065	1.000
6	0.087	0.191	0.367	0.723	1.078	1.000
7	0.100	0.220	0.426	0.837	1.052	1.000
8	0.113	0.248	0.486	0.936	1.024	1.000
9	0.125	0.277	0.548	1.015	1.006	1.000
10	0.137	0.307	0.610	1.072	0.998	1.000
11	0.150	0.337	0.674	1.105	0.995	1.000
12	0.162	0.368	0.737	1.119	0.996	1.000
13	0.175	0.400	0.799	1.119	0.997	1.000
14	0.187	0.428	0.855	1.108	0.999	1.000
15	0.206	0.467	0.909	1.092	1.000	1.000

Variable modifier imaginary part

d/δ	t/d 0.01	0.025	0.05	0.1	0.25	0.5
1	0.003	0.007	0.012	0.021	0.024	0.000
2	0.012	0.026	0.048	0.081	0.092	0.000
3	0.026	0.057	0.103	0.173	0.168	0.000
4	0.043	0.092	0.165	0.270	0.181	0.000
5	0.058	0.124	0.222	0.340	0.113	0.000
6	0.072	0.153	0.267	0.372	0.035	0.000
7	0.084	0.178	0.304	0.370	−0.009	0.000
8	0.096	0.202	0.336	0.343	−0.021	0.000
9	0.108	0.226	0.363	0.298	−0.018	0.000
10	0.120	0.249	0.383	0.241	−0.012	0.000
11	0.133	0.271	0.397	0.181	−0.005	0.000
12	0.145	0.292	0.403	0.125	−0.002	0.000
13	0.157	0.312	0.403	0.077	0.000	0.000
14	0.166	0.328	0.394	0.040	0.001	0.000
15	0.170	0.343	0.376	0.013	0.001	0.000

Appendix A.5

Solution of $\int W(mr)dr$

The integral required for the solution of eqn. 18.29 is

$$\int W(mr)dr = \int (\text{ber } mr \text{ bei}' mr - \text{bei } mr \text{ ber}' mr)\, dr$$

This is not a standard form (see Section 18.3.2), so a general solution is not possible. From section 4.1.2, we have:

$$\text{ber } 2\theta = 1 - \frac{\theta^4}{(2!)^2} + \frac{\theta^8}{(4!)^2} \cdots$$

$$\text{bei } 2\theta = \theta^2 - \frac{\theta^6}{(3!)^2} + \frac{\theta^{10}}{(5!)^2} \cdots$$

$$\text{bei}' \, 2\theta = \theta - \frac{3\theta^5}{(3!)^2} + \frac{5\theta^9}{(5!)^2} \cdots$$

$$\text{ber}' \, 2\theta = -\frac{2\theta^3}{(2!)^2} + \frac{4\theta^7}{(4!)^2} - \frac{6\theta^{11}}{(6!)^2} \cdots$$

These three terms give very accurate answers for small values of θ. If we include a fourth term, we cover a large part of the range we need to a sufficient accuracy for our purposes. Expanding our integral:

$$\int \text{ber}2\theta \text{ bei}'2\theta \, d\theta = \int \left[1 - \frac{\theta^4}{(2!)^2} + \frac{\theta^8}{(4!)^2} - \frac{\theta^{12}}{(6!)^2}\right]$$

$$\times \left[\theta - \frac{3\theta^5}{(3!)^2} + \frac{5\theta^9}{(5!)^2} - \frac{\theta^{13}}{(7!)^2}\right] \delta\theta$$

$$= \int [\theta - \theta^5 (1/4 + 1/12) + \theta^9 (1/48 + 1/576 + 1/2880)$$
$$- \theta^{13} (1/518\,400 + 1/3\,629\,000 + 1/11\,520 + 1/6912)] \, d\theta$$

$$\int \text{ber}'2\theta \text{ bei}2\theta \, d\theta = \int \theta^2 \left[1 - \frac{\theta^4}{(3!)^2} + \frac{\theta^8}{(5!)^2} - \frac{\theta^{12}}{(7!)^2}\right] (-2\theta^3)$$

$$\times \left[\frac{1}{(2!)^2} - \frac{2\theta^4}{(4!)^2} + \frac{3\theta^8}{(6!)^2} - \frac{4\theta^{12}}{(8!)^2}\right] d\theta$$

$$= \int [-\theta^5/2 + \theta^9\,(1/72 + 1/144)$$
$$- \theta^{13}\,(1/5184 + 1/28\,800 + 1/86\,400)]\ d\theta$$

Combining these gives $\int W(2\theta)$:

$$\int [\theta + \theta^5/6 + \theta^9/480 + \theta^{13}\,(5.5/10^6)]\ d\theta$$
$$= \theta^2/2 + \theta^6/36 + \theta^9/4800 + \theta^{13}\,(3.93/10^7)$$

Hence, substituting for $2\theta = mr$ or mR, as appropriate,

$$W(mR) = (mR/2) + (mR/2)^5/6 + (mR/2)^9/480 + (mR/2)^{13}\,(5.5/10^6)$$
$$= (mR/2) + (mR)/192 + (mR)^9/245\,760 + (mR)^{13}\,(6.71/10^{10})$$

and

$$\int W(mr)\ d(mr/2) = (mr/2)^2/2 + (mr/2)^6/36$$
$$+ (mr/2)^{10}/4800 + (mr/2)^{14}\,(3.93/10^7)$$

So

$$\int W(mr)\ dr = \frac{2}{m}\,[(mr)^2/8 + (mr)^6/2304 + (mr)^{10}/4.92 \times 10^6$$
$$+ (mr)^{14}/4.17 \times 10^{10}]$$

References

1 DAVIES, E. J., 'Depth of penetration in electroheating', *Elektrowärme International*, **46**(B1), 29–35, 1988.

2 WARREN, A. G., *Mathematics Applied to Electrical Engineering*, Chapman and Hall, London, 1949.

3 CARTER, G. W., *Electromagnetic Field in its Engineering Aspects*, Longmans, London, 1967.

4 MCLACHLAN, N. W., *Bessel Functions for Engineers*, Oxford University Press, Oxford, 1934.

5 HAMMOND, P., *Applied Electromagnetism*, Pergamon, 1971.

6 DWIGHT, H. B., 'Bessel functions for AC problems', *AIEE Trans.*, 812–820, July 1929.

7 DAVIES, E. J. and P. G. SIMPSON, *Induction Heating Handbook*, McGraw-Hill, 1979.

8 BROWN, G. H., C. N. HOYLER, and R. A. BIERWIRTH, *Theory and Application of Radio-frequency Heating*, Van Nostrand, New York, 1947.

9 SUNDBERG, Y., *Induction Heating, with Special Reference to Bodies inside Metallic Shells*, Vastra Aros Tryckeri AN, Västerås, 1965.

10 LAMMERANER, J. and M. STAFL, *Eddy Currents*, Illiffe, London, 1966.

11 DAVIES, E. J., and A. L. BOWDEN, 'Direct resistance heating of ferromagnetic billets', paper presented at the 7th UIE Congress, Warsaw, 1972, Paper N628.

12 DREYFUS, L. A., 'High-frequency heating and temperature distribution in surface hardening of steel', *Acta polytechnica*, **4** (5), 1952.

13 BAKER, R. M., 'Design and calculation of induction heating coils', *AIEE Trans.*, **57** (4), 31–40, 1957.

14 DAVIES, E. J., 'Two experimental studies in dynamo-electric machinery', PhD thesis, University of Bristol, 1962.

15 DAVIES, E. J., 'An experimental and theoretical study of eddy-current couplings and brakes', *IEE Trans.*, **PAS-82**, 401–417, 1963.

16 DAVIES, E. J., 'General theory of eddy-current couplings and brakes', *Proc. IEE*, **113**(5), 825–837, 1966.

17 DAVIES, E. J., B. JAMES, and M. T. WRIGHT, 'Experimental verification of the generalized theory of eddy-current couplings', *Proc. IEE*, **122**(1), 67–73, 1975.

18 DAVIES, E. J., M. T. WRIGHT, and R. C. JOHNSON, 'Transient performance of eddy-current couplings', *Proc. IEE*, **122**(10), 1128–1135, 1975.

19 BOWDEN, A. L., 'A study of magnetic non-linearity and finite-length effects in solid iron subjected to a travelling m.m.f. wave', PhD thesis, University of Aston in Birmingham, June 1973.

20 THORNTON, C. A. M., 'Resistance heating of mild-steel containers at power frequencies', *Proc. IEE Part II*, **99**(68), 85–93, 1952.

21 ELLETT, R. C., 'Induction heating of mild steel vessels in a pulsating field', MPhil thesis, University of Aston in Birmingham, 1974.

22 HAMMOND, P., 'Calculation of magnetic field of rotating machines – eddy currents induced in a solid slab by a circular current loop', *IEE Monograph 514S*, pp. 508–515, April 1962.
23 EDE, A. J., *An Introduction to Heat Transfer*, Pergamon Press, London, 1967.
24 FISHENDEN, M., and O. A. SAUNDERS, *An Introduction to Heat Transfer*, Oxford University Press, Oxford, 1950.
25 DAVIES, E. J., *Heat Transfer for Induction Heating*, Teaching Information Note No. 5, Electricity Council, 1986.
26 CARSLAW, H. S. and H. C. JAEGER, *Conduction of Heat in Solids*, Oxford University Press, Oxford, 1959.
27 BAKER, R. M., 'Classical heat flow problems applied to induction billet heating', *AIEE Trans.*, **77**, 106–112, 1958.
28 Institution of Electrical Engineers, 'Units and symbols for electrical and electronic engineering', IEE, Stevenage, 1986.
29 British Standards Institution, 'Letter symbols, signs and abbreviations', BS 1991: 1960 Part I.
30 SUNDBERG, Y., *Electrougnar och induktiva omrörare*, ASEA, Västerås, 1969.
31 TUDBURY, C. A., *Basis of Induction Heating* (Vols. 1 and 2), Chapman and Hall, London, 1960.
32 WALSHAW, A. C., and D. A. JOBSON, *Mechanics of Fluids*, 2nd ed., Longmans, London, 1972.
33 STANSEL, N. R., *Induction Heating*, McGraw-Hill, New York, 1949.
34 SMITHELLS, C. J. (ed.), *Metals Reference Book*, 5th ed., Butterworths, London, 1976, pp. 960–968.
35 LAUSTER, F., *Manuel d'electrothermie industrielle*, Dunod, Paris, 1968, p. 240.
36 *V.D.I. – Arbeitsblatt* 5-3131.
37 KASPER, R. J., 'Temperature distribution for induction through heating of magnetic materials to forging temperatures', ASME publication 68HT30, August 1968.
38 MAY, E., *Industrial High-frequency Electric Power*, Chapman and Hall, London, 1949.
39 DICKSON, D. J., 'Thermal insulation for energy storage', *Proc. IEE*, **118**(10), 1399–1407, 1971.
40 KOREY, W. J., 'Induction heating of slabs – today'.
41 ROSS, N. V., 'A system for induction heating of large steel slabs', *IEEE Trans.*, **IGA-6** (5), 449–454, 1970.
42 British patents 1513241/2; US patent application 774 242.
43 DAVIES, E. J. and A. L. BOWDEN, 'Travelling wave induction heaters', paper presented at Universities Power Engineering Conference 1975 (not freely available).
44 DELLACASA, M., de LAMBEXTERIE, B., MICHELET, M., EYGLUMENET, B., CHEDAILLE, J., 'Four prototype a induction pour le rechauffage des brames d'acier avant laminage'. Paper presented at the 9th UIE Congress, Cannes, 1980.
45 HARVEY, I. G., 'A method of improving the energy transfer in induction heating processes and its application in a 1 MW billet heater', in *IEE Conference Publication* 149: *Electricity for Materials Processing and Conservation*, 1977, pp. 16–20.
46 HARVEY, I. G., 'The theory of multi-layered windings for induction heating and their application to a 1 MW, 50 Hz longitudinal flux billet heater', paper presented at the 8th UIE Congress, Liège, Belgium, 1976, Paper II(a)4.
47 GIBSON, R. C., 'SLEDDY, a computer program for calculating the induction and other heating of metal slabs and long cylindrical billets', *Electricity Council Research Centre, Capenhurst, Report MM16*, 1973.
48 HARVEY, I. G., R. H. JOHNSON, and M. G. GIBBS, 'The commercial exploitation of the multi-layer coil: a final report on an EEC demonstration project ref. EE/147/6B', Electricity Council Report R1708, May 1983.
49 BAKER, R. M., Scale Model Testing as Applied to Induction Heating, AIEE, New York, 1955.

50 CALVIN, R. A., L. W. CRANE, and E. J. DAVIES, 'Physical properties of sintered low alloy steels', *Powder Metallurgy*, **30**(2), 111–117, 1987.

51 CALVIN, R. A., L. W. CRANE, and E. J. DAVIES, 'Induction heating of sintered, low alloy steel billets', *Power Metallurgy*, **32**(1), 1989.

52 American Society of Metals, *Metals Handbook*, 1947.

53 EILENDER, W. and L. MINTROP, 'Influence of rate of heating steel on the pearlite-austenite transformation, with special reference to surface hardening', *Stahl v. Eisen*, **68**, 83, 1948.

54 LOZINSKII, M. G., 'The continuous progressive surface hardening of steel using high-frequency current', *Vestnik Metalloprom.*, 3, 50–60, 1940.

55 LOZINSKII, M. G., *Industrial Applications of Induction Heating*, Pergamon Press, London, 1969.

56 WARBURTON-BROWN, D., *Induction Heating Practice*, Odhams, London, 1956.

57 CURTIS, F. W., *High-frequency Induction Heating*, 2nd ed., Mcgraw-Hill, New York, 1950, pp. 226–246.

58 American Society of Metals, *Metals Handbook, 1955 Supplement*, 1955, pp. 107–123.

59 BURDETT, M. R., 'Polymer quenchants', *Heat Treatment of Metals*, **1**, 4–6, 1977.

60 SAUVEUR, A., *The Metallography and Heat Treatment of Iron and Steel*, 4th ed., McGraw-Hill, New York, 1935.

61 BABAT, G. and M. G. LOZINSKII, 'The heat treatment of steel by high frequency currents', *JIEE*, **86**, 161, 1940.

62 KUGLER, A. N., 'Brazed joints', *Machine Design*, pp. 116–121, 23 Feb. 1956.

63 LIPSCH, J. and P. CAPOLONGO, 'High-frequency induction brazing and soldering', *Lepel High-frequency Heating Review*, **1** (5), undated.

64 BAKER, R. M., 'Manufacture of seam-welded tubes by low frequency induction', US Patent 2 647 982.

65 Inductron Ltd, *Notes on Tube Welding*.

66 ANGEL, R. F., 'The place of standard frequency (50/60 Hz) induction heating in the technology of electric surface heating', *Elektrowärme International*, June 1973, 31.B3.

67 British Electrical Development Association, *Induction and Dielectric Heating*, Electricity and Productivity Series No. 6, 1962.

68 BAKER, R. M., 'Induction heating of moving magnetic strip', *AIEE Trans.*, **64**, 184–189, 1945.

69 SIMPSON, P. G., *Induction Heating*, McGraw-Hill, New York, 1960.

70 BAKER, R. M., 'Transverse flux induction heating', *AIEE Trans.*, **69**, 711–719, 1950.

71 BAKER, R. M., US Patents 2 448 009, 2 448 011, 2 448 012.

72 BAKER, R. M., 'Transverse flux induction heating for continuous heat treatment of non-ferrous strip', paper presented at the AIEE Heating Conference, May 1953.

73 LAMOURDEDIEU, M., 'Continuous heat treatment of aluminium alloy strip', *Metal Progress*, pp. 88–92, Oct. 1951.

74 LACKNER, R. V., 'Continuous flash annealing produces fine-grained aluminium sheet', *Modern Metals*, pp. 62–64, Sept. 1960.

75 JACKSON, W. B., 'Transverse flux induction heating of flat metal products', paper presented at the 7th UIE Congress, Warsaw, 1972, paper N206.

76 WAGGOTT, R., 'Spatial temperature distribution of induction heated workpieces', paper presented at the Tempcon Conference, Brighton, Mar. 1974.

77 JACKSON, W. B., R. C. GIBSON, and R. H. JOHNSON, 'Improved methods for induction heating of thin slabs', *Electricity Council Research Centre, Capenhurst, Report M499*, May 1972.

78 IRESON, R. C. J., 'Induction heating with transverse flux in strip-metal process lines', *Power Engineering Journal*, **3**, (2), 83–91, 1989.

79 PFANN, W. G., *Zone Melting*, Wiley, New York, 1958.

80 BUEHLER, E., 'Contribution to the floating zone refining of silicon', *Rev. Scient. Instrum.*, **28**(6), 453–460, 1957.

81 BARDSLEY, W., 'The floating zone process', *Research*, **12**, 183–187, 1959.
82 PFANN, W. G., 'Principles of zone melting', *J. Metals*, **194**, 147, 1952.
83 NYLANDER, G. A., 'Applications of induction heating to the manufacture of semiconductors', paper presented at the AIEE Conference on Electric Heating, Toledo, Ohio, 23/24 April 1957.
84 ROBIETTE, A. G. E., 'Electrical melting/holding supplement', *Foundry Trade Journal*, Sept. 1976.
85 — 'Electric furnaces for melting iron', *Iron and Steel*, Jan./Feb. 1968.
86 — 'Electric melting of non-ferrous metals', *Electricity Council, EC2797R*, 1974.
87 — 'Electric melting for iron and steel foundries', *Electricity Council, EC3128*, 1973.
88 HARRISON, W. L., 'Melting by induction', paper presented at the BCIRA Conference on Electric Melting and Holding Furnaces in the Iron Foundry Industry, University of Technology, Loughborough, 1967, p. 269.
89 CHESTNUT, F. T., 'Melting by induction', *Induction Heating*, pp. 1377–1382, July 1956.
90 CLAMER, G. H., 'Development of the coreless induction furnace', *Metals and Alloys*, **6**, 119, 1935.
91 ROBIETTE, A. G. E., *Electric Melting Practice*, Charles Griffin, London, 1972.
92 STEFANO, R., '20 ton stainless steel melting', *IEEE Conference Record, Twelfth Electric Process Heating in Industry Conference*, 1975, pp. 20–25.
93 SOMMER, R. A., 'Channel furnace melting with jet-flow inductors', *IEEE Conference Record, Twelfth Electric Process Heating in Industry Conference*, 1975, pp. 20–25.
94 WORMLEY, K. G., 'Electrical safety of induction melting-furnaces: legislation, earthing, earth-leakage and lining-wear', *British Cast Iron Research Association, Report 1267*, 1977.
95 BAILEY, D., 'Cooling water systems for use with electric induction furnaces', *British Foundryman*, pp. 34–40, Feb. 1977.
96 ROWAN, H. M., 'Relating coreless induction furnace size to frequency', *Foundry*, Sept. 1971.
97 OLAUSSON, B., 'ASEA quick-change inductors for channel furnaces', paper presented at the BCIRA Conference on Electric Melting and Holding Furnaces in the Iron Foundry Industry, University of Technology, Loughborough, 1967.
98 DWIGHT, H. B., *Electrical Coils and Conductors*, McGraw-Hill, New York, 1945.
99 RUSSELL, A., *Theory of Alternating Currents*, Vol 1, 2nd Ed., Cambridge University Press, 1914.
100 HEAVISIDE, O., *Electrical Papers*, Vol. 2, p. 69.
101 BOWDEN, A. L., DAVIES, E. J., 'Analytic separation of the factors contributing to the eddy-current loss in magnetically nonlinear steel', *IEE Proc. B*, **130**, 364–372.
102 MARTIN, D. H., 'High efficiency induction billet heater', Paper presented at the 10th UIE Congress, Stockholm, 1984.
103 EDGERLEY, C. J., 'Electrical characteristics of coreless furnaces and their relevance to practical operating conditions', BCIRA Conference on Developments for Future Founding Prosperity, April 1984.
104 WILFORD, C. F., 'Refining a coreless furnace using reusable formers: A description of two techniques', ECR Report M1620, January 1983.
105 EDGERLEY, C. J., SMITH, L., and WILFORD, C. F., 'Electric metal melting - a review', *Power Engineering Journal*, **2**, (2), 83–94.
106 STOLL, R. L., *Analysis of Eddy Currents*, Oxford 1974.
107 HEAVISIDE, O., 'The induction of currents in cores', *The Electrician*, pp. 583 *et seq*, May 1884.
108 THOMSON, J.J., 'On the heat produced by eddy currents in an iron plate exposed to an alternating magnetic field', *The Electrician*, April 1892.
109 Copper Development Association 'Copper for Busbars', Publication No. 22, 1984.

Index

www.ingramcontent.com/pod-product-compliance
Lightning Source LLC
Chambersburg PA
CBHW071306180926
51613CB00001B/135